Springer Tracts in Modern Physics 84

Ergebnisse der exakten Naturwissenschaften

Springer Tracts in Modern Physics

Volume 66 **Quantum Statistics in Optics and Solid-State Physics** With contributions by R. Graham, F. Haake

Volume 67 **Conformal Algebra in Space-Time and Operator Product Expansion** By S. Ferrara, R. Gatto, A. F. Grillo

Volume 68 **Solid-State Physics** With contributions by D. Bäuerle, J. Behringer, D. Schmid

Volume 69 **Astrophysics** With contributions by G. Börner, J. Stewart, M. Walker

Volume 70 **Quantum Statistical Theories of Spontaneous Emission and their Relation to Other Approaches** By G. S. Agarwal

Volume 71 **Nuclear Physics** With contributions by J. S. Levinger, P. Singer, H. Überall

Volume 72 **Van der Waals Attraction:** Theory of Van der Waals Attraction By D. Langbein

Volume 73 **Excitons at High Density** Edited by H. Haken, S. Nikitine. With contributions by V. S. Bagaev, J. Biellmann, A. Bivas, J. Goll, M. Grosmann, J. B. Grun, H. Haken, E. Hanamura, R. Levy, H. Mahr, S. Nikitine, B. V. Novikov, E. I. Rashba, T. M. Rice, A. A. Rogachev, A. Schenzle, K. L. Shaklee

Volume 74 **Solid-State Physics** With contributions by G. Bauer, G. Borstel, H. J. Falge, A. Otto

Volume 75 **Light Scattering by Phonon-Polaritons** By R. Claus, L. Merten, J. Brandmüller

Volume 76 **Irreversible Properties of Type II Superconductors** By H. Ullmaier

Volume 77 **Surface Physics** With contributions by K. Müller, P. Wißmann

Volume 78 **Solid-State Physics** With contributions by R. Dornhaus, G. Nimtz, W. Richter

Volume 79 **Elementary Particle Physics** With contributions by E. Paul, H. Rollnick, P. Stichel

Volume 80 **Neutron Physics** With contributions by L. Koester, A. Steyerl

Volume 81 **Point Defects in Metals I:** Introduction to the Theory By N. Breuer, G. Leibfried

Volume 82 **Electronic Structure of Noble Metals, and Polariton-Mediated Light Scattering** With contributions by B. Bendow, B. Lengeler

Volume 83 **Electroproduction at Low Energy and Hadron Form Factors** By E. Amaldi, S. P. Fubini, G. Furlan

Volume 84 **Collective Ion Acceleration** With contributions by C. L. Olson, U. Schumacher

Volume 85 **Solid Surface Physics** With contributions by J. Hölzl, F. K. Schulte, H. Wagner

Collective Ion Acceleration

With Contributions by
C. L. Olson U. Schumacher

With 63 Figures

Springer-Verlag
Berlin Heidelberg New York 1979

Craig L. Olson, PhD
Plasma Theory Division - 4241, Sandia Laboratories
Albuquerque, NM 87115, USA

Dr. Uwe Schumacher
Max-Planck-Institut für Plasmaphysik, Bereich Beschleunigung
8046 Garching b. München, Fed. Rep. of Germany

Manuscripts for publication should be addressed to:

Gerhard Höhler
Institut für Theoretische Kernphysik der Universität Karlsruhe
Postfach 6380, D-7500 Karlsruhe 1, Fed. Rep. of Germany

Proofs and all correspondence concerning papers in the process of publication should be addressed to:

Ernst A. Niekisch
Institut für Grenzflächenforschung und Vakuumphysik der Kernforschungsanlage Jülich GmbH
Postfach 1913, D-5170 Jülich 1, Fed. Rep. of Germany

ISBN 3-540-09066-5 Springer-Verlag Berlin Heidelberg New York
ISBN 0-387-09066-5 Springer-Verlag New York Heidelberg Berlin

Library of Congress Cataloging in Publication Data. Olson, C. L. Collective ion acceleration. (Springer tracts in modern physics; v. 84) Includes bibliographies and index. CONTENTS: Olson, C. L. Collective ion acceleration with linear electron beams. — Schumacher, U. Collective ion acceleration with electron rings. 1. Ion accelerators. I. Schumacher, Uwe, 1938-. II. Title. III. Series. QCI.S797 vol. 84 [QC787.L5] 539'.08s [539.7'3] 78-11116

Offset printing and bookbinding: Brühlsche Universitätsdruckerei, Lahn-Giessen
2153/3130 — 543210

Contents

Collective Ion Acceleration with Linear Electron Beams

By *C.L. Olson*. With 32 Figures

Collective Ion Acceleration with Electron Rings

By *U. Schumacher*. With 31 Figures

Collective Ion Acceleration with Linear Electron Beams

Craig L. Olson

1. Introduction

Methods for the production of intense energetic ion beams are receiving increasing interest due to the growing number of applications for these beams. Existing applications involve such diverse fields as nuclear physics research, materials research, medical radiography, and cancer therapy. In addition, potential applications that have recently arisen include use of energetic light ion beams in spallation breeders to produce fissile material from fertile material /1/, use of energetic light ion beams for possible military applications /2/, and the use of energetic heavy ion beams to produce inertially confined fusion /3/.

Collective field accelerators show promise of producing a new breed of energetic ion accelerators. In collective ion accelerators, the collective field effects of a large number of electrons are used to accelerate a smaller number of ions to high energies. In conventional ion accelerators, the accelerating and focusing fields are produced by charges and currents in structures external to the ion beam path. Because of this, the accelerating field is ultimately limited by electrical breakdown at the accelerating gap(s), and large multipole magnet systems must typically be used for focusing. In collective accelerators these restrictions are removed, and net charge and net current densities are allowed directly in the ion beam path. This renders possible the production of a compact, very-high-gradient, strongly-focused, high energy ion accelerator.

The earliest accounts suggesting use of collective fields to accelerate charged particles were by ALFVEN and WERNHOLM /4/, HARVIE /5/, and RAUDORF /6/ in the early 1950's. However, substantial interest in the possibility of collective acceleration was initiated with the works of VEKSLER /7 - 9/, BUDKER /10/, and FAINBERG /11/ in the late 1950's. Since that time, research on collective acceleration has split into two main areas of development - collective acceleration with electron rings and collective acceleration with linear electron beams. Electron ring accelerator (ERA) research began with the introduction of this concept by VEKSLER, SARANTSEV, et al. /12/ in the USSR in 1967. As discussed by SCHUMACHER in the second part of this book, the ERA concept has received widespread study at many laboratories throughout the world. Collective ion acceleration in a special linear geometry (plasma-filled diodes) was reported by PLYUTTO /13/ in 1960. However, substantial interest in collective acceleration with linear electron beams began when GRAYBILL and UGLUM /14, 15/ in the USA in 1968 discovered that injecting an intense relativistic electron beam (IREB) into a neutral gas produced collectively accelerated ions with energies greater than the injected IREB energy. Collective ion acceleration with linear electron beams is the subject of this part of this book.

The existence of naturally occurring linear beam collective acceleration processes lends strong support to the linear beam approach to collective acceleration. The existing process that occurs for IREB injection into neutral gas has already produced very large accelerating fields (~ 100 MV/m) over short distances (~ 10 cm). [These fields should be compared with the 1 - 1.5 MV/m average accelerating fields of conventional ion linacs, such as LAMPF, the ZGS injector at Argonne, or the AGS injector at Brookhaven.] Since the discovery of this acceleration process, many experiments have been performed to verify and characterize it. During the same time, many theoretical explanations were proposed to account for this process. A theory developed by OLSON /16 - 18/ in 1973, and subsequently confirmed by the numerical simulations of POUKEY and OLSON /19/, has been successful in explaining this process. The historical development of these experiments and theories is contained in a number of timely review papers /20 - 34/. Here we will offer extensive and recent comparisons of experiments, theories, and computer simulations for this acceleration process.

At the same time the naturally occurring acceleration processes were being investigated, several linear beam collective accelerator concepts were being proposed. We will summarize these concepts here. From these, three scalable collective accelerator concepts have emerged which show promise of producing and controlling large accelerating fields over large distances (many meters). These are the ionization front accelerator (IFA) of OLSON /35 - 38/, the autoresonant accelerator (ARA) of SLOAN and DRUMMOND /39, 40/, and the converging guide accelerator (CGA) of SPRANGLE et al. /41/; we will investigate these in some detail. Experiments to demonstrate feasibility of these three accelerator concepts are currently in progress.

Essentially all of these linear beam concepts were made possible through the advent of intense relativistic electron beams. IREB's have their origins in the pulsed power technology pioneered by MARTIN /42/ in the early 1960's. Today, a typical IREB machine consists of a Marx generator, which charges a transmission line or Blumlein, which when switched, applies a high voltage pulse to a diode (see, e.g., /43, 44/). A typical diode consists of a cold, field emission cathode, and a thin foil anode through which the IREB emerges. Typical IREB parameters are:

electron energy	100 keV - 10 MeV
electron current	1 kA - 1 MA
pulse length	10 nsec - 100 nsec
beam radius	1 cm - 10 cm
beam energy	10 J - 1 MJ

These parameters typically yield electron densities in the range 10^{11} - 10^{13} cm^{-3}, which is ideally suited for linear beam collective accelerators. Here we simply note that such machines are readily available today; our concern in this article is to examine how to use these beams to collectively accelerate ions to high energies.

A comparison of the peak currents of conventional ion accelerators and the expected development trend of linear beam collective ion accelerators is given in Fig. 1 for a light ion (proton), and in Fig. 2 for a heavy ion (uranium), following OLSON /45/. In both figures, the peak instantaneous particle current (including macroscopic and microscopic duty cycle effects) is plotted against the ion energy. In Fig. 1, it is shown that although very high ion energies are attainable with conventional accelerators, the peak currents are less than about 10 A. On the other hand, the development of ion diodes using pulsed power technology has led to very high ion currents but energies that are limited by technology considerations to less than 10 MeV /45/. A summary of experimental data points for ions collectively accelerated by IREB injection into neutral gas is also shown. The proposed development trend for linear beam collective accelerators is also indicated; this development should open up the new parameter space of high energy (>> 10 MeV), high current (>> 10 A) ion beams. In Fig. 2, for uranium, the results are even more dramatic. The conventional heavy ion accelerators shown are all proposed or under construction, except for the Unilac, which has not yet produced a beam intensity up to the design goal indicated. The projected linear beam collective accelerator development is also indicated.

Thus the development of linear beam collective acceleration research may ultimately result in the formation of a compact, high-gradient, high energy collective ion accelerator. In addition, as shown in Figs. 1 and 2, linear beam collective accelerator development should open up a new high power parameter space for ion beams. In fact, such high power beams are actually required for certain future applications such as heavy ion fusion.

Basic linear electron beam properties are discussed in Sect. 2, and collective ion acceleration mechanisms that use linear electron beams are discussed in Sect. 3. These mechanisms are of interest since some of them have been used to explain the naturally occurring collective acceleration processes, and others have been used to formulate linear beam collective ion accelerator concepts. Naturally occurring collective ion acceleration processes are discussed in Sect. 4, and several linear beam collective accelerator concepts are discussed in Sect. 5. The main scalable collective ion accelerators (IFA, ARA, CGA) are then described in detail in Sect. 6. A brief summary is given in Sect. 7.

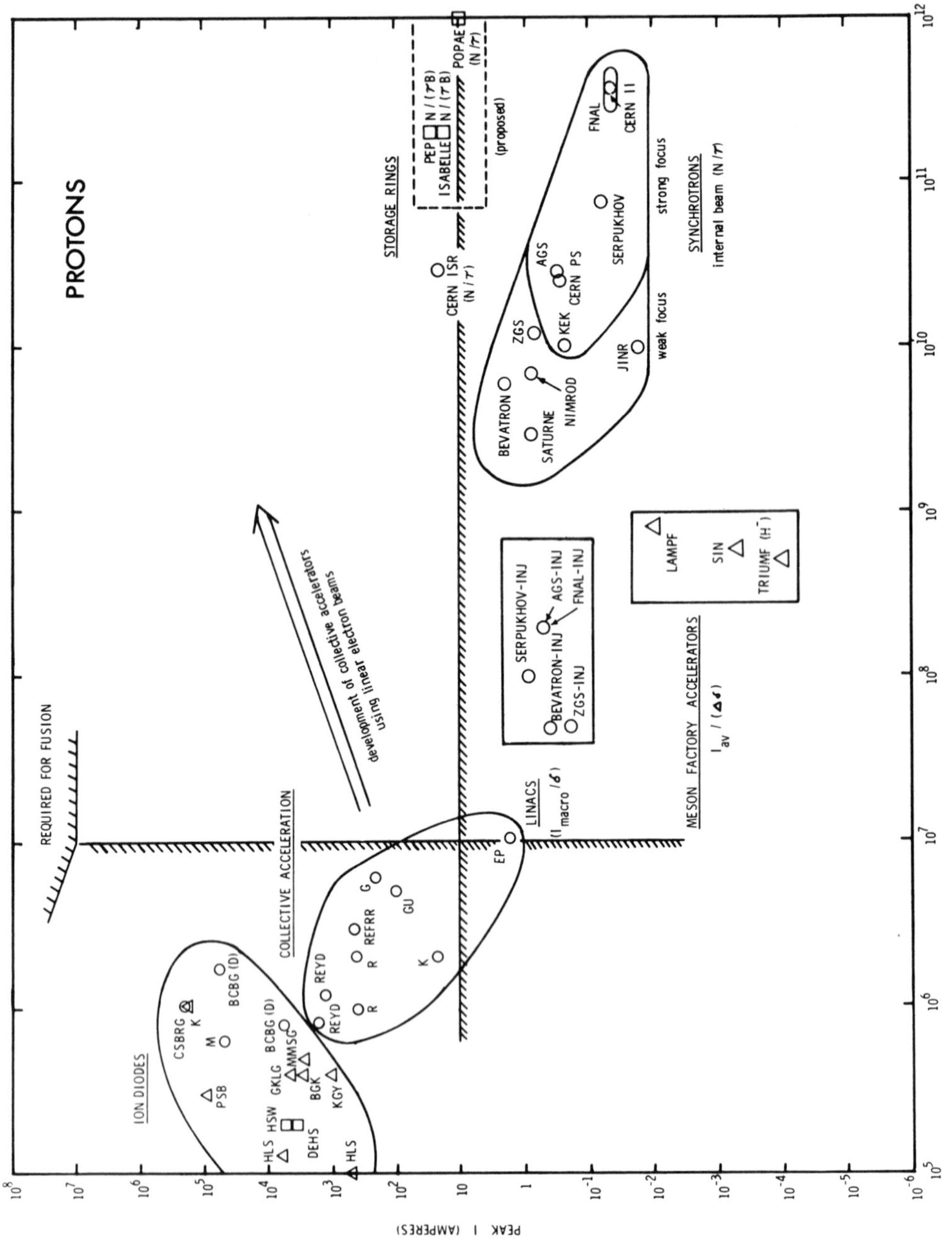

Fig. 1. Peak instantaneous currents of proton accelerators, showing the expected development trend of linear beam collective accelerators

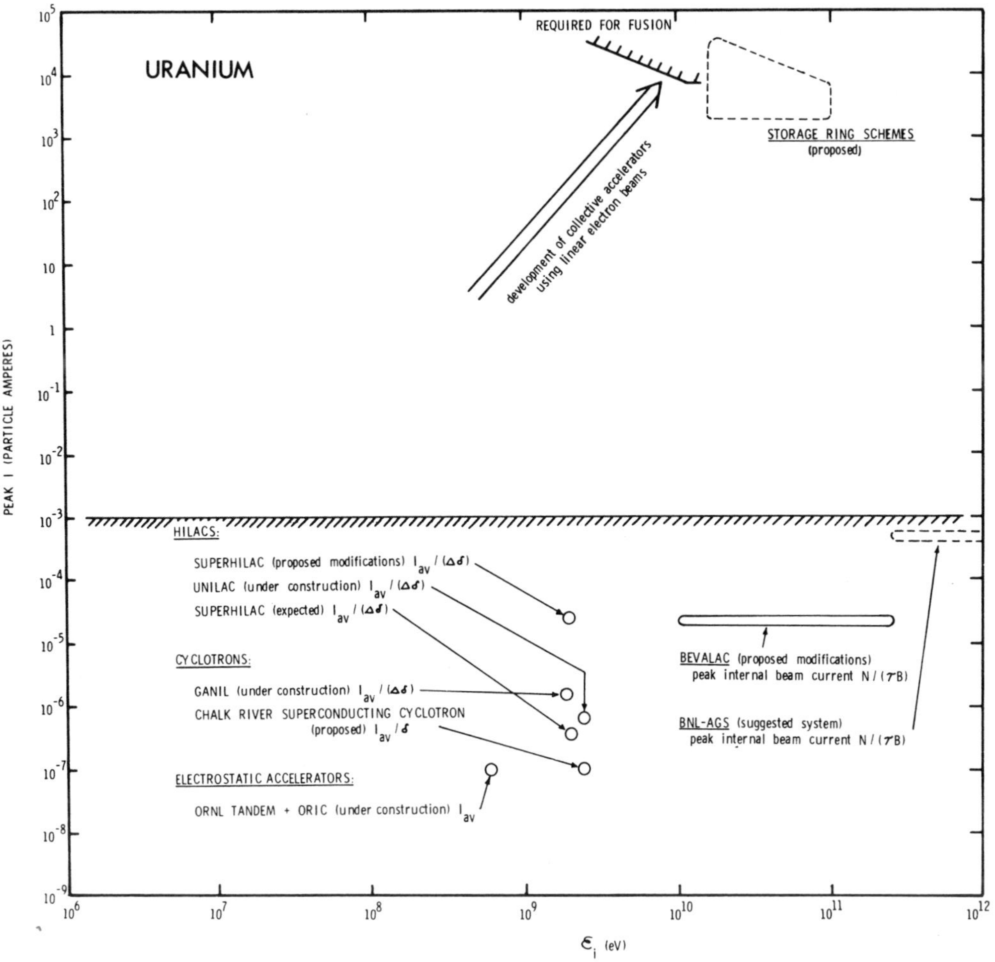

Fig. 2. Peak instantaneous currents of heavy ion accelerators for uranium (proposed and under construction), showing the expected development trend of linear beam collective accelerators

2. Basic Linear Electron Beam Properties

A brief discussion of linear electron beam properties that are significant for their use in collective ion acceleration is given here. It should be noted that for linear beam collective acceleration, we are typically interested in currents much larger than those used for electron ring accelerators. The existence and basic equilibrium properties of linear beams were first investigated by BENNETT /46, 47/ in 1934. Magnetic stopping effects were first discussed by ALFVEN /48/ in 1939, and later by BUDKER /10/ in 1956. LAWSON /49 - 53/ has since classified and explored these beams in detail.

Basic properties of propagating IREB's may be demonstrated by imagining a section of a propagating beam in which the injected electrons have energy $\mathscr{E}_e = (\gamma_e - 1)\, mc^2$, where $\gamma_e = (1 - \beta_e^2)^{-1/2}$, $\beta_e = v_e/c$, v_e is the injected electron speed, m is the electron rest mass, and c is the speed of light. If the beam is cylindrically symmetric and has uniform density n_b (in the laboratory frame), then in cylindrical coordinates (r,θ,z) the beam self-electric field (E_r) and self-magnetic field (B_θ) are given by

$$\begin{aligned} E_r(r) &= [2I_e/(\beta_z c r_b)]\, r/r_b && 0 \le r \le r_b \\ &= 2I_e/(\beta_z c r) && r \ge r_b \end{aligned} \qquad (1)$$

$$\begin{aligned} B_\theta(r) &= [2I_e/(c r_b)]\, r/r_b && 0 \le r \le r_b \\ &= 2I_e/(cr) && r \ge r_b\, , \end{aligned} \qquad (2)$$

where the beam current $I_e = \pi r_b^2 n_b e \beta_z c$, r_b is the beam radius, e is the magnitude of the electron charge, and $\beta_z c$ is the average speed of the electrons in the z direction ($\beta_z < \beta_e$). In practical units, the electric field at the beam edge is

$$E_r = 60\; I_e(A)[\beta_z r_b(\mathrm{cm})]^{-1}\; \mathrm{V/cm}. \qquad (3)$$

For a modest IREB (I_e = 30 kA, r_b = 1 cm, $\beta_z \approx 1$) this gives $|E_r| \approx 1.8 \times 10^6$ V/cm.

The radial force equation for a single electron at the beam edge is, for $\beta_z = \beta_e$,

$$d^2r/dt^2 = [2I_e e/(\beta_e c r_b \gamma_e m)][(1-f_e) - \beta_e^2(1-f_m)]. \qquad (4)$$

Here we imagine a charge-neutralizing background of ions (charge +Ze) of uniform density n_i in the IREB channel, so that there is a charge neutralization factor $f_e \equiv n_i Z/n_b$. Similarly we imagine a uniformly distributed reverse current (I_r) in the beam channel, so that there is a magnetic neutralization factor $f_m \equiv |I_r/I_e|$. A variety of trajectories is possible depending on the values of f_e and f_m. For $f_e = 0$, $f_m = 0$, the beam will blow up radially. For $f_e = \gamma_e^{-2}$, $f_m = 0$, the beam is said to be radially "force-neutral" and straight trajectories should occur. For $f_e = 1$, $f_m = 0$ (a charge neutralized beam), the electrons will oscillate radially due to the Lorentz force. For $f_e = 1$, $f_m = 1$ (a charge and current neutralized beam), there are no forces and the beam particles will again have straight trajectories.

By assuming the existence of the propagating beam in (1) - (4), we have ignored the problem of the creation of the beam and its injection into the drift space. Depending on the beam parameters, it is possible that the electrostatic potentials that would be set up by such a beam may be larger than $\mathscr{E}_e/e$ and therefore preclude the existence of such a beam: this is the space charge limiting current problem. Similarly, for a charge neutral beam, it is possible that the Lorentz force is so large that it would actually reverse the direction of the electron flow: this is the magnetic stopping problem.

2.1 Space Charge Limiting Currents

The space charge limiting current problem can be envisioned by considering the beam described by (1) - (4) to be inside a conducting cylindrical cavity of radius R. If the beam injection plane (anode foil) is at z = 0, then for axial distances $z \gg 2R$, the beam electric field is mainly radial and given by (1). The electrostatic potential difference ϕ_0 between the drift tube wall (taken to be at zero potential) and the center of the beam is then

$$\phi_0 = [I_e/(\beta_z c)][1 + 2 \ln (R/r_b)] (1-f_e). \qquad (5)$$

We consider the case $\beta_z = \beta_e$. Then setting the electrostatic energy $e\phi_0$ equal to the electron energy $(\gamma_e - 1)mc^2$, and solving for I_e yields for the space large limiting current (for $f_e = 0$)

$$I_\ell = \beta_e(\gamma_e - 1)(mc^3/e)[1 + 2 \ln (R/r_b)]^{-1}. \quad \text{(Olson-Poukey)} \qquad (6)$$

Beams with $I_e \gtrsim I_\ell$ will not propagate due to the large electrostatic potential, whereas beams with $I_e \ll I_\ell$ will propagate freely, not being hindered axially by the electrostatic potential. The space charge limiting current formula (6) was given by OLSON and POUKEY /54/, who were the first to note its fundamental importance to the problem of collective ion acceleration with linear beams. The result (6) is relativistically correct but does not include self-consistent γ effects; it should however be a reasonable approximation for beams injected into neutral gases with no externally applied magnetic field.

According to NEZLIN /55/, the space charge limiting current problem was first considered in the one-dimensional (planar beam), nonrelativistic case by BURSIAN and PAVLOV /56/ in 1923. In 1940, SMITH and HARTMAN /57/ considered the nonrelativistic, two-dimensional problem of a beam propagating inside a conducting, cylindrical cavity, with an externally applied, infinitely large, axial magnetic field. Including the effects of self-consistent radial shear in the axial velocities, they found (for $R = r_b$)

$$I_\ell = 3.0 \times 10^{-5}\ V^{3/2}, \quad \text{(Smith-Hartman)} \qquad (7)$$

where V is the accelerating voltage applied to the diode. For the relativistic case, BOGDANKEVICH and RUKHADZE /58, 59/ included self-consistent γ effects, but did not include radial shear effects in γ [i.e., they examined the limit $\ln(R/r_b) \gg 1$]. They reported the interpolation result

$$I_\ell = (\gamma_e^{2/3} - 1)^{3/2}\ (mc^3/e)\ [1 + 2\ \ln\ (R/r_b)]^{-1}. \quad \text{(Bogdankevich-Rukhadze)} \qquad (8)$$

This result was also derived by FESSENDEN /60/ and RYUTOV et al. /61, 62/; it is obtained by maximizing the self-consistent current that is allowed to flow as a function of $\gamma(r = 0)$ [which must be less than the injection γ_e due to the presence of the electrostatic potential (5)]. The three expressions for the limiting currents (6) - (8) are plotted in Fig. 3. We note that all three are approximations to the true limiting current (which should include relativistic, self-consistent, and radial shear effects). Continuing with the case of an infinite axial magnetic field (to insure straight particle trajectories), NECHAEV /63/ reports numerical results that show (8) underestimates the true limiting current by factor up to 25 %. Similarly, THOMPSON and SLOAN /64/ report that (8) may differ as much as 20% from the correct numerical results. It should also be noted that space charge limiting currents have also been calculated for other beam geometries, such as annular beams /65 - 68/.

In any event, it should be noted from Fig. 3 that the various results differ at most by a factor of 2 in the nonrelativistic limit, and by less in the relativistic limit. Since we will be interested in the case of beam injection into gases with no external magnetic field (where radial beam spreading may occur), we note here that none of the existing treatments have studied this case exactly. Therefore, for this case, we will consider (6) to be a reasonable approximation, which has in fact been demonstrated by numerical simulations /54/.

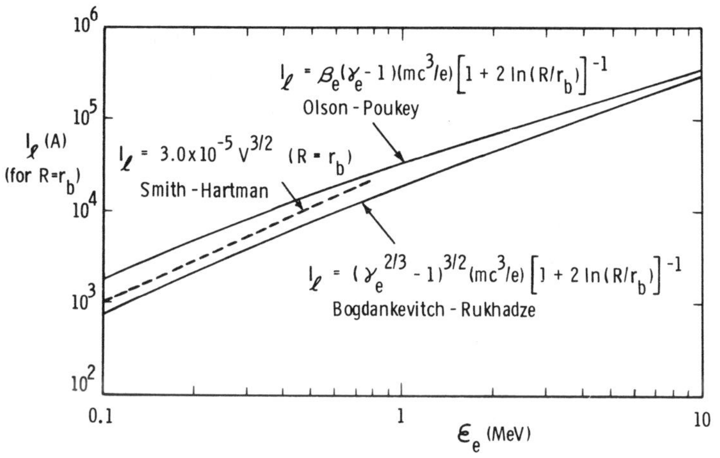

Fig. 3. Space charge limiting currents

2.2 Magnetic Stopping Currents

For the case of charge neutralized beams ($f_e = 1$), we are still left with the beam's self-magnetic field effects. To demonstrate these effects we note that the cyclotron radius r_{ce} of a beam electron at the edge of the beam is

$$r_{ce} = \beta_e\gamma_e mc^2/(eB_\theta), \tag{9}$$

where B_θ is given by (2) for $r = r_b$. When $r_{ce} \approx r_b/2$, beam stopping will occur; for this case, using (2) and (9) and solving for I_e yields for the magnetic stopping current

$$I_A = \beta_e\gamma_e mc^3/e, \tag{10}$$

which is the well-known Alfven-Lawson magnetic stopping current /48 - 50/. In practical units, this is

$$I_A = 17\ \beta_e\gamma_e\ (\mathrm{kA}). \tag{11}$$

A related parameter for linear beams is ν_e/γ_e,

$$\nu_e/\gamma_e = I_e/I_A, \tag{12}$$

where the Budker parameter $\nu_e = \pi r_b^2 n_b e^2 (mc^2)^{-1}$ is the number of beam electrons contained in an axial length of beam equal to the classical electron radius $r_o = e^2(mc^2)^{-1}$, i.e., $\nu_e = \pi r_b^2 n_b r_o$ /10/. Beams with $I_e \gtrsim I_A$ ($\nu_e/\gamma_e \gtrsim 1$) should not propagate due to magnetic stopping effects, whereas beams with $I_e \ll I_A$ ($\nu_e/\gamma_e \ll 1$) should propagate freely.

The current limit imposed by (10) is a fundamental limit for the case considered - a uniform beam, charge-neutralized ($f_e = 1$), with no magnetic neutralization (f_m=0), no rotational motion ($v_\theta = 0$), and no externally applied magnetic field ($B_z = 0$). By relaxing these restrictions, it becomes possible to propagate currents in excess of I_A, and many studies have been made of such equilibrium propagating beam configurations /30, 49, 50, 69 - 79/. A brief listing of some of these configurations is as follows:

(1) *partially charge-neutralized beam:* $\gamma_e^{-2} \leq f_e < 1$, $f_m = 0$ /30, 49, 50, 69/. Here the charge neutralization is relaxed to allow the net electric repulsive force to counteract some of the attractive magnetic force on the beam electrons and permit $I_e > I_A$. Note, however, that $I_e < I_\ell$ is still required.

(2) *partially current-neutralized beam:* $f_e = 1$, $0 < f_m < 1$ /30, 69, 73/. This is the typical configuration for IREB propagation in neutral gases at moderate pressures (~ 1 Torr air). This allows $I_e > I_A$, but of course, the net current $I_n = I_e - I_r$ must still be less than I_A.

(3) *annular beam* /69 - 71, 74/. By allowing the beam to be bunched near its outer radius, so the electrons experience the magnetic force in only a thin layer, it is possible to have $I_e > I_A$.

(4) *rotating beam:* $v_\theta \neq 0$ /74 - 77/. Here the centrifugal force of the rotating electrons is used to counteract the magnetic pinch force to permit $I_e > I_A$. The cases with all particles rotating in one direction /74 - 76/, or evenly rotating in both directions /77/ have been investigated separately.

(5) *magnetically confined beam:* $B_z \neq 0$ /69, 78/. A strong B_z has been shown to permit $I_e > I_A$.

In actual useage, configurations (2), (3), and (5) have proved most useful. For collective ion acceleration applications, we wish to simply note that the limit $\nu_e/\gamma_e = 1$ is a real limit for the beam considered, but that special configurations can be devised to permit $\nu_e/\gamma_e > 1$ beams to propagate.

Magnetic field effects can also cause pinching inside the diode that generates the electron beam. For a uniform cathode of radius r_c, with an anode-cathode gap distance d, and with $d < r_c/2$, beam pinching will occur inside the diode for $I_e \gtrsim I_c$, where /80, 81/

$$I_c = I_A r_c (2d)^{-1} \tag{13}$$

$$I_c = 8.5\ \beta_e \gamma_e r_c/d\ (\mathrm{kA}). \tag{14}$$

It should be noted that the space charge limiting current I_ℓ and the magnetic stopping current I_A are fundamental limiting currents based on zero-order effects. For "compensated beams", other critical currents may be defined that may also affect beam propagation; these critical currents are generally based on instabilities /82 - 86/ caused by the beam streaming through a charge and/or current neutralizing plasma background /55, 58, 59/. For example, for the case of a charge-neutralized beam ($f_e = 1$, $f_m = 0$), the electron-ion two stream instability of BUNEMAN /85 - 86/ may develop. The threshold current for this instability is known in the Soviet literature as the "critical current for a compensated beam"; such critical currents have been investigated by NEZLIN et al. /55, 87 - 89/ and RAIZER et al. /90/. This critical current is typically many times larger than I_ℓ. Here we will consider this phenomenon in the category of linear waves and instabilities, which we will discuss in Sect. 3.4.

3. Collective Ion Acceleration Mechanisms Using Linear Electron Beams

There are many conceivable mechanisms by which the collective fields of an IREB could be used to accelerate ions. Here we have organized these mechanisms into several distinctly different categories, and we present a discussion of the general properties of each category. This classification is useful for several reasons. First, some of these mechanisms have been used to explain naturally occurring collective acceleration processes (as will be discussed in Sect. 4). Second, several of these mechanisms form the basis for linear beam collective ion accelerator concepts (as will be discussed in Sect. 5).

The starting point for each mechanism is an IREB for which we permit

$$\nabla \cdot \underset{\sim}{E} \neq 0$$
$$\nabla \times \underset{\sim}{B} \neq 0 \,,$$

i.e., we allow net charge and net current to be directly in the acceleration region (as distinct from conventional accelerators where these quantities must be zero). As we have noted in Sect. 2, the electric field at the edge of an unneutralized IREB may reach very large values ($|E_r| \sim 10^6$ V/cm) even for modest IREB parameters ($\gamma_e = 3$, $I_e = 30$ kA, $r_b = 1$ cm). This large space charge field is, of course, in the radial direction. The problem is to devise a mechanism for creating a large accelerating field in the axial direction, and then use it to trap and accelerate ions in a carefully controlled manner with an ever increasing phase velocity. Thus while it is true that IREB's can produce large fields, it is also true that some ingenuity is required to control these fields and use them to accelerate ions over substantial distances.

The acceleration mechanisms to be discussed include: (1) net space charge mechanisms, (2) induced field mechanisms, (3) inverse coherent drag mechanisms, (4) linear waves and instabilities, (5) nonlinear waves and solitons, (6) stochastic acceleration, and (7) impact acceleration. All of these mechanisms are distinctly different, yet together they represent essentially all of the linear beam collective acceleration concepts discussed in the literature to date. A simplified sketch of each of these mechanisms is given in Table 1, as is an estimate of the maximum axial field E_z possible (these results occur in the following discussions). It is useful to note even at this time that the most important category, in regard to explaining the acceleration processes that occur during IREB injection into neutral gas or vacuum, is "net space charge mechanisms". Also present scalable linear beam collective ion accelerator concepts are based on "net space charge mechanisms" and "linear waves and instabilities".

Table 1. Linear beam collective ion acceleration mechanisms

MECHANISM	PICTURE	E_z
net space charge		large
inductive		small
inverse coherent drag	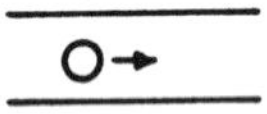	small
linear waves		moderate
nonlinear waves		large
stochastic		very small
impact		-

3.1 Net Space Charge Mechanisms

Net space charge mechanisms are those in which the electric field created by the space charge of a bunch of electrons (which may be partially charge-neutralized) is used to accelerate a cluster of ions. They are the simplest, yet perhaps the most useful, of all collective ion acceleration mechanisms. They produce very large electric fields which are relatively easy to create, and which, under special conditions, should be amenable to control. The net space charge comes from an unneutralized ($f_e = 0$), or partially neutralized ($0 < f_e < 1$), portion of an IREB. The net space charge is arranged so as to produce an axial potential gradient. There are various ways of controlling f_e, or the boundary conditions, to produce such gradients.

Some simple estimates of the accelerating fields that may be produced with this mechanism are as follows. Consider an infinitely long, unneutralized IREB that is constrained to propagate with a fixed radius r_b. The radial electric field at the beam edge is then $E_r = 2\,\phi_0/r_b$, where the potential difference between the center and edge of the beam is $\phi_0 = \pi r_b^2 n_b e' = I_e/(\beta_z c)$. If we were free to specify f_e everywhere in the beam, then the net charge could be made to produce an axial electric field E_z,

$$E_z = 2\alpha\phi_0/r_b \,, \tag{15}$$

where α is a dimensionless factor which depends on the detailed shape of the net charge region. For example, if we had $f_e = 1$ for $z < z_0$, and $f_e = 0$ for $z > z_0$, then the net space charge would appear as a cylindrical rod, and at $z = z_0$ we would find $\alpha = 1$. If the neutralization transition occurred over a distance ξ, with $f_e = 1$ for $z < z_0$, $f_e = 1 - [(z - z_0)/\xi]$ for $z_0 < z < (z_0 + \xi)$, and $f_e = 0$ for $z > (z_0 + \xi)$, then at $z = z_0$ we would find

$$\alpha = 2^{-1}[1 + (\xi/r_b)^2]^{1/2} + [r_b/(2\xi)]\ \ln\{(\xi/r_b) + [1 + (\xi/r_b)^2]^{1/2}\} - \xi/(2r_b). \tag{16}$$

This result gives $\alpha = 0.65$ for $\xi = r_b$, and $\alpha = 0.15$ for $\xi = 10\ r_b$. Thus α decreases as ξ increases, as one would expect. At the other extreme, consider $f_e = 1$ everywhere, except in the region $z_0 < z < (z_0 + \Delta\xi)$ where $f_e = 0$ and $\Delta\xi \ll r_b$. Then the charge region is disc shaped and $\alpha \approx \Delta\xi/r_b$. From these examples, we conclude that net space charge can be used to produce the large E_z given by (15), where $\alpha \lesssim 1$ for the simple geometries considered. For a short charge section ($\Delta\xi \ll r_b$), or for a long neutralization transition section ($\xi \gg r_b$), we find $\alpha \ll 1$. However, for a steep charge neutralization gradient ($\xi \lesssim r_b$), we find $\alpha \approx 1$ and the axial field produced is roughly equal to the peak radial electric field of the unneutralized beam.

In general, net space charge mechanisms produce closed, three-dimensional potential wells. However, to demonstrate ion trapping and acceleration effects, it is convenient to consider the simplest case of a one-sided, one-dimensional potential well with

$$\begin{aligned} E_z &= \phi_0/\ell \qquad z_w \leq z \leq (z_w + \ell) \\ &= 0 \qquad z < z_w, \quad z > (z_w + \ell). \end{aligned} \tag{17}$$

Ion trapping and acceleration effects are then demonstrated by considering the energy $\mathscr{E}_i$ attained by an ion of mass M and charge Ze as it is accelerated by this well. For nonrelativistic ion motion we consider the well to be stationary ($z_w = 0$), translating with constant velocity v_0 ($z_w = v_0 t$), or accelerating with constant acceleration a_0 ($z_w = a_0 t^2/2$). At $t = 0$, the ion is taken to be at $z = 0$ for the stationary case, and at $z = z_w + \ell$ for the other cases. The ion energy attained, as well as the acceleration time T, have then been calculated by OLSON /91/ to be as shown in Table 2. Note that several cases emerge depending on the relative size of $Ze\phi_0$.

3.2 Induced Field Mechanisms

Induced electric fields caused by a time-varying beam current I(t), or by a time-varying beam inductance L(t), may in principle be used to collectively accelerate ions. To demonstrate these inductive effects, we assume $f_e = 1$ and $f_m = 0$, and consider a uniform IREB of radius r_b inside a metallic drift tube of radius R.

The induced electric field on axis produced by a time-varying beam current is /92/

$$E_z(\dot{I}) = - L\, \partial I/\partial t = -c^{-2}\,[1 + 2 \ln (R/r_b)]\; \partial I/\partial t. \tag{18}$$

For $I(t) \approx I_e t/t_r$, where t_r is the current risetime, we find

$$E_z(\dot{I}) = -I_e(c^2 t_r)^{-1}\,[1 + 2 \ln (R/r_b)]. \tag{19}$$

For typical IREB parameters (I_e = 30 kA, R/r_b = 2, t_r = 10 nsec), this electric field is $\lesssim$ 10 kV/cm, and is therefore relatively small compared to the fields that can be produced by the net space charge.

The induced electric field produced by a radial time variation of the same beam is

$$E_z(\dot{L}) = - I\, \partial L/\partial t = + 2\, I_e c^{-2}\, [r_b(t)]^{-1}\; \partial r_b(t)/\partial t. \tag{20}$$

Table 2. Potential well ion acceleration

Well	Restriction	$\mathscr{E}_i$	T	$\mathscr{E}_i$ scaling
Stationary	none	$Ze\phi_0$	$\ell\left[\frac{2M}{Ze\phi_0}\right]^{1/2}$	$\sim Z$
Translating ($\dot{z}_w = v_0$)	(i) ion lags $Ze\phi_0 < \frac{1}{2}Mv_0^2$	$\frac{1}{2}Mv_0^2\left[1-(1-\frac{2Ze\phi_0}{Mv_0^2})^{\frac{1}{2}}\right]^2$	$\frac{Mv_0\ell}{Ze\phi_0}\left[1-(1-\frac{2Ze\phi_0}{Mv_0^2})^{\frac{1}{2}}\right]$	$\sim Z^2/M$ (for $Ze\phi_0 \ll \frac{1}{2}Mv_0^2$)
	(ii) border case $Ze\phi_0 = \frac{1}{2}Mv_0^2$	$\frac{1}{2}Mv_0^2$	$\frac{Mv_0\ell}{Ze\phi_0}$	$\sim M$
	(iii) ion shot ahead $Ze\phi_0 > \frac{1}{2}Mv_0^2$	$4\,(\frac{1}{2}Mv_0^2)$	$\frac{2Mv_0\ell}{Ze\phi_0}$	$\sim M$
Accelerating ($\ddot{z}_w = a_0$)	(i) ion lags $Ze\phi_0 < Ma_0\ell$	$\frac{Ze\phi_0}{\{[(Ma_0\ell)/(Ze\phi_0)] - 1\}}$	$\left[\frac{2\ell}{a_0 - [(Ze\phi_0)/(M\ell)]}\right]^{\frac{1}{2}}$	$\sim Z^2/M$ (for $Ze\phi_0 \ll Ma_0\ell$)
	(ii) ion trapped $Ze\phi_0 \geq Ma_0\ell$	$\frac{1}{2}M(a_0t)^2$	-----	$\sim M$

For $r_b(t) = r_b(0) \exp(-t/t_p)$, where t_p is the characteristic pinch time, we find

$$E_z(\dot{L}) = -2\, I_e(c^2 t_p)^{-1}. \qquad (21)$$

The effectiveness of this field in accelerating ions has been calculated by OLSON /91/ for a number of cases as shown in Table 3. For the stationary pinch case, a uniform beam is assumed to contract radially at all z, and the ion energy is computed at time $t = t_p$. For the translating case, the beam envelope is assumed to contract from r_b to r_b/e in a distance δ, and this structure is assumed to propagate with velocity $\dot{z}_p = v_0$. For the synchronized pinch case, the pinch velocity $\dot{z}_p$ is set equal to the accelerated ion velocity. In all of these cases, the acceleration effectiveness of the induced fields is still relatively small for typical IREB parameters.

Some useful comparisons of the induced fields with the net space charge field can be made using (15), (19), and (21). We find

$$|[E_z(\dot{I})]/[E_z(\text{space charge})]| = \beta_z r_b (2\alpha c t_r)^{-1}\, [1 + 2 \ln (R/r_b)] \qquad (22)$$

$$|[E_z(\dot{L})]/[E_z(\text{space charge})]| = \beta_z r_b (\alpha c t_p)^{-1}. \qquad (23)$$

These expressions show that for the induced field effects to be important as compared to the space charge effects, requires an extremely small current risetime ($\alpha t_r \lesssim r_b/c$), or an extremely small pinch time ($\alpha t_p \lesssim r_b/c$). Similarly, the effective total potential difference produced by the induced fields is

$$\phi(\dot{I}) = E_z(\dot{I})\, \beta_f c t_r = -\beta_f I_e c^{-1}\, [1 + 2 \ln (R/r_b)] \qquad (24)$$

$$\phi(\dot{L}) = E_z(\dot{L})\, \beta_f c t_p = -\beta_f 2 I_e c^{-1}, \qquad (25)$$

where $\beta_f c$ is the beam front (or pinch) velocity, and in (25) we integrated $E_z(\dot{L})$ over one pinch e-fold time. The induced field potentials (24) and (25) may be compared with the electrostatic potential (5) for $f_e = 0$ to give

$$|[\phi(\dot{I})]/[\phi(\text{space charge})]| = \beta_f \beta_z \qquad (26)$$

$$|[\phi(\dot{L})]/[\phi(\text{space charge})]| = \beta_f \beta_z\, 2[1 + 2 \ln (R/r_b)]^{-1}. \qquad (27)$$

These results show that the induced field potentials become significant only when $\beta_f \beta_z$ approaches unity.

Table 3. Pinch ion acceleration

Pinch type	Pinch time	$\mathscr{E}_i$	$\mathscr{E}_i$ scaling
Stationary	t_p	$\frac{2I_e^2 Z^2 e^2}{Mc^4}$	$\sim \frac{Z^2 I_e^2}{M}$
Translating $\dot{z}_p = v_0$	$\frac{\delta}{v_0}$	$\frac{1}{2}Mv_0^2 \left[1 - \left(1 - \frac{4Ze\lvert I_e\rvert}{Mv_0 c^2}\right)^{\frac{1}{2}}\right]^2$ $(4Ze\lvert I_e\rvert \leq Mv_0 c^2)$	$\sim \frac{Z^2 I_e^2}{M}$ (for $4Ze\lvert I_e\rvert \ll Mv_0 c^2$)
Synchronized $\dot{z}_p = v_i(t)$	$\frac{\delta}{v_i(t)}$	$\frac{1}{2}M\left[v_i(0)\right]^2 \; e^{2t/\tau}$ $\tau = Mc^2\delta/(2Ze\lvert I_e\rvert)$	depends on M, Z, $v_i(0)$, and time cutoff

3.3 Inverse Coherent Drag Mechanisms

Drag acceleration refers to a process whereby the intense beam, as a result of "frictional forces", literally drags a stationary ion (or bunch of N ions) from rest and accelerates it. The process is best viewed in the beam frame, where the ion bunch appears to have a velocity $-v_z$. In the beam frame, the ion bunch slows down due to stopping power effects, and hence experiences an effective electric field E_z that decelerates the ion in the beam frame. Since the relativistic transformation back to the laboratory frame does not alter the axial electric field, the effective accelerating field in the laboratory frame is given by the same E_z. In the laboratory frame, the particle accelerates in the same direction as the IREB propagates. This mechanism is interesting in that if certain coherence conditions are fulfilled, $E_z \sim N$. The acceleration mechanism is then called an "inverse coherent" drag effect.

VEKSLER /7, 8/ originally proposed inverse coherent Cerenkov radiation as a potential collective acceleration mechanism. Later, LAWSON /23/ noted the inverse coherent Bohr-Bethe effect as a potential acceleration mechanism. In other work, DOGGETT and BENNETT /93 - 96/ claimed to have explained the Cerenkov effect in terms of the Bohr-Bethe effect. In fact, the two effects refer to different physical processes, and have different functional dependences as noted by OLSON /97/; in this work, it was also shown that the total drag force experienced by an ion bunch includes contributions due to Cerenkov radiation and due to Bohr-Bethe stopping, and that each of these contributions can in turn have both coherent and incoherent parts. We will summarize this general case briefly here.

The Bohr-Bethe effect describes the effective energy loss rate of an ion due to small-angle Coulomb collisions with electrons of the medium through which the ion propagates. The effect is derived by considering a collision between the incident ion and a single beam electron; the effective energy loss is then found by summing over all beam electrons. We consider an ion of charge Ze and velocity $-v_z$ immersed in a plasma of electron density n_b/γ_z (in our final result, n_b will correspond to the IREB density in the laboratory frame). The effective BOHR /98/ stopping power is then /99, 100/

$$-d\mathscr{E}_i/dz = 4\pi n_b e^4 Z^2 (\gamma_z m v_z^2)^{-1} \ln(b_{max}/b_{min}) , \tag{28}$$

where $\gamma_z = (1 - \beta_z^2)^{-1/2}$ and $\beta_z = v_z/c$. The maximum impact parameter b_{max} is roughly equal to the Debye length $\lambda_d = v_{T_e}(\sqrt{2}\omega_{pe})^{-1}$ where v_{T_e} is the beam thermal velocity, and $\omega_{pe} = [4\pi n_b e^2 (\gamma_z m)^{-1}]^{1/2}$. The minimum impact parameter is given by the largest of the classical value (b^c_{min}), or the BETHE /101/ quantum mechanical value (b^{QM}_{min}),

$$b^c_{min} = Ze^2(\beta_z^2 \gamma_z mc^2)^{-1} \tag{29}$$

$$b_{min}^{QM} = \hbar(\beta_z\gamma_z mc)^{-1}, \tag{30}$$

where $\hbar = h/2\pi$ and h is Planck's constant. The effective electric field is found by noting $-d\mathscr{E}_i/dz = ZeE_z$. Now imagine the "ion" to be a cluster of N ions uniformly contained within a spherical volume of radius ℓ. Then if $\ell < b_{min}$, it is readily seen that $E_z \sim NZ$, which demonstrates inverse coherent Bohr-Bethe stopping. If $b_{min} < \ell < b_{max}$, the impact parameters in the range $\ell < b < b_{max}$ ($b_{min} < b < \ell$) will produce a coherent (partially coherent) effect /97/, as we will summarize shortly.

The Cerenkov effect /102 - 104/ is a collective wave effect. It depends intrinsically on the ability of the background beam to support waves with phase velocities $v_\phi < v_z$. For EM waves, this is possible only if the plasma is in a magnetic field /105 - 110/. In the absence of an external magnetic field, however, it is still possible to support longitudinal plasma waves, if thermal effects are included /110 - 118/. These waves have phase velocities in the range $v_{T_e} \lesssim v_\phi < \infty$, so Cerenkov radiation is possible for $v_z \gtrsim v_{T_e}$. For this case, we find /91, 97/

$$-d\mathscr{E}_i/dz = 4\pi n_b e^4 Z^2(\gamma_z m v_z^2)^{-1} \; (1/2) \ln \left[1 + 2(v_z/v_{T_e})^2\right] \tag{31a}$$

$$\approx 4\pi n_b e^4 Z^2(\gamma_z m v_z^2)^{-1} \ln \left[\beta_z(c/\omega_{pe})/\lambda_d\right]. \tag{31b}$$

Numerous results of this type have been quoted in the literature /7, 8, 111 - 122/. In obtaining this result, we note that the density n_b entered the calculation at the outset through the dispersion relation for electrostatic waves. All contributing wave states were then summed (by transform techniques) to arrive at (31). For our ion cluster of radius ℓ, it is evident that for $\ell \ll \lambda_d$, (31) predicts an inverse coherent Cerenkov effect with $E_z \sim NZ$. For $\ell \gtrsim \lambda_d$, this Cerenkov effect will no longer be completely coherent.

In comparing results (31) and (28), we note that small angle Coulomb collisions depend on impact parameters from b_{min} to λ_d, whereas Cerenkov radiation effects depend on "impact parameters" from λ_d to $\beta_z c/\omega_{pe}$. For the general case of $\ell < \lambda_d$, the relative amounts of Bohr-Bethe stopping and Cerenkov radiation will depend on the relative sizes of ℓ and the various impact parameters. For this case, we find /97/

$$E_z = NZe\omega_{pe}^2 (\beta_z c)^{-2}\{\ln[\beta_z(c/\omega_{pe})/\lambda_d] + \ln(\lambda_d/\ell) + \\ + [1.75 - 2\ln 2 + O(R_0^2/\ell^2)] + N^{-1}(3/2)(R_0^2/\ell^2)\ln(R_0/b_{min})\}, \tag{32}$$

where the four terms on the right hand side correspond to coherent Cerenkov radiation, coherent Bohr-Bethe stopping, partially coherent Bohr-Bethe stopping, and

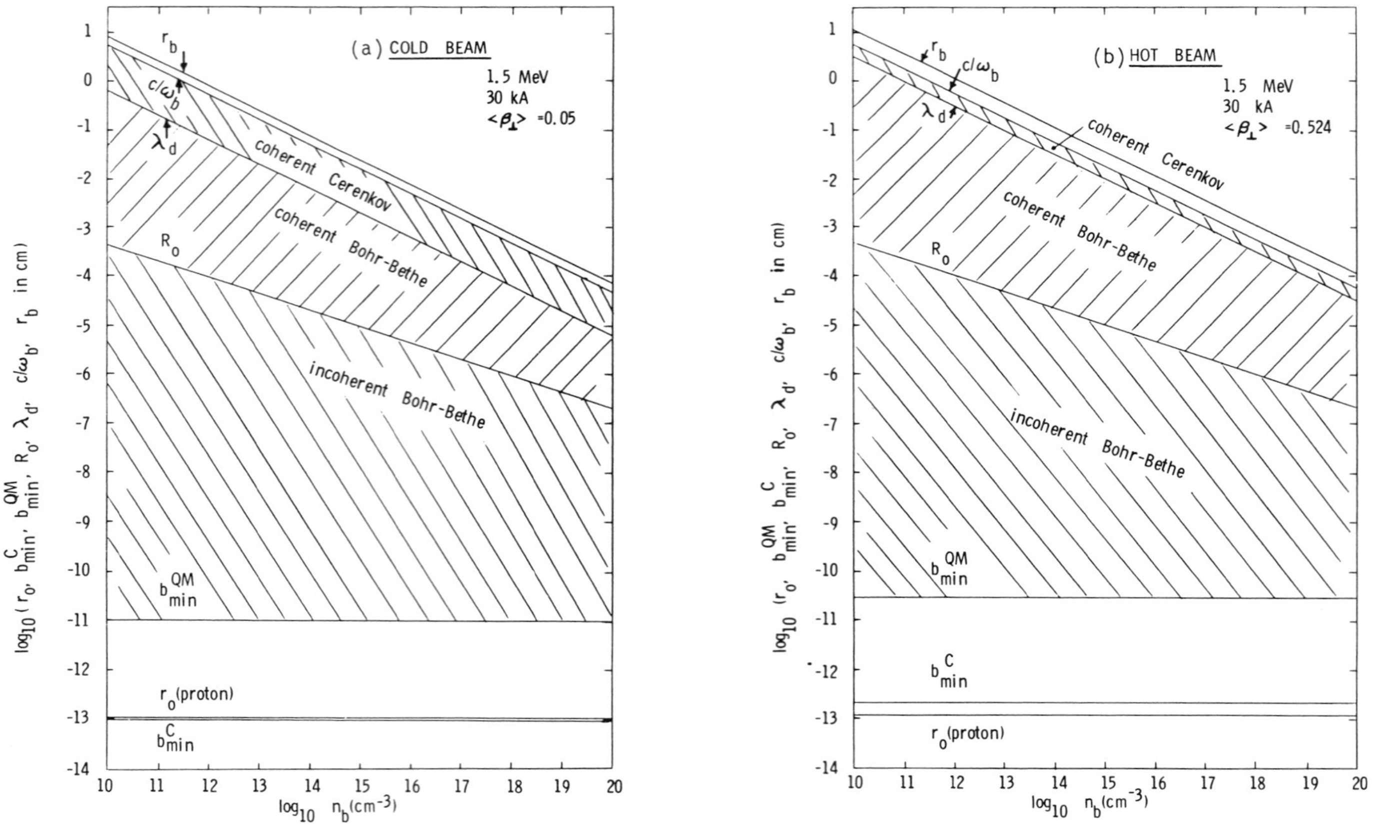

Fig. 4a and b. Impact parameters and characteristic lengths for inverse coherent drag acceleration: (a) cold beam, (b) hot beam incoherent Bohr-Bethe stopping. Here R_0 is the interparticle electron spacing ($R_0 = n_b^{-1/3}$). To calculate E_z, we imagine that the beam and an ion bunch can coexist with

$$NZ = (4/3)\pi \ell^3 a n_b \qquad a \leq 1$$
$$\ell/\lambda_d = b \qquad b \ll 1, \tag{33}$$

where $a \leq 1$ is required to insure the ion charge density is $< n_b$, and $b \ll 1$ is required to permit coherence. To demonstrate the various effects we consider an IREB with $\mathcal{E}_e = 1.5$ MeV and $I_e = 30$ kA, and either $\langle \beta_\perp \rangle = 0.05$ (cold beam) or $\langle \beta_\perp \rangle = 0.524$ (hot beam), where $\langle \beta_\perp c \rangle = v_{T_e}$. The various impact parameters are plotted in Figs. 4 a, b for various values of n_b. The various mechanisms may be operative in the shaded areas as indicated. For an ion bunch with a = 1 and b = 1/3, the coherent drag fields produced are as shown in Figs. 5 a, b. Note that in this example, coherent Cerenkov radiation dominates for the cold beam while coherent Bohr-Bethe stopping dominates for the hot beam. If we consider the extreme case of a = 1, b = 1, we find the only coherent term is that of coherent Cerenkov radiation (there is also an incoherent Bohr-Bethe term); for this case the coherent Cerenkov field would be larger than those in Fig. 5 by a factor of $(3)^3 = 27$. In all cases, however, note that the electrostatic space charge field at the beam edge drastically dominates all coherent and incoherent field effects /97/.

It should also be noted that we had to "imagine" a rigid ion bunch, and a propagating IREB in the above discussion. In general, some neutralizing uniform background ($\gamma^{-2} < f_e < 1$) is desirable, and the bunch must preserve its size (by some form of focusing) if the coherence effects are to be maintained. The bunch focusing aspect of this problem has been investigated by IRANI and ROSTOKER /123, 124/ who found that for a cyclic geometry, bunch focusing does become possible. The accelerating fields, however, do remain small.

3.4 Linear Waves and Instabilities

Linear waves in systems involving electron beams have frequently been considered as potential collective ion acceleration mechanisms. To demonstrate the existence of such waves, one must begin with a system that possesses an equilibrium configuration. Then the effects of small linear wave perturbations (of frequency ω and wavenumber k) about this equilibrium may be investigated. This leads to the derivation of a dispersion relation $\varepsilon(\omega,k) = 0$ which characterizes the propagation of the wave. For collective ion acceleration, the basic goal is to isolate and produce a *single* wave with the following properties. (1) The wave should grow (as an instability) until nonlinear effects saturate it at a value such that the resultant wave E_z is

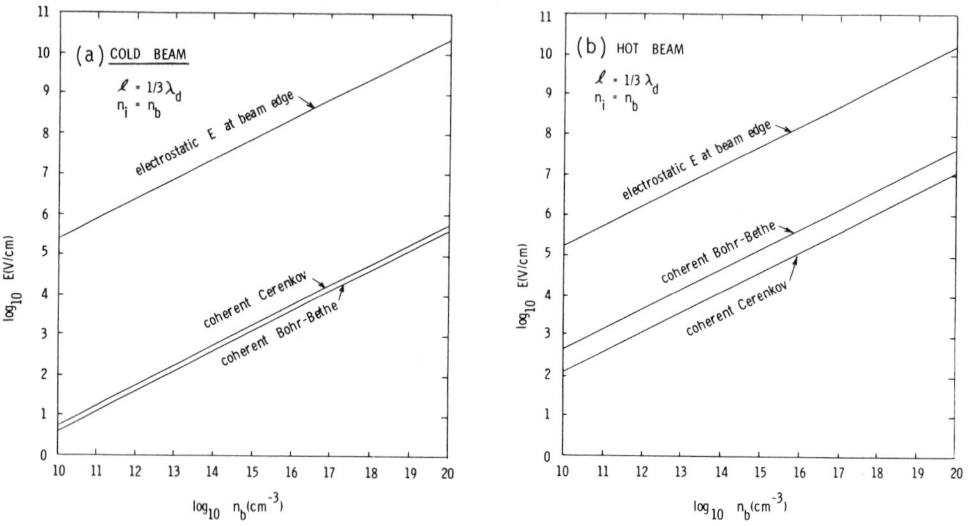

Fig. 5a and b. Coherent drag fields for (a) cold beam, and (b) hot beam

as large as possible. (2) The phase velocity (ω/k) should be amenable to axial control. (3) The wave should permit ion trapping and acceleration without destruction of the wave.

The original interest in linear wave systems as potential collective accelerators began with the work of FAINBERG /11/ in 1956, who proposed the use of waves in plasma (i.e., "plasma waveguide") for this purpose. This approach has resulted in the extensive investigation of waves in beam-plasma systems by Fainberg and his co-workers at the Physical Technical Institute in Kharkov up until the present time /125-127/. Research has been directed toward the isolation of a single wave, and the suppression of unwanted streaming instabilities that drain away the beam energy. Beam modulation or direct RF excitation has been used to improve wave excitation in low power (10^7-10^9 W) beam experiments. Ion energies of tens of keV have been achieved over distances of ~ 100 - 150 cm /125/, giving an effective $E_z \sim 10^2$ V/cm. In computer studies, phase velocity control has been attained by use of a changing plasma density. It is hoped that similar results may eventually be scaled up for use with IREB's.

Actually, beam-plasma interactions with high current IREB's have been investigated at many laboratories over the past few years for a different reason - to study plasma heating for CTR applications. Here, the electron-electron two stream instability /128, 129/, or the electron-ion instabilities of the Buneman or ion-acoustic type driven by the induced back current /130, 131/, have been used to explain the observed plasma heating /132/. These processes, which lead to a turbulent plasma state, must be avoided if a linear wave collective accelerator with a large-amplitude, narrow-wave-spectra, controllable-phase-velocity wave is to be achieved.

In the above beam-plasma systems, we note that the beam is typically well neutralized ($f_e = 1$, $0 < f_m \lesssim 1$). This means that the desired large amplitude wave must be created and grow large enough to gain back fields of the order of the large fields that were originally present with the "bare" beam (but were not in a form useful for collective ion acceleration).

An alternative approach is to consider a bare beam ($f_e = 0$, $f_m = 0$) equilibrium that is maintained by a large axial magnetic field B_z. Then both the large space charge and magnetic fields of the beam are present in the zero-order configuration. Accordingly it may be possible, by considering linear wave perturbations of this equilibrium, to create wave accelerating fields with E_z a sizeable fraction of the radial bare beam electric field. To achieve phase velocities in the range $0 \leq \omega/k < c$ with this equilibrium, two natural wave modes presently appear relevant - the slow cylotron wave, and the slow longitudinal space charge wave. The cyclotron wave is the basis for the auto-resonant accelerator concept /39, 40/, and the slow space charge wave is the basis for the converging guide accelerator /41/. Both of these modes are negative energy modes and can be made to grow automatically at the expense of the beam energy. We will discuss these modes in Sects. 5 and 6.

Another, yet different, approach is to start with a radially force neutral beam equilibrium ($f_e = \gamma_e^{-2}$, $f_m = 0$, $B_z = 0$). With this equilibrium, a beam envelope focusing instability has been derived by TSYTOVICH and KHODATAEV /133 - 136/, who proposed use of this wave to collectively accelerate ions. This concept will be discussed in Sects. 4 and 5.

A further instability that we will have occasion to consider is the electron-ion two stream instability caused by fast beam electrons (velocity v_e) streaming through a cold ion background. Here we briefly summarize its basic features. This instability was originally investigated by PIERCE /84/ and BUNEMAN /85, 86/, with subsequent analysis by BERNSTEIN and KULSRUD /137/. The relativistic case has been treated by BLUDMAN et al. /138/, YOSHIKAWA /139/, NEZLIN /55/, and BOGDANKEVICH and RUKHADZE /59/. The dispersion relation for these electron-ion, electrostatic, waves in an unbounded beam-plasma system is

$$\omega_b^2(\omega - k_z v_e)^{-2} + \omega_i^2\,\omega^{-2} = 1, \tag{34}$$

where $\omega_b^2 = 4\pi n_b e^2(\gamma_e^3 m)^{-1}$, $\omega_i^2 = 4\pi Z^2 n_i e^2/M$, and $0 < Zn_i \leq n_b$ corresponds to $0 < f_e \leq 1$. For

$$k_z v_e < (\omega_b^{2/3} + \omega_i^{2/3})^{3/2}, \tag{35}$$

unstable waves result, the maximum growth rate being /138/

$$\mathrm{Im}\{\omega\} = 3^{1/2}\,2^{-4/3}\,\omega_b^{1/3}\,\omega_i^{2/3}, \tag{36}$$

which occurs for waves with

$$\mathrm{Re}\{\omega\} = 2^{-4/3}\,\omega_b^{1/3}\,\omega_i^{2/3} \tag{37}$$

$$k_z = \omega_b/v_e, \tag{38}$$

so the phase velocity is

$$v_\phi = 2^{-4/3}[\gamma_e^3 m Z^2 n_i (M n_b)^{-1}]^{1/3}\, v_e. \tag{39}$$

For a bounded beam of radius r_b inside a conducting cylinder of radius R, a threshold current $I_{e,i}$ for this instability may be derived, corresponding to (35). The analytic results in the limit $\ln(R/r_b) \gg 1$ /55, 59, 139/, and in the limit $R = r_b$ /139/, and the numerical results for the general case /139/, lead us to propose here the useful interpolation formula

$$I_{e,i} = \beta_e^3\gamma_e^3(mc^3/e)[1 + 2\ln(R/r_b)]^{-1}\{1 + [\gamma_e^2 m Z^2 n_i (n_b M)^{-1}]^{1/3}\}^{-3}. \tag{40}$$

This result is correct in the limit $\ln(R/r_b) \gg 1$, but is about 30% too small in the limit $R = r_b$ according to Yoshikawa's numerical results /139/. For reference, we note using (6) and (40)

$$I_{e,i}/I_\ell = \gamma_e(\gamma_e + 1)\{1 + [\gamma_e^3 m Z^2 n_i (n_b M)^{-1}]^{1/3}\}^{-3}. \tag{41}$$

For $\mathscr{E}_e$ = 1 MeV ($\gamma_e \approx 3$), this gives $I_{e,i} \approx 12\, I_\ell$.

We conclude this section with a comment relevant to all linear wave schemes. For collective ion acceleration, one would like to have the maximum electric field $E_z = 2\pi\phi/\lambda_z$ to be as large as possible (where ϕ represents the amplitude of the wave potential variation, and $\lambda_z = 2\pi/k_z$ is the axial wavelength). At the same time it would be desirable to maintain this E_z as the wave accelerates. However, for ω fixed, increasing the phase velocity ω/k_z means increasing λ_z. Thus to accelerate a linear wave from low phase velocity to high phase velocity, λ_z must be made to increase, and therefore the effective accelerating field of the wave ($E_z \sim \lambda_z^{-1}$) must necessarily decrease. This intrinsic drawback seems to be a fundamental property of all linear wave collective accelerator concepts.

3.5 Nonlinear Waves and Solitons

Instead of considering small amplitude waves on an initially stable, equilibrium, linear beam system, it is also possible to imagine strongly developed nonlinear waves. These waves may be periodic (producing a wavetrain of nonlinear pulses) or they may produce a singular propagating disturbance (i.e., a soliton). A soliton can be produced by using nonlinear effects to counteract dispersive effects; the result is an isolated pulse-like wave that propagates with constant velocity /140/. For collective ion acceleration, the initial theoretical problems are to (1) demonstrate that a nonlinear wave exists, and (2) show that it is stable, even in the presence of trapped ions. If these problems can be solved for a realistic (2-D, 3-D) experimental configuration, then one must investigate (3) how to create the nonlinear wave, (4) how to load ions into it, and (5) how to vary the phase velocity. Present research has been directed primarily toward goals (1) and (2), although in regard to (3) it has been suggested that certain linear instabilities may grow and saturate into a desired nonlinear wave mode /141, 142/.

One example of a periodic nonlinear wave is that which has been investigated by TSYTOVICH et al. /25, 28, 141 - 143/. Here a 1-D relativistic electron fluid is considered streaming through a uniform stationary ion background. A nonlinear wave with phase velocity v_ϕ is assumed to exist, and the problem is therefore investigated in a reference frame moving with velocity v_ϕ. In this frame, the wave is stationary, and the motion of the relativistic electron fluid is described by the coupled nonlinear equations,

$$\partial\gamma/\partial\tilde{\zeta} + \gamma(\gamma^2 - 1)^{-1/2} \, \partial\gamma/\partial\tilde{\tau} = \tilde{\varepsilon} \tag{42}$$

$$\partial\tilde{\varepsilon}/\partial\tilde{\zeta} + \gamma(\gamma^2 - 1)^{-1/2} \, \partial\tilde{\varepsilon}/\partial\tilde{\tau} = \gamma_\phi - \gamma(\gamma^2 - 1)^{-1/2} (\gamma_\phi^2 - 1)^{-1/2} , \tag{43}$$

where γ is the relativistic factor for the electrons *relative to the wave frame*, $\gamma_\phi = [1 - (v_\phi/c)^2]^{-1/2}$, $\tilde{\zeta} = z(c/\omega_{pe})^{-1}$, $\tilde{\tau} = t\omega_{pe}$, $\tilde{\varepsilon} = Ee(mc\omega_{pe})^{-1}$, $\omega_{pe} = [4\pi n_0 e^2/m]^{1/2}$, and n_0 is the ion number density in the laboratory frame. By analysis of (42) and (43), it is found /142/ that for ultrarelativistic waves ($\gamma_\phi \gg 1$),

$$\gamma_{max} - \gamma_{min} \approx \gamma_{max} \approx 2\gamma_\phi^2 \tag{44}$$

$$\phi_{max} \approx 2\gamma_\phi^2 \, mc^2/e, \tag{45}$$

where γ_{max} (γ_{min}) is the maximum (minimum) value of γ in the wave frame, and ϕ_{max} is the maximum potential well depth in the wave frame. These maximum values correspond to wave amplitudes close to the point of wave braking. Use of the wave potential well represented by (45) for collective acceleration of ions has been suggested /25, 28, 141, 142/.

The result (45) is interesting as it suggests a rapid increase in ϕ as γ_ϕ increases. However, it should be noted that the γ_{max} value in (45) used to create this well is measured in the wave frame. By a Lorentz transformation to the laboratory frame, it can be shown that for $\gamma_\phi \gg 1$,

$$\gamma_{\ell ab} \approx 4\,\gamma_\phi^3 . \qquad (46)$$

This means that extremely energetic electrons are required to make the well depth given by (45). The half period of the wave has length ζ_p (scaled to c/ω_{pe}), which using Eq. (2.29) of Ref. /142/ gives for $\gamma_\phi \gg 1$,

$$\zeta_p \approx (8/\sqrt{2})\gamma_\phi^2 . \qquad (47)$$

Thus,

$$E^{max} \sim \phi^{max}/\zeta_p \sim \text{CONST} \qquad (48)$$

as γ_ϕ increases, and $\varepsilon \equiv \tilde{\varepsilon}/\zeta_p \sim \text{CONST}/\zeta_p \sim \gamma_\phi^{-2}$, as is shown by the numerical results presented in Fig. 3 of Ref. /142/. This means that the maximum electric field for this wave has a value of order $(mc^2/e)/(c/\omega_{pe})$ even for $\gamma_\phi \gg 1$, a result which is expected from simple net space charge considerations.

The stability of this 1-D nonlinear wave has been investigated numerically, and the wave has been found to be quite stable under a variety of perturbations /142/. It was also found that multistreaming instabilities were less likely to develop as γ_ϕ increases.

A number of other nonlinear wave investigations have also been performed with the intent of using these waves for particle acceleration /144 - 148/. While these theoretical investigations /141 - 148/ are interesting, it is clear that much remains to be done before nonlinear waves can be convincingly demonstrated and controlled in the laboratory.

3.6 Stochastic Acceleration

Stochastic acceleration here concerns the statistical acceleration of particles by a random, fluctuating, plasma wave spectrum; this mechanism has been investigated by TSYTOVICH et al. /149, 150/. A turbulent wave spectrum might be created by a beam-plasma instability. As an example, we consider the statistical acceleration of an ion in a longitudinal plasma wave spectrum in the presence of collisions. For the nonrelativistic case, the energy gain rate for an ion of charge Ze, mass M, and velocity v_z is /149/

$$d\mathscr{E}_i/dz = + 4\pi n_p e^4 Z^2 \,(mv_z^2)^{-1}\, [T_{eff}(v_z)](Mv_z^2)^{-1} , \qquad (49)$$

where n_p is the plasma density, T_{eff} is the effective plasmon temperature,

$$T_{eff}(v_z) = \hbar\omega_{pe}N\Big|_{v_\phi = v_z},$$

and N is the plasmon number for waves with phase velocity equal to v_z. The energy loss rate due to collisions is given by

$$d\mathcal{E}_i/dz = -4\pi n_p e^4 Z^2 (mv_z^2)^{-1} \ln(b_{max}/b_{min}) \tag{50}$$

as discussed in Sect. 3.3. T_{eff} may be taken to be constant for $v_z \gtrsim v_e$, and decreasing for $0 < v_z < v_e$ due to Landau damping /149/. Then assuming that the turbulent wave spectrum lasts long enough, an ion should gain energy until the energy gain rate is matched by the energy loss rate. In this case, the equilibrium average ion energy would be

$$\mathcal{E}_i = Mv_z^2/2 = T_{eff}\,[\ln(b_{max}/b_{min})]^{-1}\,(2^{-1}) \tag{51}$$

which means ion energies of order T_{eff} are ultimately possible. Thus, for this particular mechanism to accelerate ions requires an initial ion velocity $v_z \gtrsim v_e$ (whereas v_z is usually $\ll v_e$), and

$$T_{eff} > 2(M/m)\,T_e \ln(b_{max}/b_{min}), \tag{52}$$

where $T_e = mv_e^2/2$ (in energy units).

Other turbulent wave spectra may also be present depending on the plasma and the means of wave excitation /149/. Here we note that the statistical acceleration mechanisms all produce relatively small accelerating fields, compared to the space charge field of the initial electron beam. For example, the electric field due to turbulent acceleration [as contained in (49)] when compared to the electric field due to inverse Cerenkov acceleration [as contained in (31b)] for a *single ion* gives

$$[E_z(\text{turbulent})]/[E_z(\text{Cerenkov})] = [T_{eff}(v_z)]\{Mv_z^2 \ln[\beta_z(c/\omega_{pe})/\lambda_d]\}^{-1} \tag{53}$$

which is of order unity. However for a cluster particle of charge NZ we find inverse *coherent* Cerenkov radiation raises the Cerenkov field by a factor of NZ [see (32)], whereas the turbulent field would typically not be similarly enhanced. Thus, for collective ion acceleration, the ordering of field strengths is apparently

$$E_z(\text{turbulent}) \ll E_z(\text{inverse coherent Cerenkov}) \ll E_z(\text{space charge}). \tag{54}$$

The stochastic acceleration mechanism has been invoked to explain the presence of ~ keV ions in some beam-plasma experiments /149, 151, 152/. However, here it is apparent that stronger acceleration mechanisms exist, and may be used, for collective ion acceleration with linear electron beams.

3.7 Impact Acceleration

Impact acceleration, as proposed by VEKSLER /7 - 9/ involves a relativistic elastic collision between a fast ($\gamma \gg 1$) heavy particle bunch of mass M_1 (e.g., a bunch composed of N_1 electrons of mass m so $M_1 = N_1 m$) and an initially stationary light bunch of mass M_2 (e.g., a bunch composed of N_2 ions of mass M so $M_2 = N_2 M$). If certain criteria are met for a head-on collision of these bunches, then every ion in the second bunch will receive the enormous energy /7, 8, 37, 153, 154/

$$\mathscr{E}_i \approx 2\gamma^2 Mc^2 . \tag{55}$$

VEKSLER believed impact acceleration would be the only method whereby ultrahigh ion energies ($\gtrsim 10^{12}$ eV) might be achieved /9, 153/.

By considering a head-on collision, we find using relativistic dynamics that following the collision the ion bunch will receive the relativistic factor γ_2 where

$$\gamma_2 M_2 = M_2 + \gamma M_1 - \gamma M_1 \left[1 + (M_2/M_1)^2(\gamma_2^2 - 1)\gamma^{-2} - 2(M_2/M_1)(\gamma^2 - 1)^{1/2} \cdot \cdot (\gamma_2^2 - 1)^{1/2} \gamma^{-2}\right]^{1/2} \tag{56}$$

which is exact. In the limit $\gamma_2 M_2 \ll \gamma M_1$, (56) reduces to $\gamma_2 \approx 2\gamma^2$, giving (55). This mechanism requires that the bunches remain intact during the collision, and that /7, 8, 37/

$$\gamma N_2 M << N_1 m \tag{57a}$$

$$p^* \le N_1 N_2 e^2 (M_0 c^2)^{-1} \tag{57b}$$

$$\ell \le p^* , \tag{57c}$$

where p^* is the impact parameter, ℓ is the characteristic bunch dimension, and M_0 is the reduced mass given by $M_0 \equiv (MmN_1 N_2 \gamma) \cdot (mN_1 + \gamma MN_2)^{-1}$. Requirements (57b) and (57c) are that p^* be small enough to effectively produce a head-on collision, and that the bunch size be smaller than this p^*. If we consider the projectile bunch $N_1 m$ to be a spherical segment (of radius ℓ) of an intense beam, then $N_1 = (4/3)\pi\ell^3 n_b$, and the peak current associated with this bunch is $I=\pi \ell^2 n_b e \beta_e c$ or $I=(3/4)\beta_e c e N_1 \ell^{-1}$. But (57 a, b, c) combine to require $\ell < N_1 e^2 (\gamma_e Mc^2)^{-1}$, or equivalently /37/

$$I \gg (3/4)\ \beta_e \gamma_e Mc^3/e \tag{58}$$

which in practical units is

$$I \gg \beta_e \gamma_e (23.4)\ \text{mega-amperes} \tag{59}$$

for M equal to the proton rest mass. Note that this is roughly equivalent to $I \gg (M/m) I_A$, where $I_A = \beta_e \gamma_e mc^3/e$ is the usual Alfven-Lawson magnetic stopping current.

The electron beam current required by (59) is clearly beyond the realm of present intense beam technology. The required beam has an impedance $\mathcal{Z} = [\mathcal{E}_e/(eI_e)] \ll m/M \approx 0.0005\ \Omega$, whereas present high-$\gamma$ IREB beams have $\mathcal{Z} \gtrsim 1\ \Omega$. Thus, with existing pulsed power technology, collective acceleration using impact acceleration is not an allowed possibility.

4. Naturally Occurring Collective Ion Acceleration Processes

There are now several experimental configurations involving IREB's that produce collectively accelerated ions by naturally occurring processes. These configurations are of interest because they produce large accelerating fields ($\sim 10^6$ V/cm) over small distances ($\sim$ 10 cm) that result in the acceleration of substantial quantities of ions. These naturally occurring processes fall into three categories:

(1) Collective acceleration during IREB injection into neutral gas.

(2) Collective acceleration in modified drifting beam geometries (vacuum drift tube, dielectric drift tube, cusp magnetic field, etc.).

(3) Collective ion acceleration in IREB diodes.

Category (1) has been the most extensively investigated to date. One reason for this is that an understanding of this process should suggest how to devise a scalable collective ion accelerator. There is now a theoretical explanation for this process that accounts for its many parametric dependences. Here we will discuss all three catagories, with extended emphasis on category (1). The natural occurrence of collective ion acceleration in all three catagories is, of course, very encouraging to the general concept of linear beam collective acceleration.

4.1 Collective Ion Acceleration During Intense Electron Beam Injection into Neutral Gas

The basic experimental configuration for this mode of collective ion acceleration is shown in Fig. 6. In the basic experiment, an IREB is formed in the diode section, and injected through a thin metallic foil anode into a metallic drift tube filled with neutral gas at a pressure of the order of 0.1 Torr H_2. Ions from the gas are observed to be accelerated in the *same* direction that the beam propagates, and attain energies *higher* than the injected electron energy. Ion energy multiplication factors up to 10 - 15 have been reported. The ion pulse length is typically much shorter than the injected electron pulse length. The basic acceleration process depends on time

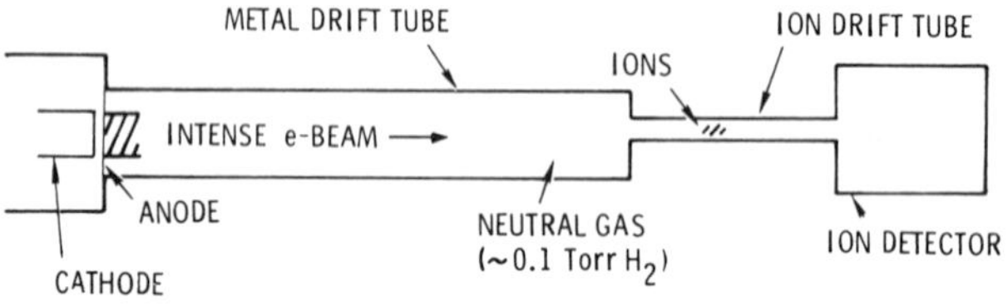

Fig. 6. Basic experimental configuration for IREB injection into neutral gas that produces collectively accelerated ions

and at least eleven other independent parameters. These include the IREB (peak electron energy $\mathcal{E}_e$, peak current I_e, voltage risetime t_v, current risetime t_r, pulse length t_b, beam radius r_b) metallic drift tube (radius R, length L), and neutral gas (pressure p, ion mass M, charge state Z). In addition, the diode parameters (cathode radius r_c, anode-cathode distance d) can affect the injected beam parameters' time dependence (e.g., r_b may decrease in time).

This acceleration process was first observed by GRAYBILL and UGLUM /14, 15/ in 1968. Since that time, it has been investigated in further detail at seven more laboratories /155 - 195/. Here we present a comprehensive summary of the large body of data that now exists. A total of thirteen theories /16 - 18, 133 - 136, 196 - 222/ have been proposed to explain this acceleration process; in reviewing these here, we will demonstrate that the theory of OLSON /16 - 18/ is uniquely capable of accounting for the acceleration process with its many parametric dependences. We will also summarize the results of four computer simulation studies /19, 223 - 228/ of this process. Detailed comparisons of experiments, theories, and simulations will then be given. Lastly, the natural limitations of the process will be discussed, as will be the results of attempts to overcome them.

4.1.1 Experiments

A large amount of data now exists /14, 15, 155 - 195/ which characterizes the basic collective acceleration process for IREB injection into neutral gas. For proton acceleration, this data is summarized in Table 4, with ordering according to the injected electron energy. Shown are the basic independent parameters ($\mathscr{E}_e$, I_e, t_r, t_v, t_b, r_c, d, R, L, type of gas, p) which control the acceleration process. Next, seven relevant derived parameters are given (I_ℓ, I_A, I_c, I_e/I_ℓ, ν_e/γ_e, I_e/I_c, and the beam impedance $\mathscr{Z} \equiv \mathscr{E}_e I_e^{-1} e^{-1}$). In calculating I_ℓ, (6) was used and r_b was taken to be equal to r_c, since r_b was usually not measured in the experiments. The ions reported are then summarized (ion type, ion energy $\mathscr{E}_i$, ion number N, ion pulse length τ, and ion current I_i). The acceleration efficiency $\eta \equiv (N\mathscr{E}_i)(\mathscr{E}_e I_e t_b/e)^{-1}$ is also given. Except for the derived quantities, any quantity calculated here and not contained in the original publications is given in parenthesis. The ion energy distribution function $f(\mathscr{E}_i/\mathscr{E}_e)$ is plotted, which shows the relative distribution of ion energies scaled to the beam energy. The number of ion pulses, whether or not IREB pinching was reported, and the authors of the various experiments, are then given. Note that there are 30 independent data sets spanning a large range of all of the basic parameters. Also note that in some experiments there is a lack of basic information in the publications so that the table can not be completely filled in. Nonetheless, Table 4 contains an extensive amount of data that describes the basic acceleration process for protons. With minor added complications due to the range of charge states possible, the same acceleration process occurs for other ions; the basic ion pulse characteristics for these cases (D^+, $He^{+1,2}$, N^{+3-7}, A^{+6}) have been summarized earlier /229/. We will use all of this data and more (e.g., data of $\mathscr{E}_i$ as a function of p) in our comparisons with theories in Sect. 4.1.4.

Considering a typical case in Table 4, note that an IREB with $\mathscr{E}_e \sim 1$ MeV and $I_e \sim 30$ kA when injected into ~ 0.1 - 0.3 Torr H_2 can produce an intense ion bunch with energy $\mathscr{E}_i \sim 3$ - 5 MeV and current ~ 100 - 200 A in a 3 - 5 nsec pulse. To complement the data summary in Table 4, we now present some brief comments on the main experiments performed at the various laboratories where the acceleration process has been observed.

GRAYBILL and UGLUM /14, 15, 155/ at Ion Physics Corporation (IPC) discovered the process in 1968 with a $\nu_e/\gamma_e < 1$ beam. Ion diagnostics included TOF (time of flight) current screens, magnetic analysis, and neutron production by the $B_e(p,n)$ reaction (for protons). Detailed ion pulse shapes were given for IREB injection into H_2, D_2, He, and N_2. The data showed $\mathscr{E}_i \sim Z$. The data was interpreted to suggest $\mathscr{E}_i \sim I_e^2$, that the beam waited at the anode until the force neutralization time τ_{FN} (the time at which $f_e = \gamma_e^{-2}$), and that $\beta_i \ll \beta_f$ so the acceleration must occur behind the beam front (here $\beta_i c$ is the final ion velocity and $\beta_f c$ is the beam front velocity). In subsequent investigations, GRAYBILL /156 - 158/ measured the effect of varying L, and also found that $\mathscr{E}_i$ had a pressure dependence.

RANDER et al. /160 - 166/ at Physics International Company (PI) first observed the acceleration process in 1969 with a $\nu_e/\gamma_e > 1$ beam. Ion diagnostics used by this group included TOF screens and nuclear emulsion plates. With three Rogowski coils

Table 4. Collective acceleration of protons by IREB injection into neutral gas

DATA SET	IREB						
	$\mathcal{E}_e$(MeV)	I_e(kA)	t_r(nsec)	t_v(nsec)	t_b(nsec)	r_c(cm)	$\mathcal{E}_e$(J)
1	0.2	200			80	3.81	3,200
2	0.25	200				3.8?	
3	0.5	160	15		50	3.8?	4,000
4	0.5-1.1	20-48		30	55	6	550-2,900
5	0.5	> 40		30	55	6	~ 1,100
6	0.6	7			3		13
7	0.6	150-160			50	3.8-2.54[a]	4,650
8	0.65	15-20	15		50		570
9	0.65	145	10			3.8?	
10	0.75	100	10		90	1.27,2.54,3.81	6,750
11	1.0	110			50	3.8	5,500
12	1	115	10	15	90	2.54,1.27	10,350
13	1	115	10	15	90	2.54	10,350
14	1-1.4	160			80	3.81?	12,800
15	0.45-1.35	32-150	8		55	1.27,2.54,3.81	790-11,140
16	1.3	35	10	12	40	1.25	1,820
17	1.3	50	35		50	2.5-2[a]	3,250
18	1.3	50	35		50	2.5-2[a]	3,250
19	1.5	20			20		600
20	1.7	30	10		50	1.25	2,550
21	1.8	75	60	70	90	0.5	12,150
22	2	6			20		240
23	2-2.3	15-20	6	10	45	0.15,0.64,1.25[a]	~ 1,600
24	2-2.3	15-20	6	10	45	0.15,0.64,1.25[a]	~ 1,600
25	2	15	6		50	0.7[a]	1,500
26	2	80	20		125	5.1[a]	20,000
27	3	55,80	20-25		125	5.1,2.55[a]	20,600-30,000
28	4.5	40			65		11,700
29	5	38-40	20		125	0.65,2.55[a]	25,600
30	8	230	70		200	15	368,000

[a] Hollow beam

Table 4 (part 2)

DATA SET	DIODE	DRIFT				
	d(cm)	R(cm)	L(cm)	gas	p(torr)	B_z(kG)
1		1.59	> 40	H_2	0.200	0
2		3.8		H_2	0.2	0
3		3.8	58	H_2	0.03-0.96	0
4		7,9	>60,varied	H_2		0
5		9		H_2	0.026	0
6			1-30	D_2	0.001-15	0
7	0.60	3.8	58	air,He		0
8		5	varied	H_2	0.005-0.4	0
9		3.8?		H_e	0.200	0
10	0.95	3.81	82	H_2	0.25-0.7	0
11		3.8	58	H_2	0.03-0.96	0
12	0.95	3.81	73	H_2	0.15-0.65	0
13	0.95	3.81	73	H_2	~ 0.3	0.1-10
14		1.59	> 40	H_2	0.200	0
15		3.81	82	H_2	0.3, 0.55	0
16	2.0	7.6	50, varied	H_2	0.05-0.30	0
17		10.5	> 100	H_2	0.300	0
18		10.5	> 100	H_2	0.300	0-3
19		15.9	20	H_2	0.005-0.015	0
20		7.6	20-100	H_2	0.005-0.5	0
21	0.1-1.0	2.5	> 70	H_2	0.150	0
22				D_2	0.001-15	0
23	3.18,2.54,1.43	2.5,5.0,12.7	63	D_2	0.05-0.6	0
24	3.18,2.54,1.43	2.5,5.0,12.7	63	D_2	0.05-0.6	0
25	2	7.5	50	D_2	0.03-0.5	7.8
26	5	32	117	H_2	0.02-0.30	0
27	3.1,4.8-5.1	32	117	H_2	0.02-0.30	0
28				H_2	0.08-0.16	0
29	6-6.7,15	32	117	He	0.02-0.30	0
30		60	120	He	0.1	0

Table 4 (part 3)

DATA SET	DERIVED PARAMETERS						
	I_ℓ(kA)	I_A(kA)	I_c(kA)	I_e/I_ℓ	ν_e/γ_e	I_e/I_c	$\mathcal{F}(\Omega)$
1	4.6[b]	16.4		43.5	12.2		1.0
2	6.2	18.8		32.3	10.6		1.3
3	14.4	29.1		11.1	5.5		3.1
4	8.0-27.3	29-52		0.7-6.0	0.4-1.7		10.4-55
5	8.0	29.1		> 5.0	> 1.4		< 12.5
6	17.6[b]	32.7		0.4	0.2		86
7	17.6	32.7	104	8.8	4.7	1.5	3.9
8	4	34.6		4	0.5		37
9	19.4[b]	34.6		7.5	4.2		4.5
10	7.1,12.7, 22.9	38.4	25.7-77	4.4,7.9,14.1	2.6	1.3-3.9	7.5
11	31.4	47.4		3.5	2.3		9.1
12	9.8,17.3	47.4	31.7-63.4	6.6,11.7	2.4	1.8-3.6	8.7
13	17.3	47.4	63.4	6.6	2.4	1.8	8.7
14	32-44.5[b]	48-61		3.6-5.0	2.6-3.3		6.3-8.8
15	4.1-42.6	27.5-59		~ 1-37	0.5-5.5		3-42
16	9.0	57.7		3.9	0.6		37.1
17	11.6	57.7		4.3	0.9		26
18	11.6	57.7		4.3	0.9		26
19	48.4[b]	64.8		0.4	0.3		75
20	11.9	71.6		2.5	0.4		57
21	13.8	74.9		5.4	1.0		24
22	65.1[b]	81.7		0.092	0.073		333
23	varied	82-92		0.6-1	~ 0.2		100-153
24	varied	82-92		1-2.5	~ 0.2		100-153
25	11.3	81.7		1.3	0.2		133
26	13.9	81.7		5.8	1.0		25
27	16.4 21.1	116		2.6,4.9	0.5,0.7		55,38
28	149[b]	166		0.3	0.2		113
29	18.9,27.4	183		1.4,2.4	~ 0.2		125
30	70.5	283		3.3	0.8		34.8

[b] For $R = r_b$ (because $R < r_b$, or R & r_b not given)

Table 4 (part 4)

DATA SET	IONS Type	$\mathcal{E}_i$(MeV)	N	τ(nsec)	I_i(A)	$\mathcal{E}_i$(J)	$\eta = \frac{\mathcal{E}_i(J)}{\mathcal{E}_e(J)}$
1	p	0.8	10^{13}-10^{14}	5-8	(200-3,300)	(1.3-13)	(4 x 10^{-4} - 4 x 10^{-3})
2	p	0.1-2.1	10^{12}-10^{14}			(0.3-30)	
3	p	~ 1	10^{13}	3-5	(~ 400)	(1.6)	(4 x 10^{-4})
4	ions observed only above threshold at $I_e \approx I_\ell$						
5	p	2.5-3.5?		~ 6			
6	no ions						
7	?		10^{11}-10^{12}	20	(0.8-8)		
8	p	1-3,8	10^{12}	5-10	(16-32)	(0.3)	2 x 10^{-4}
9	p	~ 1.8	10^{12}-10^{14}			(0.3-30)	
10	p	3-7	10^{12}-10^{13}			(0.5-11)	(7 x 10^{-5} - 2 x 10^{-3})
11	p	1.8-2.2	10^{13}	3-5	(400)	(~0.3)	(5 x 10^{-4})
12	p	2-12	3x10^{10} c	~ 4^c	~1.4^c	(0.06)	(6 x 10^{-6})
13	N reduced 10^2 - 10^3 for 0.5 kG: no ions for B > 0.5 kG						
14	p	~ 1.5	10^{13}-10^{14}	5-8	(200-3,200)	(2.4-24)	(2 x 10^{-4} - 2 x 10^{-3})
15	p	2-14					
16	p	4.8	2x10^{12}	3	100	(1.5)	(8 x 10^{-4})
17	p	<4.5	~10^{13}	~5	(320)	(~ 1.6)	(5 x 10^{-4})
18	N reduced to minimum at 0.8 kG: ions peak ~ 2 kG due to cusp						
19	no ions						
20	p	4.5-6.5	2.8x10^{12}	~ 3	100-200	(3)	(1 x 10^{-3})
21	p	1-5	10^{11}-10^{14}			(0.05-50)	
22	no ions						
23	no ions below threshold at $I_e \approx I_\ell$						
24	p(in H_2)	2-5				~ 0.4	(3 x 10^{-4})
25	N reduced ~ 10^2 and ion energy down for 7.8 kG						
26	p	4-14					
27	p	3-8					
28	no ions						
29	p	4-16.5	2x10^{8d}				
30	H_e	18-40	10^{10}			(0.03)	(8 x 10^{-8})

[c]At p = 0.4 Torr

[d]With $\mathcal{E}_i$ > 16.5 MeV

Table 4 (part 5)

DATA SET	IONS f($\mathscr{E}_i/\mathscr{E}_e$)	IONS # Pulses	IREB Pinching?	Authors	References
1	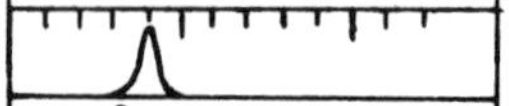	2		Rander et al.	163,165
2		2		Rander et al.	161,165
3				Rander	164,165
4		1		Bystritskii et al.	193,194
5		2	yes	Bystritskii et al.	193,194
6				VanDevender	174
7		$> 1^e$		Yonas et al.	160
8		1		Kolomensky et al.	29, 186-190
9		1		Rander et al.	161
10		sometimes >1		Drickey et al.	168
11		"initial pulse"		Rander	161,164
12				Ecker et al.	167,168,173
13				Ecker et al.	167
14		generally 2		Rander et al.	163,166
15		1		Ecker et al.	169-172
16		1		Graybill,Uglum	14, 15
17			stationary	Roberson et al.	191
18			stationary	Roberson et al.	191
19				Rander et al.	165
20		1		Graybill	156-158
21			stationary	Kuswa et al.	175-177
22				VanDevender	174
23				Straw, Miller	178-182
24				Straw, Miller	178-182
25				Straw, Miller	185
26				Miller, Straw	181, 183
27				Miller, Straw	179, 183
28				Rander et al.	165
29				Miller, Straw	179, 183
30				Graybill, Young	195

[e]Species unknown

at various distances from the anode plane, β_f was measured. The data indicated $\beta_f \sim \beta_i$ and $\mathscr{E}_i \sim Z$. The data was interpreted to suggest that the beam stopped at the anode foil until the time τ_{FN}, and then began propagation. In some instances, multiple pulses were reported. Later, ECKER et al. /167, 168/ reported a strong dependence of β_i and β_f on p, and high ion energies ($\mathscr{E}_i/\mathscr{E}_e$ up to 10 - 15). The data suggested that the ions moved with the front, but some distance behind it. Application of an axial magnetic field (B_z) appeared to stop the acceleration process. More recently, ECKER and PUTNAM /169 - 172/ have investigated the effect of varying the derived parameter $\mathscr{F}$. Also for high energy ions ($\mathscr{E}_i/\mathscr{E}_e \sim 10$), ECKER et al. /173/ have now found that the number of accelerated ions is a small fraction ($\sim 10^{-2}$) of the usual number of ions in the "low" energy case ($\mathscr{E}_i/\mathscr{E}_e \sim 3$).

VANDEVENDER /174/ at the Lawrence Livermore Laboratory in 1970 used two relatively low current, short pulse IREB's and did not find any ions, even though sensitive ion diagnostics were employed, and a wide range of pressures (0.001 - 15 Torr) was investigated.

KUSWA et al. /175-177/ at Sandia Laboratories observed ions only at low pressures ($\lesssim$ 0.15 Torr). Diagnostics employed included a Thompson parabola mass spectrometer, and a series of eight Rogowski coils to measure β_f. The presence of radially accelerated ions was reported.

MILLER and STRAW /178 - 185/ at the Air Force Weapons Laboratory (AFWL) began a series of detailed experiments in 1974. They detected, for the first time, the threshold for this process which was predicted by the Olson theory (see Sect. 4.1.2 and Sect. 4.1.4). Using nuclear activation analysis, the ion energy spectra in both the radial and axial directions was mapped as a function of z. The effects of varying L were reported, and in other experiments, a high energy ion tail ($\mathscr{E}_i/\mathscr{E}_e$ up to 7) was observed. With an applied B_z, accelerated ions were still found, but at lower energy ($\mathscr{E}_i \lesssim \mathscr{E}_e$) and much lower intensity (N down 2 to 3 orders of magnitude) than for the $B_z = 0$ case.

KOLOMENSKII et al. /186 - 190/ at the Lebedev Institute in Moscow reported the first Soviet experiments for IREB injection into neutral gas in 1974. They observed the process with a $\nu_e/\gamma_e < 1$ beam, and investigated the effects of varying L and p. Ion diagnostics included nuclear activation, magnetic analysis, use of Faraday cups, and neutron production. High energy ions ($\mathscr{E}_i/\mathscr{E}_e$ up to 12) were observed, and measurements of β_f were reported.

ROBERSON et al. /191, 192/ at the University of California at Irvine (UCI) observed the basic acceleration process in 1976 with a hollow beam. Severe modifications of the process (sudden change in R, gradient in p, and a B_z cusp field) were also examined; we will discuss these briefly in Sect. 4.2.

BYSTRITSKII et al. /193, 194/ at the Tomsk Polytechnic Institute reported a series of experiments in 1976. By varying parameters, they were able to also verify the theoretical threshold for the acceleration process. The effects of varying L and p were also observed. Multiple pulses were seen sometimes, and a threshold for multiple pulses was reported for the first time.

GRAYBILL and YOUNG /195/ at Harry Diamond Laboratories in 1976 observed the acceleration process with the very large IREB machine Aurora ($\mathcal{E}_e$ = 8 MeV, I_e = 230 kA). In this experiment, nuclear activation analysis was used to deduce the He ion energies in the range $\mathcal{E}_i \approx 18 - 40$ MeV.

4.1.2 Theories

Many theories have been advanced to explain the observed acceleration process /16 - 18, 133-136, 196 - 222/. It is clear from the above data summary, that a theory must account for a diversity of features and parametric dependences. At one time it was thought that there may be more than one acceleration mechanism, and that a particular mechanism might operate only in a particular regime. The earliest theories appeared in 1969 - 1972 /196 - 209/. It has been shown that these early theories could not account for the basic acceleration process /91/, yet it was concluded that the basic process must involve a net space charge mechanism. In 1973, the 2-D net space charge theory of OLSON /16 - 18/ first appeared. This theory demonstrated that a single mechanism could be used to explain the observed acceleration in all cases. It was believed that this theory correctly gave the lowest order description of the observed collective acceleration process; we will show in Sect. 4.1.4 that this is indeed true, even with the addition of several years of new data. However, even more theories have continued to appear /133 - 136, 214 - 222/. We will show here, that with the exception of the 1-D theory of ALEXANDER et al. /214 - 218/, there are fundamental reasons why these theories cannot account for the observed acceleration process.

Early Theories

The acceleration mechanism in four of the first theories /196 - 201/ involves a 1-D ionization wave at the front of the IREB. The basic concept of ionization front motion is that the beam has a front region of axial length $\mathcal{L}$ that cannot propagate until the neutral gas in that region is sufficiently ionized. If the beam itself can sufficiently ionize the beam front region in the time τ, then the beam front will advance with the velocity

$$v = \mathcal{L}/\tau. \tag{60}$$

Ionization waves propagating with velocity (60) were first discussed by CRAVATH and LOEB /230 - 232/ in 1935 in the context of lightning discharges. In 1968, TURCOTTE and ONG /233/ first investigated ionizing waves with $\mathcal{L}$ equal to the Debye length, and τ equal to the effective electron impact ionization time. In this context, the Debye length ($\lambda_d \sim v_e/\omega_{pe}$) is roughly the distance (at the edge of a plasma) over which the electrons stop, creating a potential drop roughly equal to the electron energy ($\phi \sim \mathcal{E}_e/e$). The choices $\mathcal{L} = c/\omega_{pe}$ and τ = (electron impact ionization time) produce the characteristic 1-D ionization wave velocity v_{1-D}

$$v_{1-D} = (c/\omega_{pe})/\tau \tag{61}$$

which enters most of the 1-D ionization wave theories. We will now briefly discuss these theories, with the aid of Table 5.

First, we must note two overriding problems with all 1-D wave theories. One problem is that the front velocity v_{1-D} produces $\beta_f < 0.001$ for typical experimental parameters /211/; this is more than a factor of 100 too small to account for the data. In some of the theories, efforts were made to produce higher β_f's. The other problem is that there is no experimental evidence to suggest 1-D front motion in the first place. Such motion should produce a current front risetime $(c/\omega_{pe})/(\beta_f c) \approx 0.3$ nsec for typical parameters ($c/\omega_{pe} \approx 1$ cm, $\beta_f \approx 0.1$); the observed current risetimes are always of order ~ 10 nsec.

ROSTOKER /196, 197/ applied the 1-D ionization wave concept (61) to the collective acceleration problem in 1969. Realizing that v_{1-D} was too small to explain the data, Rostoker proposed an artificial "precursor" beam (density n_b^*) to speed up the ionization rate, and thereby speed up the front velocity. Including this effect produces a well acceleration

$$\ddot{z}_w(t) = (2c/\omega_b)(n_b^*/n_b)\{\tau_i(1 - (n_b^*/n_b)[(t/\tau_i) - 1]) + \tau_e\}^{-2}, \tag{62}$$

where ω_b is the plasma frequency corresponding to the beam density n_b, τ_i is the electron impact ionization time, and τ_e is a radial escape time ($\tau_e \approx R/c \ll \tau_i$). Result (62) predicts that the beam front suddenly jumps ahead /91/ to a velocity $(c/\omega_{pe})/\tau_e \approx c$, and that the ions slip out the well back with energy

$$\mathcal{E}_i = (n_b/n_b^*)\ Z\ \text{MeV}, \tag{63}$$

where n_b/n_b^* is a fitted parameter. It should be noted that ion slip-out is required in this model to obtain the scaling $\mathcal{E}_i \sim Z$. Also, for typical experimental parameters, both the beam front jump-ahead time, and the ion acceleration time, are subnanosecond /91/, and the final beam front current risetime $(c/\omega_{pe})/(\beta_f c) \approx 0.03$ nsec. None of these features correspond to the observed data.

UGLUM et al. /156, 198, 199/ started by assuming a propagating force neutral beam ($f_e = \gamma_e^{-2}$) exists. Then a potential well (depth ϕ, length L) was considered to accelerate due to ionization by electron avalanching, with the acceleration rate $a_0 \approx 2L\tau^{-2}$, where τ is the electron avalanche time. This model also predicts ion slip-out, with ion energy

Table 5. Ionizing wave theories

#	THEORY	$\mathscr{L}$ meaning	$\mathscr{L}$ expression	τ meaning
1	Lightning discharge	E field region length	d	electron avalanche time
2	1-D ionizing wave	Debye length	L_d ($\approx V_e/\omega_p$)	electron impact ionization time
3	1-D space charge wave	stopping length	c/ω_b	electron impact ionization time + radial escape time
		stopping length	c/ω_b	same, but with artificial precursor beam (density n_b^*)
4	1-D accelerated potential well	well length	L_o	electron avalanche time
5	1-D beam propagation in neutral gas	stopping length	c/ω_b	collisional ionization time
6	1-D ionization front	slowing down length	$(c/\omega_{pe})(2/\beta)^{1/2}(\gamma_b^{2/3}-1)^{3/2}$	collisional ionization time
7	2-D potential well in "high pressure" regime	length of region where electron impact, ion impact, and ion avalanching are important	$(R/2.4)[(\ln\alpha) + \ln(\mathscr{E}_e/\mathscr{E}_x)] + d$ $\approx 2R + (2.4)^{-1}\ln[\mathscr{E}_e(\mathrm{MeV})]$	charge neutralization time, including ion avalanching effects
8	1-D potential well	deceleration length	$d_o \approx (c/\omega_{pe})(\gamma-1)^{1/2}$	ionization time, including ion impact effect
		include effect of partial neutralization	$d = \dfrac{d_o}{[1-(n_i/n_b)]^{1/2}}$	include effects of ions accelerated past front (n_i^+)

Table 5 (continued)

#	τ expression	Comment on $\mathscr{L}/\tau$ for ion acceleration	Authors	References
1	t_e	------	Cravath, Loeb (1935) Schonland (1956) Loeb (1965)	230 231 232
2	τ_{reac}	------	Turcotte, Ong (1968)	233
3	$\tau_i + \tau_e$ $(\tau_e \ll \tau_i)$	~ 100 times too small	Rostoker (1969)	196, 197
	$\tau_i \left\{1 - \frac{n_b^*}{n_b}\left[\frac{t}{\tau_i} - 1\right]\right\} + \tau_e$	velocity jumps to ~ c (all ions slip out)		
4	τ	won't avalanche ($\lambda_{mfp} \gg L$)	Uglum et al. (1969)	198, 199
5	ν^{-1}	~ 100 times too small	Poukey, Rostoker (1971)	200
6	ν_i^{-1}	~ 100 times too small	Rosinskii et al. (1971)	201
7	$\tau_N^{e,i} \approx t_i \ln\left[\frac{\tau_e}{\tau_i} + 1\right]$	agrees with data	Olson (1973 - 1976)	16 - 18, 31, 34, 210 - 213
8	$\left[\nu_o(1 + B\frac{n_i}{n_b})\right]^{-1}$	> 10 times too small	Alexander et al. (1973 - 1976)	214 - 218
	$\frac{1 - (n_i^+/n_b)}{\nu_o\,[1 + B(n_i/n_b)]}$	can fit to some data		

$$\mathscr{E}_i = Ze\phi_0\{[(Ma_0L)/(Ze\phi)]-1\}^{-1} \tag{64}$$

as in Table 2. The initial problem with this model is that a propagating force neutral beam cannot exist for typical experimental parameters: as shown by OLSON and POUKEY /54/, this force neutral beam would set up a potential depression several times larger than the electron energy, which is impossible. Also, it has been shown that the electron avalanche process cannot occur because the mean free path between ionizing collisions (λ_{mfp}) is much too large ($\lambda_{mfp} >> L$) /91, 211/. Thus these problems actually preclude use of this model in comparison with the data.

POUKEY and ROSTOKER /200/ investigated IREB injection into vacuum for the 1-D case, and found that a potential well is created whose depth could exceed the injected electron energy by a factor of 2 to 3; the location of the density maximum, and the potential there, were found to oscillate about their mean values (by about $\pm$ 10%) in time. Beam injection into neutral gas was also investigated, and the characteristic velocity v_{1-D} was observed numerically; no mechanisms were suggested for increasing this velocity. Nonetheless, this model did establish the new feature of a potential well depth greater than the electron energy for vacuum injection.

ROSINSKII et al. /201/ also calculated the characteristic 1-D velocity v_{1-D}; however, no methods for accelerating the front to a velocity above v_{1-D} were given.

While the ionization wave approach did not appear to be getting any closer to an explanation of the observed acceleration process, there were other concepts under investigation as well. One of these concepts concerns the envelope motion of a radially-force-neutral beam ($f_e = \gamma_e^{-2}$). As discussed in Sect. 2, such a beam should have particle trajectories that are straight, and that are maintained at essentially the same angles at which the particles entered the drift chamber. Ideally, this force-neutral beam is thought of as having exactly parallel trajectories (i.e., zero emittance). If such a beam could be made, then clearly it would be in a delicate equilibrium. A slight enhancement in f_e above $f_e = \gamma_e^{-2}$ should induce envelope pinching, and the resultant pinch should have axial electric fields (both induced and electrostatic) that might be useable for collective ion acceleration.

The pinch mechanism of a force-neutral beam ($f_e = \gamma_e^{-2} \ll 1$) should be carefully distinguished from the usual magnetic pinch phenomena of partially or completely charge-neutralized beams ($\gamma_e^{-2} < f_e \lesssim 1$). For example, a charge-neutralized IREB with low divergence at injection will exhibit a series of envelope pinches beginning near the anode, which gradually wash out as the trajectories cross and become more phase mixed downstream. This type of pinching results in the (relatively) stationary pinches seen in computer simulations and in open shutter photographs of beams in experiments. Also, as the current increases (e.g., during the risetime), these pinches will move *toward* the anode. On the other hand, for collective ion acceleration with a force-neutral beam, it is desired to have the pinch propagate *away* from the anode

Apparently the first pinch acceleration mechanism was proposed by VEKSLER /202/ in 1956, as shown in Table 6. There, a beam with straight trajectories was shown passing an ion. The presence of the ion causes the beam to pinch just past the ion, and the enhanced electron density in the pinch exerts an accelerating force on the ion in the same direction that the beam propagates. (Veksler used this picture to explain the drag effect for a nonrelativistic beam.)

PUTNAM /203-206/ proposed a localized pinch model in 1970, in which a propagating force-neutral beam exists both before and after a "localized pinch". The pinch is caused by the presence of an ion bunch which enhances f_e to $f_e > \gamma_e^{-2}$. The origin of the ion bunch was not discussed. The pinch is centered on the ion bunch, and it produces an axial electric field. As the bunch moves, the constriction should follow the bunch, so the process should be self-synchronized. The envelope motion was described by using the linearized envelope equations of KAPCHINSKII-VLADIMIRSKI /20/. Both inductive and electrostatic axial fields (E_z) were calculated. The electrostatic field may be written as

$$E_z = (2^{3/2}/\pi)(\beta_e\gamma_e mc^3/e)(\nu_e/\gamma_e)^{3/2}(\beta_e^2 r_b c)^{-1} \; [1 + (\nu_e/\gamma_e)]^{1/2} \cdot \\ \cdot \; [\gamma_e^{-2} + (\Delta\lambda_i/|\lambda_e|)]^{1/2}. \tag{65}$$

However, this result is actually invalid because it violates several assumptions used in its derivation /91/. Noting this, Putnam returned to the inductive term [which was neglected early in the derivation of (65)] and by assuming that the beam envelope pinches down at the speed of light ($\partial r_b/\partial t = -c$) obtained the "saturated" induced field

$$E_z^{sat} = -2I_e(cr_b)^{-1}. \tag{66}$$

A true synchronized acceleration case was not calculated (as we did do in Table 3), and the length of the acceleration region was taken to be L, a free parameter.

Since this theory is based on the existence of a force-neutral beam, it suffers from the same fundamental problem common to all theories based on a force-neutral beam equilibrium. As discussed above, this problem is that for typical experimental parameters where ion acceleration is observed, a force-neutral beam would produce a potential depression (5) that is several times larger than the injected electron energy /54, 91/. Alternatively, this problem can be avoided only if I_e is below the space charge limiting current. Since $f_e = \gamma_e^{-2}$, this requires $I_e < I_\ell\,(1 - \gamma_e^{-2})^{-1}$ or

$$I_e < I_\ell \beta_e^{-2} \tag{67}$$

Table 6. Envelope motion theories

#	Theory	Unperturbed equilibrium	Requirement for equilibrium	E_z
1	pinch	(force-neutral)	$(I_e < I_\ell \beta_e^{-2})$ *	-------
2	localized pinch	force-neutral	$I_e < I_\ell \beta_e^{-2}$ *	inductive (max): $2I_e/(cr_b)$ for $\partial r_b/\partial t = c$ electrostatic: $\frac{I_e}{\beta^2 r_b c}\left(\frac{\nu_e}{\gamma_e}\right)^{1/2}\left(\frac{1}{\gamma_e^2}+\frac{\lambda_i}{\lvert\lambda_e\rvert}\right)^{1/2}$
3	pinch overshoot	(force-neutral)	$(I_e < I_\ell \beta_e^{-2})$*	--------
4	focusing instability	force-neutral	$I_e < I_\ell \beta_e^{-2}$ *	electrostatic: $\frac{4}{\beta}\frac{mc^2}{r_b e}\left(\frac{\nu}{\gamma}\right)^2$
5	synchronized pinch	force-neutral	$I_e < I_\ell \beta_e^{-2}$ *	electrostatic: polarization effect included

* requires $\nu_e/\gamma_e < \gamma_e\{(\gamma_e + 1)[1 + 2\ln(R/r_b)]\}^{-1} < 1$

Table 6 (continued)

#	Acceleration length	Synchronized case calculated?	Authors	References
1	-----	(no)	Veksler (1956)	202
2	L (free parameter)	no	Putnam (1969 - 1971)	203 - 206
3	-----	(no)	Benford, Benford (1972)	208
4	L (free parameter)	no	Tsytovich, Khodataev (1976)	133 - 136
5	(free parameter)	no	Irani, Rostoker (1976)	222

or, equivalently, in terms of ν_e/γ_e,

$$\nu_e/\gamma_e < \gamma_e\{(\gamma_e + 1)[1 + 2 \ln (R/r_b)]\}^{-1} < 1 . \tag{68}$$

In the regime where ion acceleration is observed, (67) - (68) are violated. In the regime where (67) - (68) hold, no ion acceleration has been reported. It follows that any theories based on the existence of a force-neutral beam cannot be used to explain the observed acceleration process.

Even if the fundamental restrictions (67) - (68) are ignored (which they cannot be), there are other problems with the Putnam theory /203 - 206/. For example, the electrostatic field would accelerate ions to the center of the pinch; net acceleration of the ions is not predicted /91/. The inductive term does just the opposite in that ions at the front of the pinch should be accelerated, while ions at the back (in the "unpinch" region) should be accelerated backwards, thus causing spreading of the bunch /91/. As we will discuss in Sect. 4.1.3, the numerical simulations of GODFREY /227, 228/ demonstrate that the ion bunch does spread rather quickly, and that the ions receive little net acceleration.

A "pinch overshoot" mechanism was briefly discussed by BENFORD and BENFORD in 1972 /208/. This is essentially the same mechanism as suggested by VEKSLER, and therefore our above comments apply to it also.

A different mechanism - inverse coherent Cerenkov radiation - was proposed by WACHTEL and EASTLUND /209/ in 1969 as a possible explanation for the observed acceleration process. However, as we have shown in Sect. 3.3, the accelerating fields produced by inverse coherent Cerenkov radiation, or inverse coherent Bohr-Bethe stopping, are too small to begin to explain the data. Also, other details such as how the bunch might be created, and why it should remain intact were not addressed.

Thus, following the introduction of the above theories /196 - 209/, there was still no valid explanation for the acceleration process. The lack of an explanation at this time is evident in the early reviews /21 - 30/.

Theory of Olson

The theory of OLSON /16 - 18, 31, 34, 210 - 213/ appeared in 1973 and gave the first well-substantiated explanation of the observed acceleration process. This theory involves an electrostatic time-dependent acceleration mechanism which has explicit dependences on all of the basic eleven parameters. In this theory, ion acceleration occurs in the electrostatic fields of a 2-D potential well. A current threshold is predicted for the acceleration process. The effects of ion impact ionization, ion avalanching (a new process), and charge exchange are invoked for the first time, and are shown to play important roles in the acceleration process. The theory predicts the time history and current risetime of the beam front, the location of the bunch relative to the front, and ion trapping characteristics of the moving well. It also predicts the pressure dependences, cutoffs, and limits of the ion bunch velocity and of the beam front velocity. The effects of chamber length (L) variation, and the effects of a B_z are included. For certain parameters, the occurrence of multiple pulses is also predicted /18, 213/. Here we will summarize the key features of this theory.

The space charge limiting current I_ℓ plays a fundamental role in this theory, which splits into the two cases $I_e \gtrsim I_\ell$, and $I_e \ll I_\ell$. For $I_e \gtrsim I_\ell$, the beam should initially stop at the anode, and the collective ion acceleration process discussed below should occur. For $I_e \ll I_\ell$, the beam never stops at the anode, and no ion acceleration is predicted. Thus ion acceleration should occur for $I_e \gtrsim I_\ell$, and the threshold for the existence of the process is $I_e = I_\ell$. It should be noted that since $I_\ell < I_A$ always, it is possible to have $\nu_e/\gamma_e < 1$ or $\nu_e/\gamma_e > 1$ and still have $I_e/I_\ell > 1$ /18/. Thus the existence of the acceleration process depends fundamentally on space charge limiting effects ($I_e/I_\ell \gtrsim 1$), rather than on magnetic stopping effects.

For $I_e \gtrsim I_\ell$, the acceleration process is divided into several stages as shown in Fig. 7. Initially the beam stops near the anode, forming a deep potential well whose depth may be up to three times the beam energy. Through beam-induced ionization

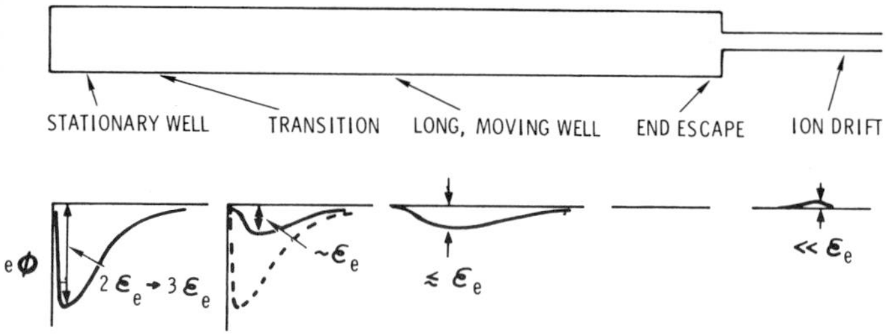

Fig. 7. The collective ion acceleration process for $I_e \gtrsim I_\ell$ in the Olson theory

processes, the beam eventually becomes almost charge-neutral near the anode. A non-adiabatic transition occurs, and the beam begins to propagate with a spreading front configuration that has a potential well depth about equal to the IREB energy. The speed at which the well moves out, the ion energy attained, and the number of ions all depend on several parameters, as discussed below. Once accelerated, the ion bunch and propagating beam front assume a quasi-equilibrium state that drifts with essentially constant velocity. At the end of the drift chamber, the ion bunch passes into an ion diagnostic section, where it undergoes spreading (mainly radial) and is finally detected.

The stationary well stage is described as follows. First it should be noted that the effects of a finite current risetime (t_r) and a finite voltage risetime (t_v) do enter into the exact stopping condition /17/. However, since the voltage turn-on usually slightly precedes the current turn-on in real beams, it is usually reasonable to assume that stopping occurs at the time $t \approx t_r I_\ell / I_e$.

The stopped beam will create a deep potential well with depth greater than the injected electron energy. The physical reason for this can be seen by imagining a stopped beam that is creating a potential depth of only $\phi \approx \mathscr{E}_e/e$. The potential minimum will occur at a small, but finite, distance from the injection plane ($z = 0$). This means that as new electrons cross the injection plane they will not be instantaneously reflected back at $z = 0$, but will actually penetrate a finite distance before being reflected. However, once these electrons are in the drift space $z > 0$, they will contribute to the total potential created there. The instantaneous potential will therefore exceed $\mathscr{E}_e/e$. Of course, this is not a steady state, and the electron density distribution will try to relax back to the $\phi \approx \mathscr{E}_e/e$ state on a fast time scale. But also, of course, new electrons will continually try to re-establish the deep potential well state on the same fast time scale. Thus a deep potential well will result with oscillations in the well depth and in the location of the well minimum. However, these oscillations are surprisingly small (~ 10%) as shown by the 1-D computer simulations of POUKEY and ROSTOKER /200/. The well depth can be solved analytically for the 1-D case up until the time of the first electron reflection. For the zero risetime case /200/, it was found that $e\phi/\mathscr{E}_e = 2.667$ in the nonrelativistic limit ($\beta \ll 1$, $\gamma \approx 1$), and $e\phi/\mathscr{E}_e = 2$ in the ultrarelativistic limit ($\beta \approx 1$, $\gamma \gg 1$). For the finite risetime case, OLSON calculated /17/

$$e\phi/\mathscr{E}_e = -2(\gamma_e - 1)^{-1} \int_0^{-(3)(2^{-1/3})\beta_e\gamma_e^{1/3}} \{[(-x^3 2^{-1}\beta_e^{-2} - 3x^2\gamma_e^{1/3}2^{-4/3}\beta_e^{-1} + \beta_e\gamma_e)^2 + 1]^{1/2} - \gamma_e\} x^{-1} dx \; . \tag{69}$$

The exact nonrelativistic limit of (69) was shown to be $e\phi/\mathscr{E}_e = 2.98125$, and the ultrarelativistic limit to be $e\phi/\mathscr{E}_e = 2.25$. In general, the 1-D requirement ($c/\omega_{pe} \ll r_b$) is frequently violated, but a deep potential well still results, as seen in 2-D numerical simulations /19/. It was concluded that a reasonable approximation is $e\phi = \alpha\mathscr{E}_e$ where $2 \lesssim \alpha \lesssim 3$.

Although the beam stops very close to $z = 0$, the potential well it creates extends a large distance into the drift tube cavity. By assuming that the stopped beam can be modelled by a charged disc of radius r_b, and axial extent from $z = 0$ to $z = d$, a Green's function calculation yielded for the axial extent $\mathscr{L}$ of the well

$$\mathscr{L}/R \approx (2.4)^{-1}[\ln \alpha + \ln (\mathscr{E}_e/\mathscr{E}_x)] + d/R \, , \tag{70}$$

where $\mathscr{E}_x$ is the well depth at which charge exchange collisions become dominant, and therefore beyond which accelerated ions could not create any additional ionization. For typical parameters ($\alpha = 2.5$, $\mathscr{E}_e = 1$ MeV, $\mathscr{E}_x = 50$ keV, and $d/R \approx 0.3$), this gives $\mathscr{L} \approx 2R$. Therefore, by explicitly keeping the $\mathscr{E}_e$ dependence, we have

$$\mathcal{L}/R \approx 2 + (2.4)^{-1} \ln[\mathcal{E}_e(\text{MeV})]. \tag{71}$$

The length $\mathcal{L}$ enters into the front velocity for a certain pressure regime (as shown below). The length $\mathcal{L}$ also requires the drift tube length L to be $L \gtrsim 2R$ if end plate effects are to be avoided. For $L \lesssim 2R$, the stopping effect and the acceleration process will be accordingly diminished /18/.

Ionization of the neutral gas in the deep potential well region involves electron and ion ionization processes /17, 210/. As shown by the cross sections in Fig. 8, when protons have energies above ~ 50 keV, they are much better ionizers of the neutral gas than the beam electrons. These considerations led to approximations /210/ for the electron impact ionization time τ_e and the ion avalanche time t_i (for H_2)

$$\tau_e \approx 5.0\ [p(\text{Torr})]^{-1}\ \text{nsec} \tag{72}$$

$$t_i \approx 0.33\ [p(\text{Torr})]^{-1}\ \text{nsec}\ . \tag{73}$$

It was noted that although the beam electrons are confined to the region next to the anode, ions created by impact ionization will traverse the entire well length and initiate an ion avalanche throughout the entire well region. Provided certain conditions are met to justify the ion avalanche /210/, the ionization growth can be estimated by the rate equation

$$\partial n_i(t)/\partial t = n_b(t)/\tau_e + n_i(t)/t_i\ . \tag{74}$$

Various initial distributions $n_i(o)$ have been considered as well as various forms for $n_b(t)$; from these, the useful estimate for the charge neutralization time including electron and ion effects $\tau_N^{e,i}$ resulted,

$$\tau_N^{e,i} \approx t_i\ \ln[(\tau_e/t_i) + 1] \tag{75}$$

which for H_2 is

$$\tau_N^{e,i} \approx 1.0\ [p(\text{Torr})]^{-1}\ \text{nsec}. \tag{76}$$

The effect of ion ionization is to speed up the charge neutralization process considerably, as can be seen by comparing (76) and (72). Indeed the early interpretation /14, 15, 155, 161, 163, 199, 204, 205/ that the beam waited at the anode until $\tau_{FN} = \tau_e \gamma_e^{-2}$, and then began propagation, was shown by OLSON and POUKEY /54/ to be incorrect. That picture also required a propagating force-neutral beam which would have $I_e > I_\ell$, a condition we have shown is impossible. For typical data ($I_e/I_\ell \gtrsim 3$),

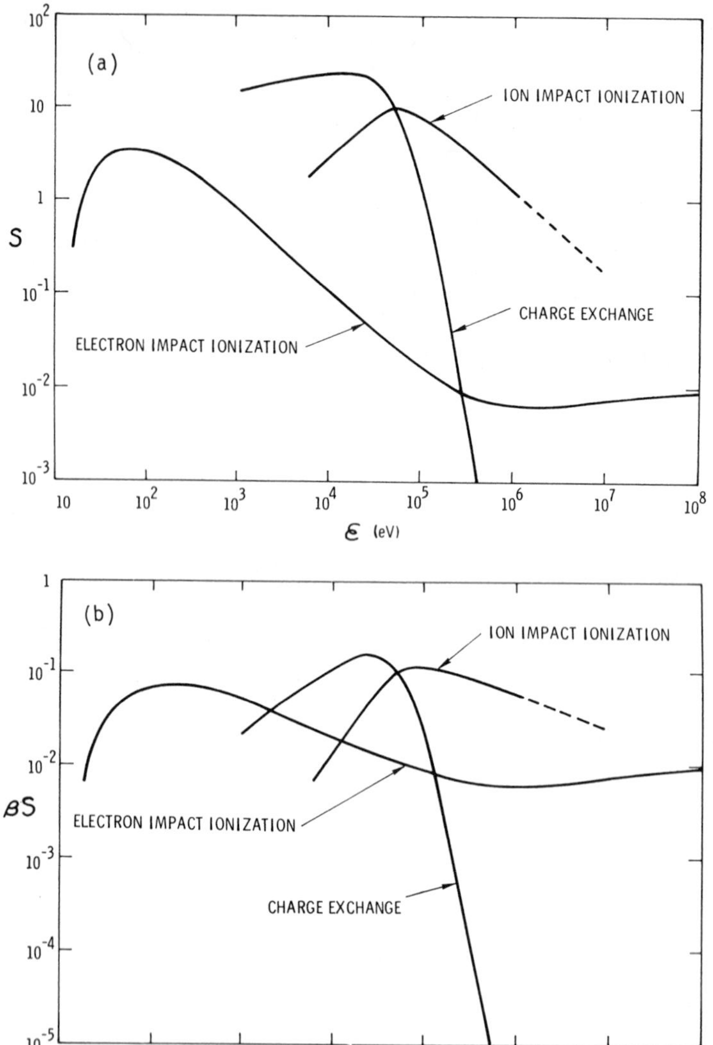

Fig. 8a and b. Cross sections for electron impact ionization, ion impact ionization, and charge exchange for electrons and protons in hydrogen /210/. In (a), S is the number of events per cm per Torr, so that the mean free path between events is $\lambda = [p(\mathrm{Torr})S]^{-1}$ cm, and the conventional cross sections are given by $\sigma = 2.82 \times 10^{-17} \times S$ cm^2. In (b), the mean time between events is given by $t = [\beta S\ p(\mathrm{Torr})c]^{-1}$ sec the correct picture is that the beam waits at the anode until it is almost charge neutral ($\tau \sim \tau_N^{e,i}$), and then it commences propagation.

Ion acceleration occurs during the deep well stage (which produces an ion distribution with energies up to $\sim 3Z\mathcal{E}_e$) and during the transition to the moving well stage. Once established, the moving well equilibrium (described below) will propagate with essentially constant velocity. Depending on the beam front motion and certain trapping criteria /17/, ions from the initial deep well ion distribution may or may not be trapped and attain the beam front velocity. The beam front seeks the fastest propagation route available, and there are three distinct pressure regimes as shown in Fig. 9. In the *low pressure regime*, the beam front velocity is determined by the fastest ions created during the deep well stage; the horizontal line $\beta = \beta_0$ corresponds to the β of an ion with energy $\alpha Z\mathcal{E}_e$. In the *high pressure regime*, the beam front velocity assumes the velocity $\approx \mathcal{L}/\tau_N^{e,i}$. The transition pressure p_T occurs where $\mathcal{L}/\tau_N^{e,i} = \beta_0 c$. In both the low and high pressure regimes, the final beam front velocity should be acquired over a time of order $\tau_N^{e,i}$. In the *runaway regime*, the beam never stops, a deep stationary well never forms, and no accelerated ions should occur. Effectively, runaway will occur if the beam, during its risetime, becomes charge-neutralized before it reaches the limiting current; then the beam will never be stopped by its own space charge. The runaway pressure p_R is therefore defined by the condition $\tau_N^{e,i} \approx (I_\ell/I_e)\ t_r$, so for H_2,

$$p_R(\text{Torr}) \approx (1.0)(I_e/I_\ell)[t_r(\text{nsec})]^{-1}. \tag{77}$$

The beam front velocity in any regime can never exceed the power balance limit as shown. However, for typical experimental parameters, this limit is important only in the runaway regime. The power balance limit was introduced into the context of collective ion acceleration by OLSON /17/, where it was noted that GRAYBILL /156/ first considered the effect for a different purpose. Power balance simply states that the injected power must equal the sum of the expended powers. For example, power must be expended to give the beam particles kinetic energy, to set up the beam's self-magnetic field, and to account for the rate of particle energy deposited in the chamber walls (etc.). The power balance *limit* occurs when all of the injected power is used to meet the *minimum* power expenses. For the relevant case of a charge-neutral (but not current-neutral) beam, power must at least be supplied to set up the beam's magnetic field, and give the particles kinetic energy /17, 156/

$$UI_e = (I_e/c)^2[1 + 4\ \ln(R/r_b)]\ \beta_f c/4 + (\gamma_f - 1)\ mc^2\ I_e/e\ , \tag{78}$$

where U is the effective diode voltage, $\gamma_f = (1 - \beta_f^2)^{-1/2}$, and $\beta_f c$ is the beam front velocity. Result (78) may be solved by iteration for β_f - this has been done to cal-

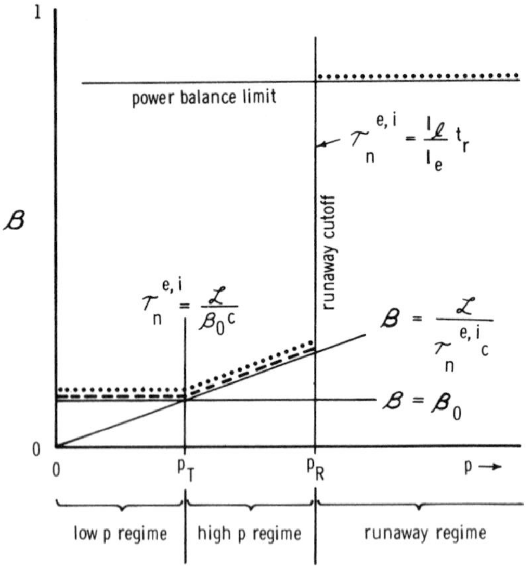

Fig. 9. Pressure regimes in the Olson theory; the final β_f (β_i) is represented by the dotted (dashed) line

culate β_f in the runaway regime /18/. Alternatively, if we assume $\beta_f \ll 1$, then the last term in (78) may be neglected to give

$$\beta_f = \{I_e(4Uc)^{-1}[1 + 4 \ln(R/r_b)]\}^{-1} \qquad (79)$$

or, in practical units,

$$\beta_f = \mathcal{Z}(\Omega)(7.5)^{-1}[1 + 4 \ln(R/r_b)]^{-1}. \qquad (80)$$

This expression produces $\beta_f \ll 1$ only for very small values of the beam impedance $\mathcal{Z}$ and/or for very large values of R/r_b.

The final moving well equilibrium is drawn in Fig. 10. Shown qualitatively are the axial dependences of (a), f_e, (b) the beam envelope and a typical electron trajectory, (c) the current enclosed by the drift tube (as would be seen by a current probe at the wall), and (d) the resultant potential well. The basic picture is that the envelope structure moves slowly ($\beta_f \sim 0.1$) while individual beam electrons stream through it and strike the walls. A description of this structure involves the self-consistent coupling of the beam dynamics with the ionization processes. If a functional form for $f_e(z)$ is assumed, then the envelope shape $r_b(z)$ can be computed; with this, the ionization rates can be computed, which in turn gives $n_i(z)$ and therefore $f_e(z)$ - this last $f_e(z)$ must be the same as our initial assumption for $f_e(z)$; i.e., the problem must be solved self-consistently as was done /17/. The time t_ℓ^*, corresponding to passage of the full well length in the laboratory frame, was shown

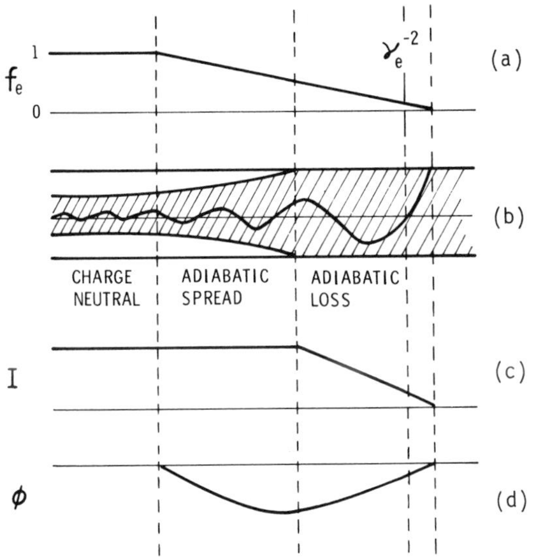

Fig. 10a - d. Self-consistent moving well equilibrium (for $\beta_f \ll \beta_e$) in the Olson theory

to be $2.0[p(\text{Torr})]^{-1} < t_\ell^* < 10[p(\text{Torr})]^{-1}$ depending on the effectiveness of ion avalanching throughout the beam front region. Note that the "knee" in the current occurs somewhere in the middle of the well, and that the central location of the accelerated ions should be at the well minimum. For $R/r_b = 2$, the well minimum should be just behind the current knee. For arbitrary values of R/r_b, graphical solution results may be used to find the location of the current knee and the well minimum /17/.

The ion pulse characteristics depend on several parameters. The ion pulse length τ should scale as $\tau \approx (1.0)\ Z[p(\text{Torr})]^{-1}$ nsec, at least for low Z. The number of ions in the low pressure regime should be $N \approx r_b^3 n_b Z^{-1}$. In the high pressure regime, only trapped ions will acquire full energy. If more than one species is present, those with the highest Z/M values will be selectively accelerated. A number of trapping criteria were derived and discussed, and it was shown that for the highest proton energies ($\mathscr{E}_i/\mathscr{E}_e \sim 10$), only a small fraction of the deep well ion distribution [ions with ($\mathscr{E}_i/\mathscr{E}_e \sim 3$)] should be trapped and accelerated to the full beam front velocity. Thus N should decrease rapidly as the pressure increases in the high pressure regime.

For comparison with data in which the derived parameter $\mathscr{Z}$ was varied for fixed values of p (in the high pressure regime) and R/r_b, the theory predicts a variation in β_f. Note that variation of $\mathscr{Z}$ means that many parameters ($\mathscr{E}_e$, I_e, t_r, t_v, r_b, etc.) are being varied, and that all of these parameters should be specified. Nonetheless, if only $\mathscr{Z}$ and the range of $\mathscr{E}_e$ is given, then the theory predictions are as shown in Fig. 11. As discussed above, note that the power balance limit (80) becomes significant only for low values of $\mathscr{Z}$. In the high pressure regime, $\beta_f c \approx \mathscr{L}/\tau_N^{e,i}$ depends on

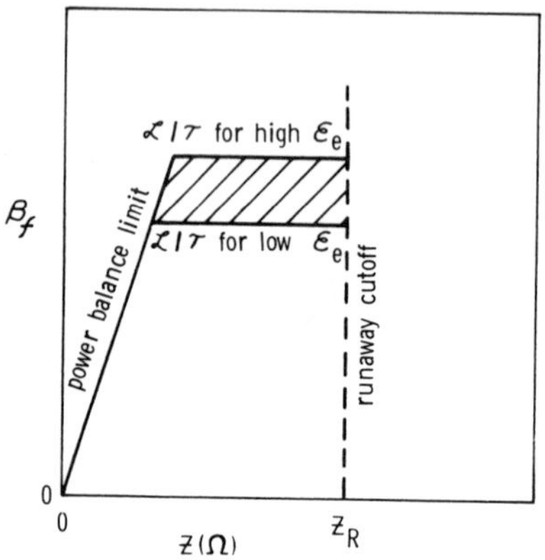

Fig. 11. Effect of variation of beam impedance $\mathcal{Z}$ on β_f for fixed values of p and R/r_b in the Olson theory

$\mathcal{E}_e$ both through $\mathcal{L}$ and $\tau_N^{e,i}$. The $\tau_N^{e,i}$ dependence means that for lower $\mathcal{E}_e$, ion avalanching can become less effective as smaller potentials are created. However, estimation of this effect would require detailed numerical calculations. Therefore, we will retain the approximation (76) for $\tau_N^{e,i}$ and use the explicit dependence on $\mathcal{E}_e$ only in $\mathcal{L}$ (71); this produces the shaded region as shown. Lastly, we note that the runaway cutoff (77) for fixed p transforms to a cutoff for $\mathcal{Z}$. Using (6), this cutoff is

$$\mathcal{E}_e I_e^{-1} e^{-1} = (\tau_N^{e,i}/t_r)[1 + 2 \ln (R/r_b)] (\beta_e c)^{-1} \tag{81}$$

or

$$\mathcal{Z}_R(\Omega) = 30[1 + 2 \ln (R/r_b)][\beta_e p(\text{Torr})\ t_r\ (\text{nsec})]^{-1} . \tag{82}$$

For $\mathcal{Z} \geq \mathcal{Z}_R$, the beam front should run away, with β_f given by (78), and no accelerated ions should be seen.

An applied axial magnetic field B_z will affect the acceleration process by allowing a portion of the injected current up to I_ℓ to propagate at all times /18/. The existence of a propagating current I_ℓ over long distances (L > 100 cm, R < 10 cm), for an injected current $I_e > I_\ell$ in the presence of a large B_z, has been demonstrated experimentally /234, 235/ for the vacuum case. For the neutral gas case here, the propagating current I ($\approx I_\ell$) will cause ionization downstream even during the stationary well stage when most of the beam is stopped near the anode. [In the $B_z = 0$ case, electrons that make it past the stopped beam will expand radially and hit the walls: no significant current will flow and cause ionization downstream.] Thus for a large B_z, an initial current ($I \approx I_\ell$) will propagate with a high beam front velocity. The stationary well stage will still produce an initial ion distribution.

However, during the transition to the moving well stage, the remaining current ($I_\ell <$ $I < I_e$) will attain a high velocity [$\beta_0, \mathscr{L}/(\tau_N^{e,i}c) < \beta < \beta$ (power balance)] because the downstream region has been continually ionized by the current I_ℓ. Ion trapping is therefore improbable, and ions from the initial stationary well distribution will find themselves in a (roughly) charge neutral beam channel. These ions will propagate with their stationary well velocities, drifting apart as they propagate (because there is no longer a potential well to focus them), and also being deflected to the walls by the beam's self-magnetic field B_θ. The main results are that some energetic ions will still be seen ($\mathscr{E}_i$ up to $\sim \alpha\mathscr{E}_e$), but their number will be drastically diminished from the $B_z = 0$ case.

Multiple pulse effects may be divided into two categories. The *first catagory* concerns a single acceleration process (as discussed above) that results in more than one ion bunch. For example, in the high pressure regime, the trapped ion bunch and the remainder of the deep well ion bunch may appear as two pulses; in this case, the second pulse would have a long pulse length (due to its large energy spread) and its peak velocity would be less than that of the first (trapped) ion bunch. Another example concerns multiple species and multiple charge states in a single shot; in either the high or the low pressure regime, these will result in the appearance of "multiple pulses". The *second category* concerns true multiple pulses caused by a repeat of the main acceleration process. For example, after the first pulse, any means by which the un-charge-neutralized current can equal or exceed the space charge limiting current, will result in well formation, possible beam stopping, and a repeat of the ion acceleration process. Note that this is an axial, electrostatic, multiple pulse effect. A generalized criterion for the appearance of well formation ($e\phi \gtrsim \mathscr{E}_e$) is from (6), /213/

$$I_e(t)[1 - f_e(t)]\{1 + 2 \ln[R/r_b(t)]\} \geq \beta_e(t)[\gamma_e(t) - 1]mc^3/e . \tag{83}$$

From this it can be seen that multiple pulses may result, depending on the detailed behavior of $f_e(t)$, $r_b(t)$, $I_e(t)$, and/or $\gamma_e(t)$:

(i) $f_e(t)$: It is possible to satisfy (83) after the first pulse leaves the anode region (and that region is momentarily charge-neutralized), if the current is still rising, and if beam-induced ionization processes produce a relatively low ionization rate in that almost charge neutral region. In the presence of the deep potential well, or the moving potential well, large potentials (ϕ) and large potential gradients (E) insure substantial ion ionization effects. However, behind the moving beam front equilibrium, ϕ and E should be small, and the dominant ionization process should be electron impact ionization; ion ionization effects should not become important again until ϕ and E become substantial again. Accounting for the acceleration process through the transition stage, it is reasonable to assume that the anode region ($z = 0$ to $z \gg r_b$) is well charge-neutralized by time $t \approx 2\,\tau_N^{e,i}$. Then if $f_e(t)$

is calculated for $2\tau_N^{e,i} \lesssim t \lesssim t_r$ for a linearly rising current ($I = I_e t/t_r$), and only electron impact ionization is assumed, we find condition (83) would be reached at about time $t \approx \tau_e$. A necessary requirement for this to occur is /213/

$$I_e/I_\ell \gtrsim \{(1/2) - (2\tau_N^{e,i}/\tau_e)[1 - (\tau_N^{e,i}/\tau_e)]\}^{-1} \approx 5.6 \; . \qquad (84)$$

If this process does occur, the time between the first and second pulses ΔT would scale as p^{-1}.

(ii) $r_b(t)$: If $\nu_e/\gamma_e \gtrsim 1$, there are at least two mechanisms which can lead to (83) being satisfied after the first pulse. One is that if $I_e/I_c > 1$, pinching will occur inside the diode; if the resultant $r_b(t)$ decreases in time fast enough, criterion (83) may be reached and another pulse would be formed /18/. This is especially true if $R < r_b$ (which was actually used in some data /164/). Even if the injected radius stays constant ($d/r_b \ll 1$), the formation of the magnetic envelope pinch(es) that occur and move toward the anode may result in (83) being satisfied again. Note that these pinches would merely provoke the axial stopping mechanism, which is the main acceleration process.

(iii) $I_e(t)$ and $\gamma_e(t)$: Large fluctuations in $I_e(t)$ and/or $\gamma_e(t)$ can lead to condition (83) being satisfied, with subsequent formation of multiple pulses /18/.

It should be noted that the second (or later) pulses could occur with $\mathcal{E}_i < \mathcal{E}_e$, if $e\phi < \mathcal{E}_e$: in this case, beam stopping is not even necessary. Even if beam stopping does occur, a high beam front velocity should subsequently result (with ion trapping unlikely) due to the downstream "preionization" caused by the first pulse. Thus typically, pulses after the first pulse should have lower energies than the first pulse. For the first two pulses, the observed pulse separation time at a distance ℓ from the anode should therefore be

$$(t_2 - t_1) = \Delta T + \ell[(\beta_2 c)^{-1} - (\beta_1 c)^{-1}] \; , \qquad (85)$$

where ΔT is the pulse separation time near the anode.

For $\nu_e/\gamma_e > 1$, and for typical beams, only a portion of the beam corresponding to $(\nu/\gamma)_{net} < 1$ should propagate, and the remainder of the beam should be lost near the anode. However, for specialized beam distributions as discussed in Sect. 2.2 (e.g., $v_\theta \neq 0$, annular beams, $f_e \neq 1$), currents in excess of $(\nu/\gamma)_{net} = 1$ may propagate.

Additional features, such as the presence of radially accelerated ions, occur in the theory also /18/. As summarized above, this theory involves many ideas not considered in previous theories, that are actually required to account for the observed acceleration process. Space charge limiting current effects (I_e/I_ℓ) were shown to define the acceleration threshold, enter into B_z effects, and enter into one multiple pulse mechanism. Two-dimensional effects were shown to influence the acceleration process through, e.g., I_ℓ, $\mathscr{L}$, p_R, and the power balance limit for β_f. Ion ionization effects were shown to play a dominant role in the charge neutralization process. It should also be noted that there are no fitted parameters in this theory. In Sect. 4.1.4, it is shown that this theory is in substantially good agreement with the extensive amount of experimental data that now exists.

More Theories

ALEXANDER et al. /214 - 218/ have independently proposed a 1-D well theory, which like the Olson theory, takes into account the effects of ion impact ionization on the front motion. Realizing that the usual 1-D front velocity (61) is too small to account for the data, they obtained the front velocity

$$\beta_f c = d_0[1 - (n_i/n_b)]^{-1/2} \, \nu_o[1 + B(n_i/n_b)][1 - (n_i^+/n_b)]^{-1}, \tag{86}$$

where d_0 is the usual 1-D slowing down length, and ν_o^{-1} is the electron impact ionization time. Three factors were included to increase the front speed well above v_{1-D}. The factor $[1 - (n_i/n_b)]^{-1/2}$ accounts for the increase in stopping length due to partial charge neutralization in the front region. The factor $[1 + B(n_i/n_b)]$ accounts for ion impact ionization in the front region, and the factor $[1 - (n_i^+/n_b)]^{-1}$ accounts for ionization by ions (n_i^+) shot ahead of the well; together, these factors can reduce the ionization time to a value well below ν_o^{-1}. A series of coupled equations relating β_f, n_i, and n_i^+ were derived and solved numerically. The ion energies which result are very similar to the "translating well" results in Table 2, as discussed earlier. These results are that a portion of the ions slip out the back of the well with relatively low energies, and another portion is shot ahead of the well with ion energies about 4 times the electron energy $[\mathscr{E}_i \approx 4(Mv_0^2/2) \approx 4(Ze\phi) \approx 4(Z\mathscr{E}_e)]$. *No ions remain with* $\beta_i = \beta_f$. In comparing with the data, the parameters of the model were adjusted to obtain a best fit /214 - 217/.

The 1-D assumption requires $d = d_0[1 - (n_i/n_b)]^{-1/2} \ll r_b$. To insure this, the model is said to apply only to data with $\nu_e/\gamma_e \gg 1$ [this eliminates from consideration much of the data, as seen from Table 4]. However, for a $\nu_e/\gamma_e \gg 1$ beam to propagate would require substantial current neutralization behind the beam front to insure $(\nu/\gamma)_{net} < 1$, whereas only charge neutralization was considered. In fact, in the ion acceleration regime, data for $\nu_e/\gamma_e \gg 1$ beams show that only a portion of the beam, with $(\nu/\gamma)_{net} \approx 1$, actually propagates /164, 165, 167/. Also if $d \ll r_b$, then the beam front current risetime is $t_r^* \ll r_b/(\beta_f c)$; for typical parameters [r_b = 1 cm, β_f = 0.1], this gives $t_r^* \ll 0.3$ nsec, whereas current front risetimes of order 10 nsec are always observed. Thus the applicability of this model to the data is in question.

As fitted to the data /214 - 217/, this model does describe some of the ion acceleration properties in the low pressure ("plateau") regime; however, this required the ad hoc addition of the well deepening effect. In later work /218/, ions shot ahead of the front were reaccelerated when the front caught up with them again; numerical results including this effect do produce higher ion energies and higher beam front velocities, as characteristic of the high pressure regime. However, this model is still unable to account for many basic features of the data, such as the location of the ions relative to the front, $\beta_i \approx \beta_f$, the runaway cutoff, the runaway regime, the power balance limit, the beam front current risetime, the effect of B_z, and true multiple pulse effects. Nonetheless, this model does demonstrate that with considerable assumptions, the 1-D well velocity can be made to achieve the observed magnitude for a limited parameter space ($\nu_e/\gamma_e \gg 1$).

GODFREY and THODE /219 - 221/ considered the electron-ion two stream instability and suggested that perhaps some of the ion acceleration observed in drift tubes is due to this mechanism. We have already discussed this instability in Sect. 3.4 [see (34) - (41)]. To achieve low phase velocities [in (39)], the ratio Zn_i/n_b was taken to be $\approx \gamma_e^{-2}$, i.e., a force-neutral beam was considered. To attain this force-neutral condition, it was assumed that the IREB is injected into a plasma, and that it pinches down to a radius $r_b \lesssim 0.01$ cm. The final stages of some computer simulations were given, showing that accelerated ions were obtained. As we have noted elsewhere /91/, trapping at the wave phase velocity (39) would predict $\mathscr{E}_i \sim (Z^2 n_i/n_b)^{2/3} M^{1/3} \gamma_e^2$ which is not seen in the data. However, the real problems in applying this e-i instability to explain the data are even more fundamental: (1) As we have discussed earlier, a force-neutral beam can exist only if (67) - (68) hold; and in the regime where (67)-(68) hold, no accelerated ions have been observed. (2) The instability threshold [which was not considered in the theory /219 - 221/, but is given here by (40)], is so high it is only reached in the first 3 data sets out of the 30 data sets listed in Table 4; thus even if we assumed the force-neutral beam could exist (which we cannot), then we would still find that for the majority of the data the threshold condition for this instability is not met. (3) For the example chosen [$\gamma_e = 6$, $I_e = 25$ kA, $r_b = 0.01$ cm], a current density of $\sim 10^8$ A/cm^2 is required in a fine filament over a distance of $\gtrsim 5$ cm /219/. While such a beam would be interesting, it is relatively clear that none of the data sets of Table 4 would produce such a beam. Thus it appears that the drifting beam data cannot be accounted for with this model.

TSYTOVICH and KHODATAEV /133 - 136/ have proposed a focusing instability model which is based on an initially force-neutral beam ($f_e = \gamma_e^{-2}$), as shown in Table 6. The electron beam envelope is allowed to vary radially, while the ions are assumed constrained to move only axially. This configuration was investigated for stability against small amplitude wave perturbations using hydrodynamic equations for the electron and ion fluids together with Maxwell's equations. Keeping only first-order terms, and assuming $\omega \ll kV_0$, $\omega^2 \ll \omega_i^2$, and $\delta V_z^e = 0$, the dispersion relation was found /134/

$$\lambda = k^2 r_b^2 \omega_i^2 \omega^{-2} \{1 + (\gamma^2 \omega_b^2)[(\omega - kV_0)^2 - \omega_b^2]^{-1}\}, \tag{87}$$

where λ is an eigenvalue (a positive number of order unity) which depends on the radial boundary conditions, since inside the beam channel

$$r^{-1} (\partial/\partial r)[r\, \partial(\delta E_z)/\partial r] = -(\lambda/r_b^2)(\delta E_z). \tag{88}$$

Here, $\omega_b^2 = 4\pi n_b e^2(\gamma^3 m)^{-1}$, $\omega_i^2 = 4\pi Z^2 n_i e^2/M$, $n_i Z/n_b = \gamma^{-2}$, V_0 is the injected electron velocity, δV_z^e is the axial electron velocity perturbation, and δE_z is the axial electric field perturbation. We have derived and analyzed the complete set of equations assuming only $\delta V_z^e = 0$, and found the more complete dispersion relation

$$\lambda = k^2 r_b^2 [(\omega_i^2/\omega^2) - 1]\Big[1 + \{\gamma^2 \omega_b^2 [1 - (2\omega V_0)(kc^2)^{-1} + \omega^2 (kc)^{-2}] - \omega^2 (\omega - kV_0)^2 (kc)^{-2}\} [(\omega - kV_0)^2 - \omega_b^2]^{-1}\Big]. \tag{89}$$

In the limit $\omega \ll kV_0$, $\omega^2 \ll \omega_i^2$, result (89) reduces to (87). Analysis of (87) leads to unstable waves with growth rate

$$\mathrm{Im}\{\omega\} = (3^{1/2}/2^{2/3})\, \omega_b \lambda^{-1/3} (\nu/\gamma)^{1/3} (Zm\gamma/M)^{1/3} \beta^{-2/3} \tag{90}$$

and wave properties

$$\mathrm{Re}\{\omega\} = (2^{-2/3})\, \omega_b \lambda^{-1/3} (\nu/\gamma)^{1/3} (Zm\gamma/M)^{1/3} \beta^{-2/3} \tag{91}$$

$$k = \omega_b/V_0 \tag{92}$$

$$V_\phi = 2^{-2/3} \lambda^{-1/3} \beta^{-2/3} (\nu/\gamma)^{1/3} (Zm\gamma/M)^{1/3} V_0 . \tag{93}$$

Due to the restriction $\omega^2 \ll \omega_i^2$, results (90) - (93) are strictly valid only for $(\nu/\gamma) \ll 4\lambda\beta^2 (Zm\gamma/M)^{1/2}$. The nonlinear saturation of the waves for $\nu/\gamma \gg 1$ was discussed, and it was estimated that the electron density perturbation δn^e would grow to be comparable to the zero-order beam density n_b. This led to an estimate of the maximum field

$$E_z \approx 4(mc^2/e)\, r_b^{-1} (\nu/\gamma)^2\, \beta^{-1}. \tag{94}$$

The final ion energy was then assumed to be $\mathscr{E}_i = ZeE_z L$, where L is a fitted parameter. It was argued, but not convincingly demonstrated, that the envelope contrictions and ion bunches should accelerate synchronously.

Several fundamental problems occur in trying to use this theory to explain the observed acceleration. These are as follows.

(1) Since a force-neutral beam is required, this theory is applicable only for $I_e < I_\ell \beta_e^{-2}$ [see (67)], which equivalently demands ν_e/γ_e to be sufficiently less than unity to satisfy (68). In the regime of this theory's applicability, no accelerated ions have been observed. Also, in this regime [$I_e < I_\ell \beta_e^{-2}$], the self-consistent value of γ for an electron is

$$\gamma(r,z) = \gamma_e - [e\phi(r,z)](mc^2)^{-1} . \tag{95}$$

Thus γ has dependences on both r and z and it is difficult to imagine how the required, complicated, $f_e(r,z)$ would ever occur "naturally" to produce the required, propagating, force-neutral (zero-emittance) beam. Also γ varies in r, so the propagating beam is not monoenergetic (although a monoenergetic beam was assumed), and ν/γ is not clearly defined.

(2) The threshold current for the onset of this instability was never calculated in the theory /134/. We have investigated the threshold using (87), which may be rewritten as $F(\omega) = \lambda/(k^2 r_b^2)$, which is a quartic equation for ω, where

$$F(\omega) = \omega_i^2 \omega^{-2} \{1 + (\gamma^2 \omega_b^2)[(\omega - kV_0)^2 - \omega_b^2]^{-1}\} . \tag{96}$$

The function $F(\omega)$ is sketched in Fig. 12 for several values of kV_0 (for definiteness we have taken $kV_0 \geq 0$). For $kV_0 = \omega_b$, the line $F = \lambda/(k^2 r_b^2)$ intersects the function $F(\omega)$ in only two places. Thus for this case there will be two real roots and two complex conjugate roots, and one of the later roots will have $\mathrm{Im}\{\omega\} > 0$ (i.e., instability). Instability also results for the case $0 < kV_0 < \omega_b$. However for $kV_0 \gg \omega_b$, a "new branch" appears which can dip below the line $F = \lambda/(k^2 r_b^2)$, resulting in the appearance of four real roots (i.e., stability). Thus we find instability results only for

$$kV_0 < \xi \omega_b , \tag{97}$$

where $1 < \xi \lesssim 2$ typically. To demonstrate this, we have calculated the threshold condition, where the minimum of the "new branch" just touches the line $F = \lambda/(k^2 r_b^2)$; this condition, which determines ξ, is

$$\xi^2 \{1 - [1 - (8/9)(1 - \xi^{-2})]^{1/2}\}^2 \left[(16)^{-1}\{1 + 3[1 - (8/9)(1 - \xi^{-2})]^{1/2}\}^2 - \xi^{-2}\right]$$
$$= (16/9)(m/M)\, \beta^2 \gamma^3 r_b^2 \omega_b^2 V_0^{-2} \lambda^{-1} Z \tag{98}$$

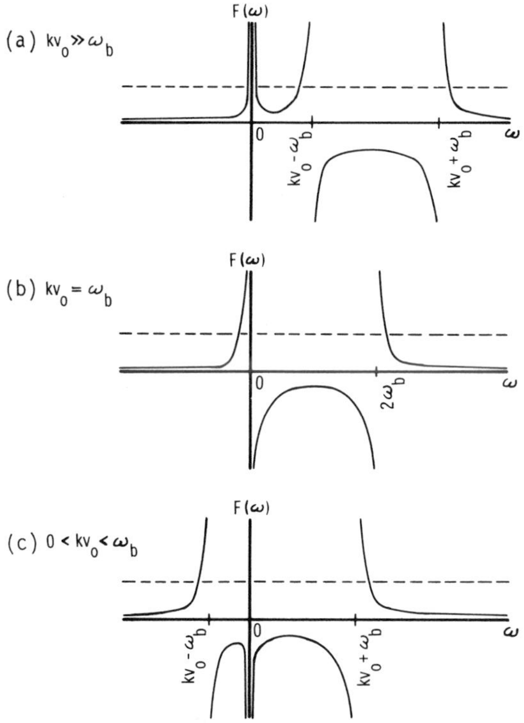

Fig. 12a - c. Analysis of the focusing instability threshold: the dashed line represents $F = \lambda/(k^2 r_b^2)$

[In deriving (98), we have neglected the 1 in (87), which is valid for $\xi^2 \ll \beta^2\gamma^2$; for the exact case, the threshold condition would be more complicated than (98)]. In analogy to the e-i instability threshold for bounded beams, (40), we therefore find for the focusing instability threshold current I_{FI}, the approximate interpolation formula,

$$I_{FI} = \xi^{-2} \beta^3 \gamma^3 (mc^3/e)[1 + 2 \ln(R/r_b)]^{-1} \tag{99}$$

$$I_{FI}/I_\ell = \gamma(\gamma + 1) \xi^{-2}. \tag{100}$$

If we could ignore the requirements (67) - (68) (which we cannot), we would find that the instability criterion $I_e \geq I_{FI}$ would be reached in only the first few of the 30 data sets in Table 4. Thus, for the majority of the data, the threshold condition is not satisfied.

(3) The focusing instability beam/wave system is unstable to an electrostatic kink (m = 1) mode. Imagine for the moment that the theory was applicable for $I_e > I_\ell$, and that the instability threshold was exceeded, $I_e > I_{FI}$. Then although the initial configuration has $f_e(z) = (n_i Z/n_b) = \gamma^{-2}$ for all z, once the instability starts to grow,

f_e will no longer be equal to γ^{-2} everywhere. In fact, the envelope oscillations characteristic of this instability are the result of having $f_e(z) \neq \gamma^{-2}$ everywhere. For the typical experimental configuration in which the IREB is inside a conducting drift tube of radius R, it can be shown /18, 213/ that a partially neutralized beam becomes kink unstable wherever $0 \leq f_e(z) < \gamma^{-2}$. This instability is due to electrostatic image forces in the drift tube walls, which dominate the radial force equation for a transversely displaced beam with $0 \leq f_e < \gamma^{-2}$. The transverse beam displacement will grow as $r = r(t=0)\cdot\exp(\Gamma_{kink}t)$, where Γ_{kink} is the linear growth rate,

$$\Gamma_{kink} = 2^{-1/2}\,(r_b/R)\,\gamma\omega_b\,(\gamma^{-2} - f_e)^{1/2}. \tag{101}$$

For the focusing instability, with $\Gamma_{FI} \equiv \mathrm{Im}\{\omega\}$, we find from (90a) and (101)

$$\Gamma_{kink}/\Gamma_{FI} = 2^{1/6}\,3^{-1/2}\,[M/(Zm\gamma)]^{1/3}[r_b/R]\,(\nu/\gamma)^{-1/3}\,\beta^{2/3}\lambda^{1/3} \cdot[1 - (\gamma^2 f_e)]^{1/2}. \tag{102}$$

For typical parameters ($\gamma = 3$, $R/r_b = 1.5$, $\beta \sim 1$, $\lambda \sim 1$, $Z = 1$, M equal to the proton rest mass, and $\nu/\gamma \sim 1$), and $f_e = 0.9\,\gamma^{-2}$ (i.e., only a 10% deviation from force neutrality) we find $\Gamma_{kink}/\Gamma_{FI} > 1$. For a more developed case ($f_e = 0$), we find $\Gamma_{kink}/\Gamma_{FI} \approx 4$. Thus these initial calculations indicate that should the focusing instability ever begin, the beam will quickly become kink unstable and be driven to the walls.

(4) Even if we could ignore the facts that the theory is only applicable for $I_e < I_\ell\,\beta_e^{-2}$ (where no ion acceleration is observed), that the threshold for this instability is never met in most of the data, and that the beam/wave system is kink unstable, we would find that the main scaling laws predicted by this theory are not seen in the data. For example, the theory predictions $\mathscr{E}_i \sim (\nu/\gamma)^2$ and $N \sim (\nu/\gamma)^2$ are clearly not displayed in the data in Table 4. Also, this theory has no explicit pressure dependence whereas the data clearly do. A strong B_z should totally suppress this acceleration mechanism (reducing $\mathscr{E}_i$ severely), whereas energetic ions are still produced with a strong B_z (but in severely reduced numbers). Thus, in summary, it is apparent that this theory is unable to account for the observed acceleration process.

ECKER and PUTNAM /169 - 172/ have recently reconsidered the Olson-Graybill power balance limit in connection with collective ion acceleration. We note that this work is neither a theory of collective ion acceleration, nor a model, but merely an attempt to add higher-order effects to the basic power balance β_f limit. The basic limit (78) may be written as

$$\mathscr{P}_0 = \mathscr{P}_M + \mathscr{P}_{PE}, \tag{103}$$

where

$$\mathscr{P}_0 = (\mathscr{E}_e I_e/e) \qquad \text{(injected power)} \tag{104}$$

$$\mathscr{P}_M = (I_e^2/c)\, 4^{-1}[1 + 4 \ln(R/r_b)]\, \beta_f \qquad \text{(magnetic)} \tag{105}$$

$$\mathscr{P}_{PE} = (\gamma_f - 1)\, mc^2 I_e/e\,. \qquad \text{(primary electron)} \tag{106}$$

In their first work /169, 170/, ECKER and PUTNAM dropped the $\mathscr{P}_{PE}$ term and added a term to account for the secondary electron power loss,

$$\mathscr{P}_{SE} = (\mathscr{E}_e I_e/e)\, \kappa(\beta_f/\beta_{e_{||}}), \tag{107}$$

where $\kappa\mathscr{E}_e$ is the average kinetic energy with which the secondary electrons hit the walls, and $\beta_{e_{||}}$ is the injected axial electron β ($\beta \approx 1$). In the ion acceleration regime, reasonable parameter estimates are $\kappa \lesssim 0.3$, and $\beta_f/\beta_{e_{||}} \approx 0.1$; thus $\mathscr{P}_{SE}/\mathscr{P}_0 \lesssim 0.03$, so the correction obtained by adding the term (107) to (103) is typically less than 3%. The next effort by ECKER and PUTNAM /171/ was to obtain a different term for the primary electron power loss $\mathscr{P}'_{PE}$ - for this they assumed that the primary electrons are expelled radially through an average electrostatic potential given by $\phi = (I_e/\beta c)(1 - f_e)$, and using $\beta = 1$, they found

$$\mathscr{P}'_{PE} = (I_e^2/c)(1 - f_e), \tag{108}$$

where f_e is an adjustable parameter. In the power balance *limit*, this term should be minimized, and we should be left with the original $\mathscr{P}_{PE}$ term. In their latest effort /172/, ECKER and PUTNAM simply reverted back to the original power balance ($\mathscr{P}_0 = \mathscr{P}_M$) as contained in the Olson theory. Actually, if any term should be added to (103), we believe a more realistic possibility, which we propose here, is to consider the electric field power term

$$\mathscr{P}_E = (I_e^2/c)\, 4^{-1}[1 + 4 \ln(R/r_b)](\beta_f/\beta_z^2)(1 - f_e). \tag{109}$$

Of course, for absolute charge neutrality behind the beam front, $f_e = 1$ and $\mathscr{P}_E = 0$; however, for real beams, $(1 - f_e)$ may acquire a small value [or, e.g., for the special case of a propagating beam with $\nu/\gamma > 1$ whose equilibria might require $(1 - f_e) \neq 0$]. A summary of the resultant β_f's for the various cases is given in Table 7.

As we have noted earlier, the power balance limit typically does not limit β_f to much below the injected β_e. Even with the addition of higher-order effects, as in Table 7, this is still typically true. However, by forcing a 2-parameter power balance curve to fit their data, ECKER and PUTNAM /169, 170/ managed to pull the

Table 7. Power balance limit for β_f

Terms used	β_f	Authors	References
$\mathscr{P}_0, \mathscr{P}_B$	$\frac{1}{[\mathscr{F}(\Omega)]^{-1}\, 7.5[1 + 4 \ln(R/r_b)]}$	Graybill (1968) Olson (1973 - 1975) (in limit $\beta_f \ll \beta_e$)	156 16, 17 (see Eq. 80)
$\mathscr{P}_0, \mathscr{P}_B, \mathscr{P}_{SE}$	$\frac{1}{[\mathscr{F}(\Omega)]^{-1}\, 7.5[1 + 4 \ln(R/r_b)] + (\kappa/\beta_{e_{\parallel}})}$	Ecker, Putnam (1976)	169, 170
$\mathscr{P}_0, \mathscr{P}_B, \mathscr{P}_{SE}, \mathscr{P}'_{PE}$	$\frac{1 - 30[\mathscr{F}(\Omega)]^{-1} (1 - f_e)}{[\mathscr{F}(\Omega)]^{-1}\, 7.5[1 + 4 \ln(R/r_b)] + (\kappa/\beta_{e_{\parallel}})}$	Ecker, Putnam (1976)	171
$\mathscr{P}_0, \mathscr{P}_B, \mathscr{P}_E$	$\frac{1}{[\mathscr{F}(\Omega)]^{-1}\, 7.5[1 + 4 \ln(R/r_b)][1 + (1 - f_e)\, \beta_z^{-2}]}$	Olson (1978)	(this paper)

power balance limit down to very small values ($\beta_f \lesssim 0.1$). This means the power balance limit in the Olson theory (Fig. 9).was pulled down below the $\beta_f \approx \mathscr{L}/\tau_N^{e,i}$ line in the high pressure regime, and it would therefore determine β_f; in fact, if the power balance limit were low enough, it could determine β_f in all pressure regimes. However, for the Ecker-Putnam case, the "fitted parameters" were r_b and κ. To fit the data, r_b had to be taken to be less than 0.2 r_c, and κ had to be taken to be $\approx$ 3.7 /169/; this means the injected beam would have to propagate with a radius 5 times *smaller* than the injected beam radius, and that the secondary electrons would have to escape with an energy about 3.7 times *larger* than the primary electron energy (!). Further, it was assumed that the full injected beam current propagated downstream even for $\nu/\gamma > 1$ (in previous PI data with $\nu/\gamma > 1$, only a portion of the beam with $\nu/\gamma \approx 1$ propagated /164, 165, 167/). With a more recent 3-parameter fitted curve /171/, the requirements on r_b and κ are still severely unrealistic; r_b=0.37 r_c was used, as was $\kappa \approx 5.4$. Also, as we will show in Sect. 4.1.4, the 3-parameter fitted curve for $\beta_f(\mathscr{F})$ leads to gross disagreements on other experimental observables [e.g., the runaway cutoff, $\mathscr{E}_i(p)$, and $\beta_f(p)$]. Thus, based on this discussion, and the results of Sect. 4.1.4, we must still conclude that the power balance limit is not typically a severe limit for β_f.

IRANI and ROSTOKER /222/ have proposed another pinch theory, which like its predecessors, is based on the initial assumption of a force-neutral beam ($f_e = \gamma_e^{-2}$). In this theory, which most closely resembles the PUTNAM theory /203 - 206/, the polarization charge produced by the Cerenkov effect is considered, and the electrostatic fields due to the resultant envelope pinching are calculated. The fields so obtained are relatively small. Also, an attempt was made to ignore the fact that for typical data, $f_e \approx 1$ is required for the beam to propagate /54/. For most of the data where accelerated ions have been reported (see Table 4), we find $3 \lesssim (I_e/I_\ell) \lesssim 30$ so that f_e must be in the range $[1 - (I_\ell/I_e)] \leq f_e \lesssim 1$ for the beam to propagate; this requires (0.66 to 0.97) $\leq f_e \lesssim 1$ (i.e., $f_e \approx 1$). Thus this theory suffers from the problems common to all theories based on force-neutral beams. For example, in the domain of this theory's applicability as given by (67) - (68), no accelerated ions have been seen.

In summary, we have found that the Olson theory /16 - 18/, a lowest-order quasi-static field theory with no fitted parameters, is apparently able to account for the collective acceleration phenomenon observed upon IREB injection into neutral gas. For various reasons as detailed above, we have also shown that other mechanisms (envelope pinching, Cerenkov radiation, etc.) cannot account for the observed acceleration phenomenon.

4.1.3 Numerical Simulations

POUKEY and OLSON /19/ have reported two-dimensional computer simulations of the collective ion acceleration process for IREB injection into neutral gas. In these simulations, only the quasistatic fields E_r, E_z, and B_θ were calculated; inductive and radiation fields were neglected, as had been justified earlier /91/. Both electron and ion trajectories were followed in time; ion production was calculated using (74), so ion ionization effects were included. Typically, about 2000 simulation electrons were employed, together with a similar number of simulation ions. The detailed time and spatial dependence of the beam dynamics and of the ion acceleration process were observed. In Fig. 13, we show the results of one of the computer runs taken during these studies. Here we have plotted the electrostatic potential on axis $\phi(r = 0)$ at various times [to minimize computing time for this run, $t_r = 0$ was used]. Fig. 13 clearly shows the essential features of the Olson theory: I_e is $> I_\ell$ so the beam initially stops at the anode; a deep potential well forms ("stationary well stage") with $e\phi/\mathcal{E}_e = \alpha \approx 2\text{-}3$ (here $\alpha = 2.4$), and the well extends a good distance downstream ($\mathcal{L} \sim 2R$); at about the charge neutralization time ($\tau_N^{e,i} \approx 3$ nsec here), the well depth decreases to $e\phi/\mathcal{E}_e \approx 1$ and the well begins to move out (transition stage); by t = 5 nsec, the potential well has length $\mathcal{L} \sim 2R$ and the beam front is now propagating (moving well stage). Computer runs were performed for parameters of several experiments /19/; we will discuss these results in Sect. 3.1.4 with the data. It was concluded that these numerical simulation results directly substantiate the Olson theory.

KOLOMENSKII and NOVITSKII /223 - 225/ have performed numerical simulations assuming an infinite B_z, so the electrons are constrained to move only in the axial direction. Ionization by ions and by secondary electrons [which must now escape longitudinally] were included. (For the $B_z = 0$ case, secondary electrons may escape radially, with negligible ionization effects /210/). For $I_e > I_\ell$, the moving ionization wave was observed for both the zero and finite risetime cases. Ion ionization and secondary electron ionization effects were shown to considerably increase the ionization rates above those created by just beam electron impact ionization. Accelerated ions with energies several times the injected electron energy were reported.

BYSTRITSKII et al. /194, 226/ have investigated the initial deep well formation using a 1-D numerical quasistatic model. They observed well deepening ($\alpha = 2.5 - 3$) for zero and finite current risetime beams, whereas for a finite voltage risetime case ($t_r = t_v$) they reported less deepening ($\alpha \approx 1.5$). (As we have noted earlier, the voltage turn-on in real beams must actually slightly precede the current turn-on, so the real case corresponds closest to the $t_r \neq 0$, $t_v = 0$ case.)

GODFREY et al. /227, 228/ have reported 2-D numerical simulations of an envelope pinch model of collective acceleration. These simulations were performed to assess the ability of envelope pinch models such as those of PUTNAM /203 - 206/ or TSYTOVICH and KHODATAEV /133 - 136/ to accelerate ions. Ideal conditions were used to initiate the process: a force-neutral beam was assumed propagating with $\nu/\gamma = 0.5$, 1, or 2, and an ion bunch was placed in it. [Note that less than these ideal conditions must

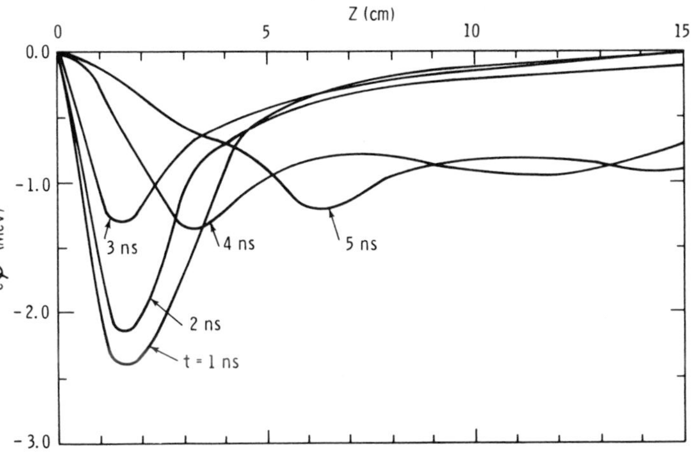

Fig. 13. Potential well formation and motion in 2-D computer simulations of IREB injection into neutral gas ($\mathcal{E}_e$ = 1 MeV, I_e = 40 kA, t_r = 0, R = 9 cm, r_b = 1.5 cm, p = 0.3 Torr H_2), including ion ionization effects /19/

occur in the drifting beam experiments, for which (67) - (68) must hold for a force-neutral beam to propagate.] However, even with these ideal conditions, the key conclusion from the simulations was that the process is phase unstable. Ions at the front of the bunch accelerate faster than those at the rear of the bunch; the bunch spreads, the envelope pinch dissipates, and acceleration eventually stops. The maximum ion energies obtained were no greater than 1.5 times the injected electron energy. It was concluded that the envelope pinch mechanism is not a useful mechanism for collective ion acceleration.

4.1.4 Comparison of Experiments, Theories, and Numerical Simulations

As concluded above, the 2-D quasistatic field theory of OLSON /16 - 18, 210 - 213/, which is a lowest-order theory with no fitted parameters, is apparently able to account for the observed collective acceleration process. Here we present some detailed comparisons of this theory with the data /14, 15, 155 - 195/ and with numerical simulations /19/. Following this, we will comment briefly on other theories in relation to the data.

Comparison of Olson Theory with Experiments and Simulations

The Olson theory predicts the acceleration process should occur only for $I_e \gtrsim I_\ell$. In Fig. 14, we have plotted I_e/I_ℓ vs $\mathcal{E}_e$ for the data in Table 4. Note that $\mathcal{E}_e$ spans almost two orders of magnitude and I_e/I_ℓ spans more than two orders of magnitude. The data clearly demonstrate that ion acceleration occurs only in the $I_e \gtrsim I_\ell$ regime. The theoretical threshold for acceleration at $I_e \approx I_\ell$ was first observed by MILLER

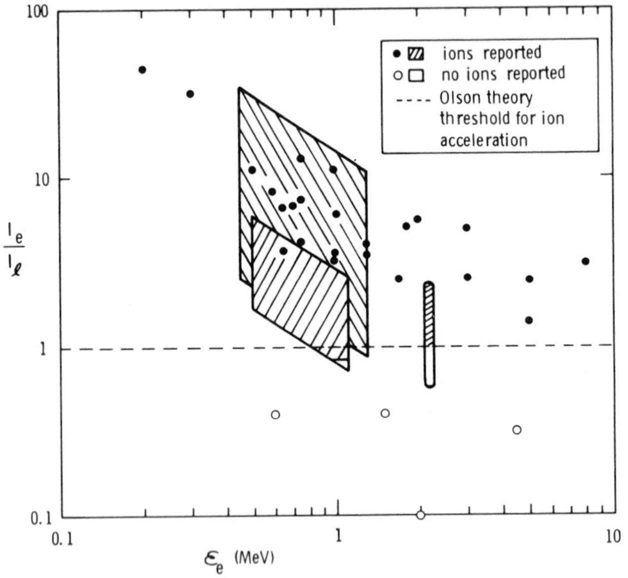

Fig. 14. I_e/I_ℓ vs $\mathcal{E}_e$ for the data in Table 4, showing the Olson theory threshold at $I_e = I_\ell$ for the acceleration process

and STRAW /178 - 182/ with $\mathcal{E}_e \approx 2$ MeV. The threshold has now also been observed by BYSTRITSKII et al. /193 - 194/ with $\mathcal{E}_e \approx 1$ MeV, as shown by the bottom tip of the diamond shaped region that crosses over into $I_e < I_\ell$ in Fig. 14. The larger diamond shaped region corresponds to the recent parameter space of ECKER and PUTNAM /169 - 172/. In Fig. 15, we have plotted ν_e/γ_e vs I_e/I_ℓ for the same data; this figure demonstrates the theoretical threshold at $I_e \approx I_\ell$, *regardless of the value of* ν_e/γ_e ($\nu_e/\gamma_e < 1$ or $\nu_e/\gamma_e > 1$). Note that much of the data from Table 4 has $\nu_e/\gamma_e < 1$ and $I_e/I_\ell > 1$, starting with the original experiments of GRAYBILL and UGLUM /14, 15/. Also note that the threshold experiments of MILLER and STRAW /178 - 182/ were performed with $\nu_e/\gamma_e \approx 0.2$, but by increasing I_e/I_ℓ to above unity by variation of R/r_b it was possible to observe the threshold at $I_e \approx I_\ell$. All of these comparisons clearly demonstrate that the acceleration process depends fundamentally on I_e/I_ℓ (rather than ν_e/γ_e).

The beam front dynamics have been compared earlier /18/. These comparisons demonstrate for a wide range of data, that the beam front stops near the anode for a time about equal to $\tau_N^{e,i}$ and then commences propagation. The propagating beam front current risetime corresponds to the distance from the beam front to the knee in the "net current" for the moving well equilibrium [see Fig. (10)]. The location of the trapped ion bunch is centered near the well minimum, whose location is near the net current knee (the precise location is given by the equations describing the beam

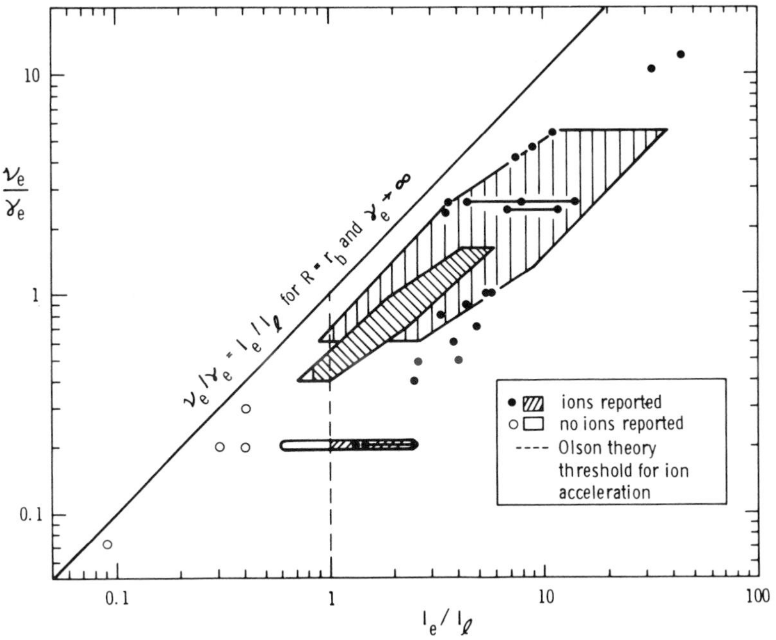

Fig. 15. ν_e/γ_e vs I_e/I_ℓ for the data in Table 4, showing the Olson theory threshold at $I_e = I_\ell$ for the acceleration process, *regardless* of the value of ν_e/γ_e

front equilibrium /17/). Evidence of the ion bunch at this location is clearly shown in the recent data of ECKER and PUTNAM /171/. The correlation of the beam front velocity with the ion bunch velocity was observed even in early investigations of the process by RANDER /164/. An early interpretation /14, 15/ that the beam front had $\beta_f \gg \beta_i$ was based on data for β_f in the runaway regime; more recent data /159/ actually indicates that $\beta_f \approx \beta_i$ for parameters in the acceleration regime (as we will show in Fig. 18a). The experimentally observed correlation in β_f and β_i has steadily improved to the point where ECKER and PUTNAM have now begun to use β_f and β_i interchangeably /169 - 172/. Thus acceleration with the beam front, but behind it a certain distance, is in agreement with the data.

The ion energies in the low pressure regime are plotted in Fig. 16 for the data in Table 4. Note that as the injected electron energy varies from 0.2 MeV to 8 MeV, the proton energies scaled to the electron energy ($\mathcal{E}_i/\mathcal{E}_e$) stay remarkably close to the theoretical prediction of $\mathcal{E}_i/\mathcal{E}_e = \alpha$, where $2 \leq \alpha \lesssim 3$. Some of the early data were originally interpreted to show $\mathcal{E}_i \sim I_e^2$ /14, 15/; however, these data were obtained by varying $\mathcal{E}_e$, and we have shown earlier that these data actually fit the scaling $\mathcal{E}_i/\mathcal{E}_e \approx \alpha$ /18/. Due to the stationary well effect, the theory predicts for a single species the scaling $\mathcal{E}_i \sim Z\alpha\mathcal{E}_e$ in the low pressure regime; this scaling has been

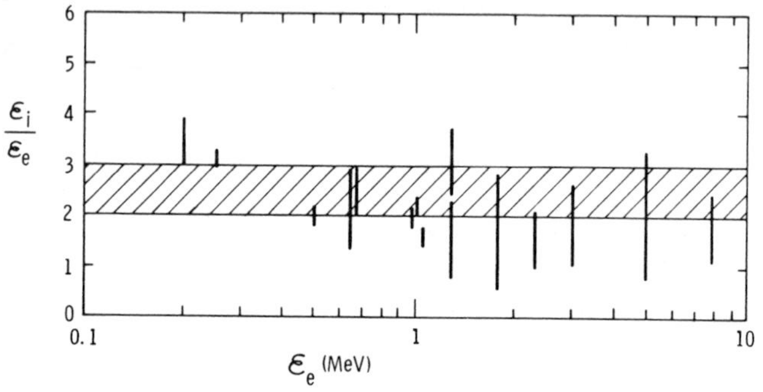

Fig. 16. Proton energies ($\mathcal{E}_i/\mathcal{E}_e$) in the *low pressure regime* for the data in Table 4, showing comparison with the Olson theory (hatched region)

demonstrated in the data /14, 15, 155 - 158/. If many species are present, the scaling $\mathcal{E}_i \sim Z$ may or may not occur depending on the parameters and the pressure regime; for example, if the beam front commences propagation before ions with the lowest Z/M values have accelerated to $\mathcal{E}_i \approx Z\alpha\mathcal{E}_e$, then these ions will slip out the well back and only ions with the highest Z/M values will attain full energy. A variety of phenomena is possible /17/, some of which have been demonstrated experimentally /18/.

The pressure dependences of β_i and β_f for all pressure regimes are shown in Figs. 17 and 18 for data from four different laboratories. Shown are the theory predictions (as in Fig. 9), the data, and the computer simulations. In Fig. 17, note the generally good agreement for β_i in the low pressure regime, for β_i in the high pressure regime, and for the location of the runaway cutoff. Note that ion energies $\mathcal{E}_i \gg 3\,\mathcal{E}_e$ may occur in the high pressure regime. In Fig. 17c, the runaway cutoff occurs at a relatively low pressure because the beam employed had a large risetime; thus it was relatively easy to charge neutralize the beam before the limiting current was reached. In Fig. 18, note the generally good agreement for β_f in all three pressure regimes, the location of the runaway cutoff, and the power balance limit for β_f in the runaway regime. In Figs. 18b and 18c, the propagating current value ($I \approx I_A$) was used in computing the power balance limit β_f. In Fig. 18b, since $I \approx I_A$ is observed downstream /164, 165/, we used I = 34 kA rather than I_e = 160 kA to compute the power balance β_f; for Fig. 18c, we used $I \approx 50$ kA as observed downstream /167/ rather than the full injected current I_e = 115 kA. If the full injected currents were used in Figs. 18b and 18c, the power balance limits would be $\beta_f = 0.39$ and $\beta_f = 0.42$ respectively - well below the observed β_f's, yet still well above the observed β_i's. Thus, as expected, the propagating current I should be used in calculating the power balance limit for β_f. Note that the computer simulations show the basic features of higher ion energies in the high pressure regime, and high beam front velocities and lack of energetic ions in the runaway regime.

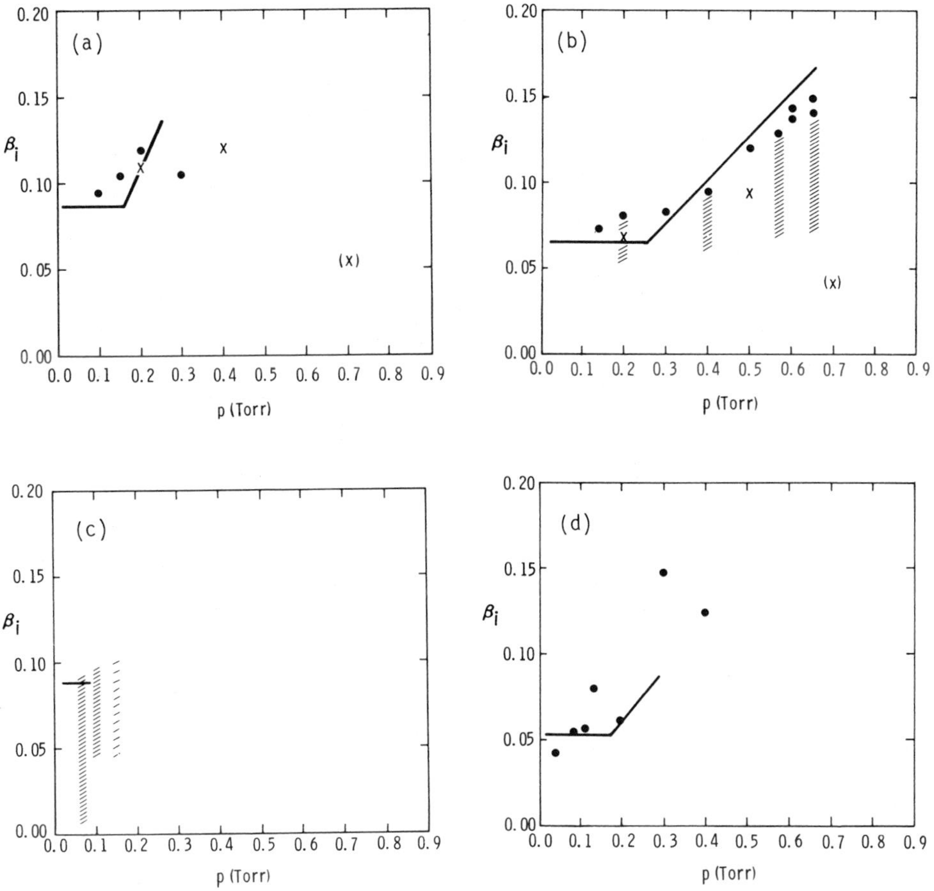

Fig. 17a - d. Proton velocities ($\beta_i c$) vs pressure for data from four different laboratories. Data (dots and hatching) of (a) GRAYBILL /156-158/, (b) ECKER et al. /167/, (c) KUSWA et al. /175-177/, and (d) KOLOMENSKII et al. /186, 189/. Theory (solid line) of OLSON /16-18/. Computer simulations (x's) of POUKEY and OLSON /19/

The theory predicts the acceleration region length should be of order ~ 2R. GRAYBILL /156 - 158/ first showed that the ion energies decrease as L decreases below ~ 2R. MILLER and STRAW /179, 182/ reported that ion energies decrease for L $\lesssim$ 2R. KOLOMENSKII et al. /186, 188, 189/ observed decreasing ion energies for L $\lesssim$ 4R, whereas BYSTRITSKII et al. /193/ found that ion currents decrease for L $\lesssim$ 3R. Thus the acceleration region length is of order 2R (although we emphasize that the actual ion acceleration occurs in the well back region over a distance which may be less than this length).

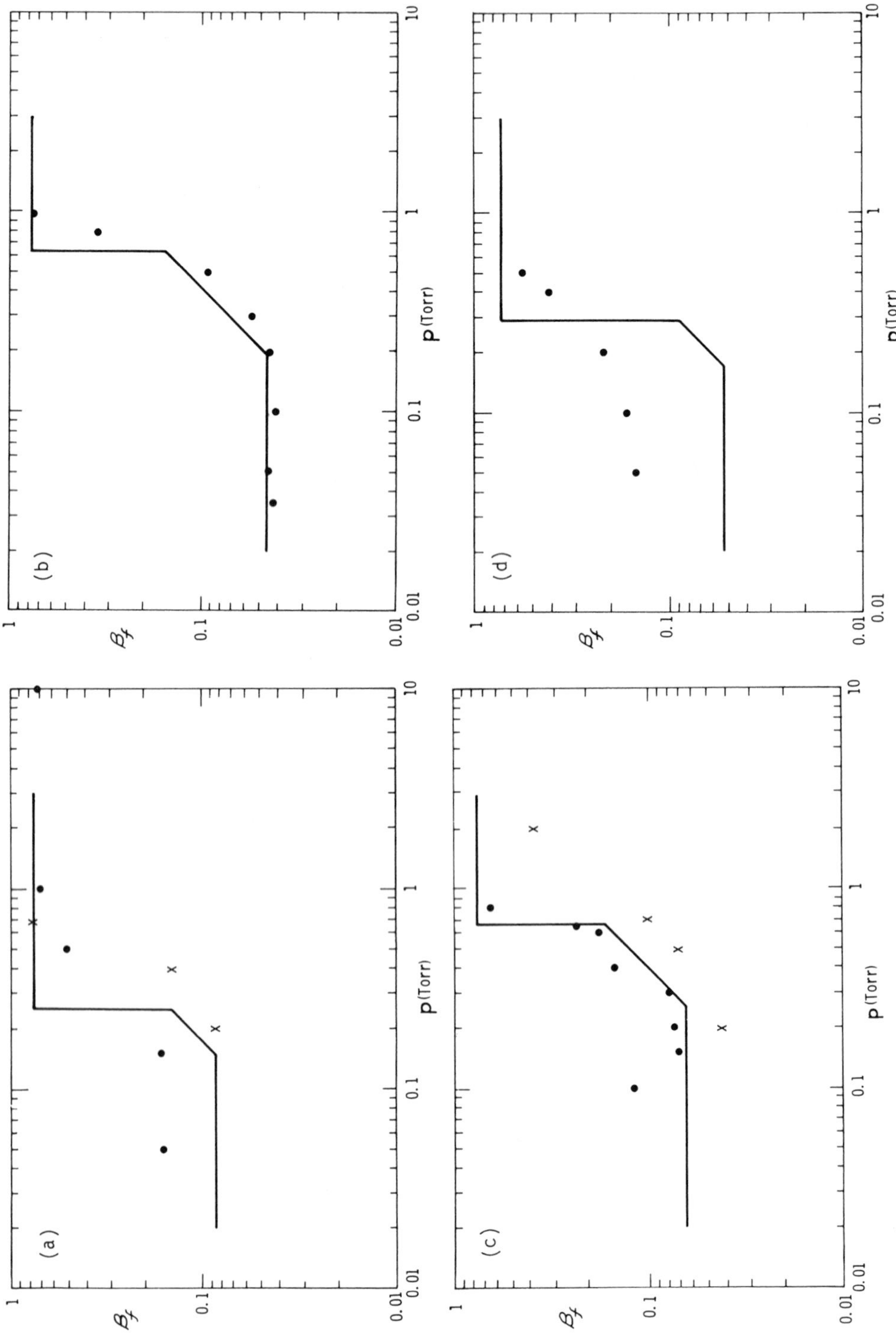
(a)
(b)
(c)
(d)
B_f
p (Torr)
0.01
0.1
1
10

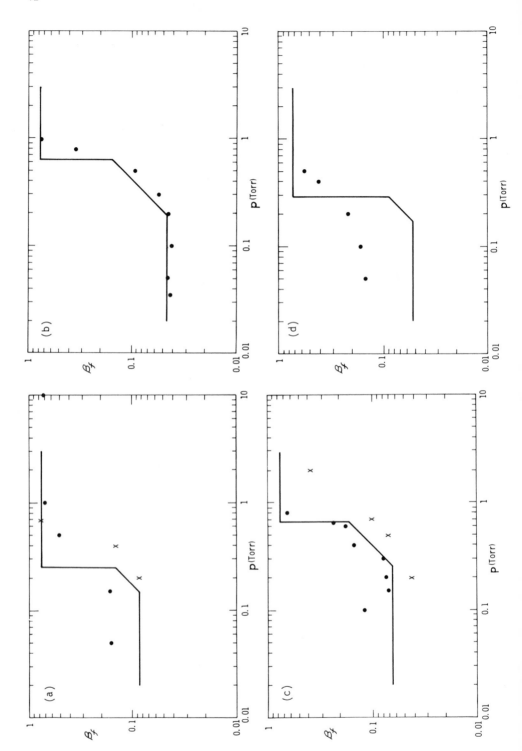

According to the theory, as discussed in Sect. 3.1.2, an applied B_z should produce high beam front velocities, lack of ion trapping, and extensive ion losses in the beam's B_θ field in the charge neutral region behind the beam front. Ion energies should typically be limited to those attained during the stationary well stage, and ion losses should severely restrict the observed ion bunch number downstream. These predictions are in agreement with the applied B_z experiments of ECKER et al. /167/ and of STRAW and MILLER /185/, and the basic effect has been seen in the 2-D numerical simulations of POUKEY and OLSON /19/.

The effects of variation of beam impedance $\mathcal{Z}$ are predicted by the theory as discussed earlier in conjunction with Fig. 11. ECKER and PUTNAM /169 - 172/ have performed experiments in which $\mathcal{Z}$ was varied, keeping p and r_b fixed. However, in their comparisons with the Olson theory, ECKER and PUTNAM did not plot the correct Olson theory curves /16 - 18/; in addition, they failed to acknowledge that the power balance limit was already in the Olson theory /16 - 18/. ECKER and PUTNAM's errors were also later published by KEEFE /32/. Here, we have plotted the correct Olson theory curves /213/. In Fig. 19, we show β_i vs $\mathcal{Z}$ for the data of ECKER and PUTNAM /169 - 172/, as well as earlier data by ECKER et al. /167, 168/. Note that the theory curves demonstrate the power balance cutoff, a hatched region (corresponding to γ_e variation in $\mathcal{L}/\tau_N^{e,i}$ for fixed $\mathcal{Z}$), and the runaway cutoff (beyond which ion acceleration should cease). The theory curves are in reasonably good agreement with the data, and show especially good agreement in regard to the runaway cutoff; note the large number of shots beyond the runaway cutoff that resulted in no accelerated ions. In their experiments, ECKER and PUTNAM /169 - 172/ reported that $\mathcal{E}_i$ decreased as r_b decreased for fixed p; however, as shown here, their data barely demonstrate this effect, and their earlier data /167, 168/ actually demonstrate the exact opposite effect. Taking all of the data into account, we find relatively little variation of $\mathcal{E}_i$ with respect to r_b. Note that there are no fitted parameters in the Olson theory, and that r_b was taken to be equal to r_c in computing the power balance limit; if r_b were really smaller than r_c, the power balance limit lines would have a less steep slope in all cases. In Fig. 20, we show β_i vs I_e for fixed p for two different values of $\mathcal{E}_e$. The theory line moves up and down with a change in $\mathcal{E}_e$ due to (71), and the line terminates at the low current end due to the runaway cutoff. Again we see reasonably good agreement between theory and experiment. Lastly, in Fig. 21, we plot β_i vs p for

---

◄ Fig. 18a - d. Beam front velocities ($\beta_f c$) vs pressure. Data (dots) of (a) GRAYBILL /159/, (b) RANDER et al. /164, 165/, (c) ECKER et al. /167/, and (d) KOLOMENSKII et al. /186, 189/. Theory (solid line) of OLSON /16 - 18/. Computer simulations (x's) of POUKEY and OLSON /19/

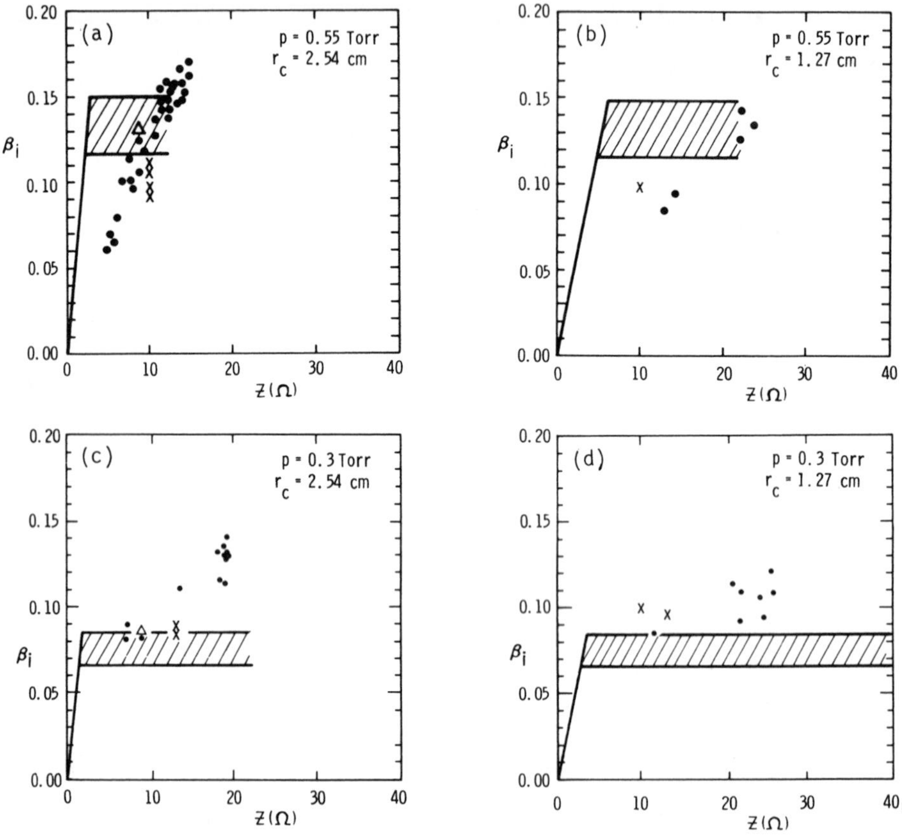

Fig. 19a - d. Proton velocities ($\beta_i c$) vs IREB impedance $\mathcal{Z}$. Data: Dots - ECKER and PUTNAM /169-172/, x's - DRICKEY et al. /168/, Δ's - ECKER et al. /167/. Theory (solid line and hatched region) of OLSON /16-18/. In (a), no ions were observed for $14.7 < \mathcal{Z}(\Omega) \leq 22.7$ in eight pulses /171/. In (b), no ions were observed for $\mathcal{Z}(\Omega) = 43$ in one pulse /171/. In (c), no ions were observed for $19.3 < \mathcal{Z}(\Omega) \leq 22.7$ in four pulses /171/

more data of ECKER and PUTNAM /172/ in which $\mathcal{Z}$ (and $\mathcal{E}_e$) were varied; the hatched region gives the range of theory predictions - the lower limit corresponds to the lowest $\mathcal{E}_e$, and the higher limit corresponds to the highest $\mathcal{E}_e$. Again reasonably good agreement is demonstrated. From Figs. 19 - 21, we therefore conclude that the theory does account for the experimentally observed values of β_i when $\mathcal{Z}$, $\mathcal{E}_e$, I_e, r_b, and/or p are varied.

The theory predicts multiple pulses for a variety of circumstances /17, 213/, as described earlier in conjunction with the multiple pulse threshold criterion (83)

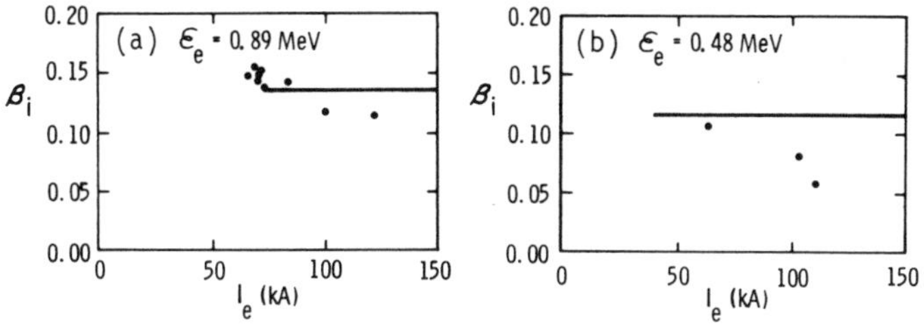

Fig. 20a and b. Ion velocities ($\beta_i c$) vs injected current I_e. Data (dots) of ECKER and PUTNAM /169, 171/. Theory (lines) of OLSON /16-18/

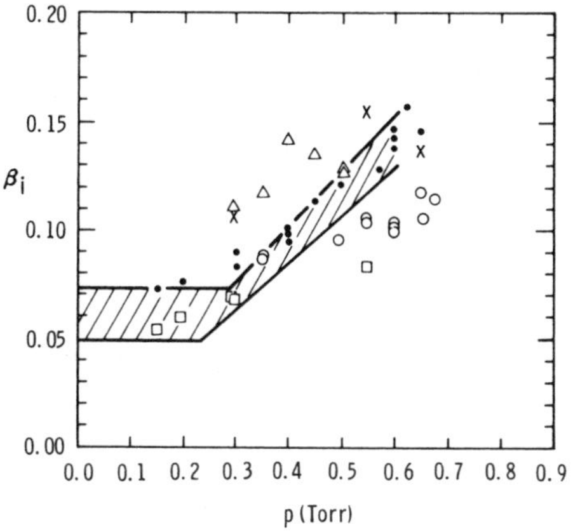

Fig. 21. Ion velocities ($\beta_i c$) vs pressure, for several values of injected electron energy $\mathcal{E}_e$. Data (dots, Δ's, □'s, o's, and x's) of ECKER and PUTNAM /172/. Theory (hatched region) of OLSON /16-18/

There it was explained how true multiple pulses may result for $I_e/I_\ell \gtrsim 5.6$, for $\nu_e/\gamma_e \gtrsim 1$, for $I_e/I_c \gtrsim 1$, and/or for large variations in $\mathcal{E}_e(t)$ and/or $I_e(t)$. It is believed that one or more of these mechanisms can account for the multiple pulses seen in the data, although multiple pulse phenomena are frequently irreproducible, and are not fully documented /18/. Nonetheless, from the data summary in Table 4, there are several cases in which there is definite experimental evidence for 0, 1, or 2 pulses. In Fig. 22 we have plotted the number of pulses vs I_e/I_ℓ for these cases. The threshold for the first pulse occurs at $I_e/I_\ell \approx 1$ (as discussed earlier). However, for $I_e/I_\ell \gtrsim 5$, the data demonstrate the appearance of a second pulse as indicated. The necessary condition ($I_e/I_\ell \gtrsim 5.6$) for the onset of a second pulse by the $f_e(t)$ mechanism is shown by the theory line in Fig. 22. Note though that since the conditions $I_e/I_\ell \gtrsim 5.6$, $\nu_e/\gamma_e \gtrsim 1$, $I_e/I_c \gtrsim 1$ frequently overlap (see Table 4), it is difficult to distinguish which of these mechanisms first leads to the multiple pulse criterion (83) being satisfied. However, the p^{-1} dependence of the pulse separation time for the $f_e(t)$ mechanism scales with the p^{-1} dependence in some of the

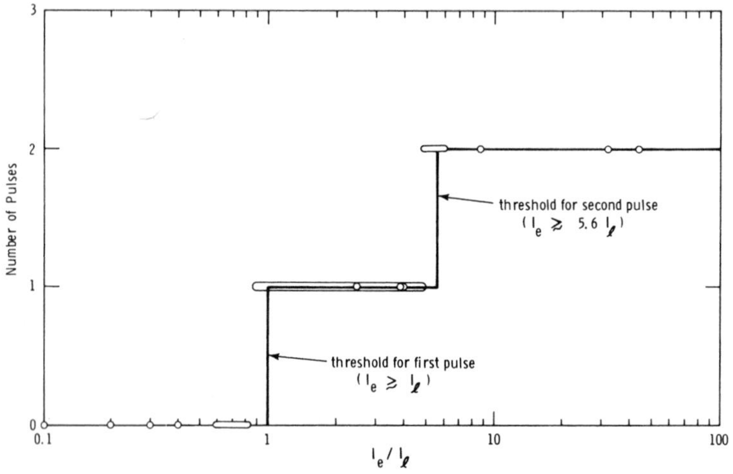

Fig. 22. Number of ion pulses vs I_e/I_ℓ. Data (circles, open blocks) of Table 4. Theory (solid line) of OLSON /16-18, 213/

early data /165/, so that this multiple pulse mechanism may in fact be the principle one.

Numerous other comparisons of the theory with the data have been made in regard to the ion number N, the ion pulse length τ, scaling of the ion pulse length with respect to Z, and the presence of radial ions /18/. In addition, scaling of β_f with R (corresponding to beam front motion in the high pressure regime) has been recently reported by BOYER et al. /236/.

We conclude that the Olson theory is indeed able to account for the large amount of data /14, 15, 155 - 195/ in which collectively accelerated ions result during IREB injection into neutral gas.

Comparison of Other Theories

The domains of applicability of the various theories discussed in Sect. 4.1.2 are summarized in Fig. 23 for the parameter space ν_e/γ_e vs I_e/I_ℓ. The domains fall naturally into three categories; (i) the 1-D theories /196, 197, 200, 201, 214 - 218/, (ii) the theories based on force-neutral beams /133 - 136, 198, 199, 203 - 206, 219 - 222/, and (iii) the Olson theory /16 - 18/. The 1-D theories may be applicable only if $c/\omega_{pe} \ll r_b$ which requires

$$(\nu_e/\gamma_e)^{1/2} \gg 1, \qquad (110)$$

whereas the theories based on force-neutral beams require (67) - (68) to hold. We note that the conditions (67), (68), and (110) are necessary requirements for

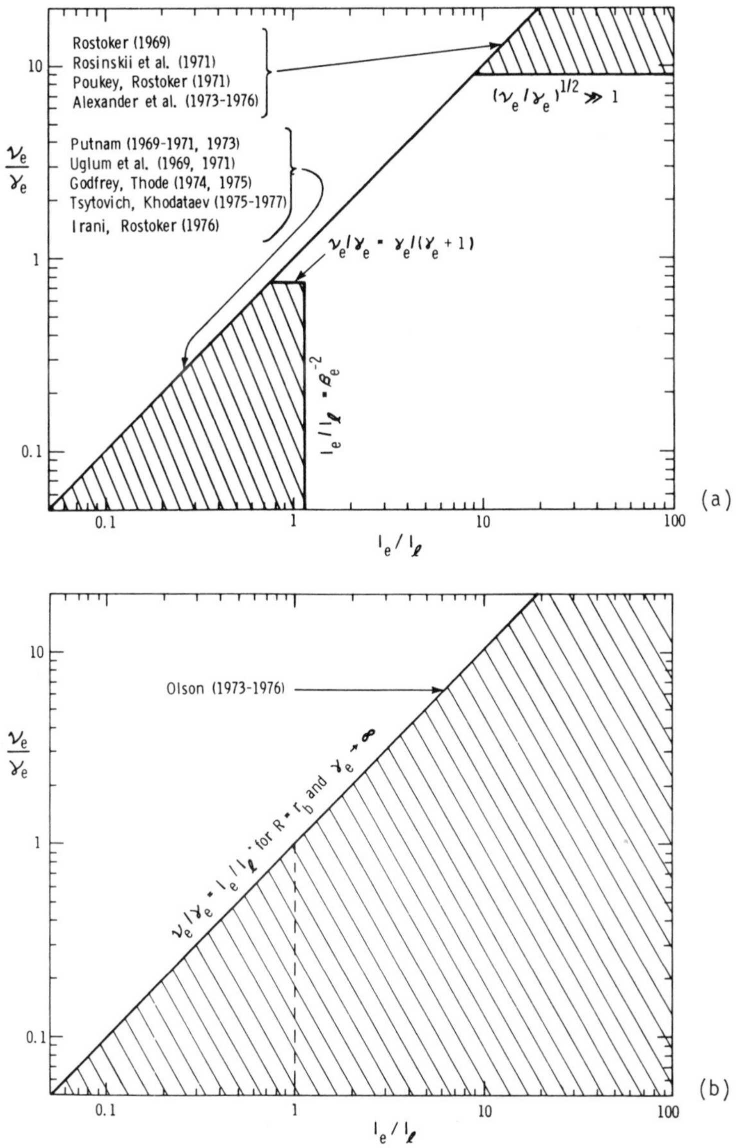

Fig. 23a and b. Domain of applicability of various theories as a function of I_e/I_ℓ and ν_e/γ_e. $\gamma = 3$ ($\mathscr{E}_e \approx 1$ MeV) was used in computing the lower hatched region boundaries in (a). This figure should be compared with the data in Fig. 15

the applicability of these theories. The Olson theory is applicable to the entire $\nu_e/\gamma_e \leq I_e/I_\ell$ space (but predicts accelerated ions only for $I_e/I_\ell \geq 1$). Note that since $I_\ell \leq I_A$ [see (6) and (10)], only the regime $\nu_e/\gamma_e \leq I_e/I_\ell$ is permitted. Comparing Fig. 23 and Fig. 15, we note that almost all of the data in which collectively

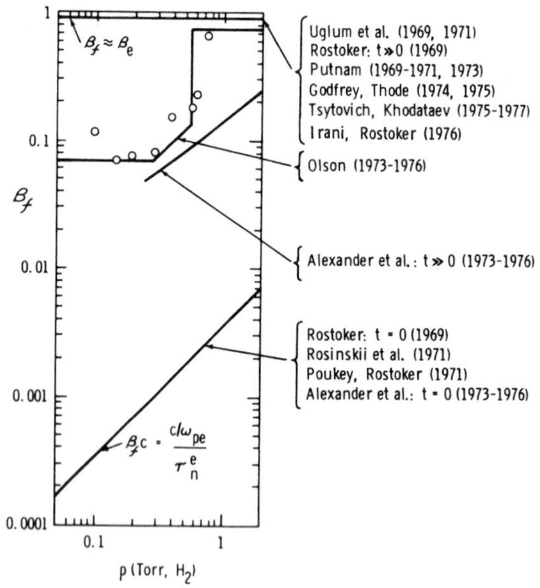

Fig. 24. Comparison of various theories for the beam front velocity ($\beta_f c$) vs pressure for the data (circles) of ECKER et al. /167/

accelerated ions have been observed lie outside the domain of applicability of the 1-D theories, and of the theories based on force-neutral beams, but fully overlap the domain of the Olson theory.

In Fig. 24, we compare the various theories for β_f vs p for the parameters of ECKER et al. /167/. As discussed earlier, the 1-D theories use the velocity $(c/\omega_{pe})/\tau_N^e$ which is much too low, whereas the force-neutral beam theories use a fast beam front. Note the difference in the theoretical predictions of ROSTOKER and ALEXANDER et al. for $t = 0$ and $t \gg 0$. The data clearly agree with the Olson theory.

In Fig. 25, we compare the 3-parameter fit for the power balance limit of ECKER and PUTNAM /171/ with the Olson theory. This β_f limit is given by the third entry in Table 7; note that the 3 "parameters" are r_b, κ, and f_e, and that there is no p dependence in this expression. [To accomodate a change in p, ECKER and PUTNAM assumed the liberty of arbitrarily varying all 3 parameter values to match any data, i.e., they reduced the 3 parameter fit to being simply a curve fit with little, if any, physical significance.] Once chosen, the values of the parameters r_b, κ, and f_e will remain fixed here; in Fig. 25a we have used the values for these parameters as chosen by ECKER and PUTNAM [r_b = 1.47 cm (whereas r_c = 2.54 cm), κ = 2.5, and f_e = 0.93]. As discussed earlier, nonphysical values of the parameters [$r_b \ll r_c$, $\kappa \gg 1$ (which means secondary electrons escape with energies $\gg \mathscr{E}_e$), etc.] had to be used to fit the data. In Fig. 25a we note the 3-parameter fit goes through these data, but has problems at both low $\mathscr{Z}$ and high $\mathscr{Z}$. For $\mathscr{Z} \approx 0$, the 3-parameter fit for β_f goes negative, which is impossible; alternatively it might be implied that for $0 \le \mathscr{Z} \le \mathscr{Z}_c$ [where $\beta_f(\mathscr{Z}_c) = 0$] the beam front cannot propagate at all, but this is also nonphysical since based on power balance arguments the beam should always be able to

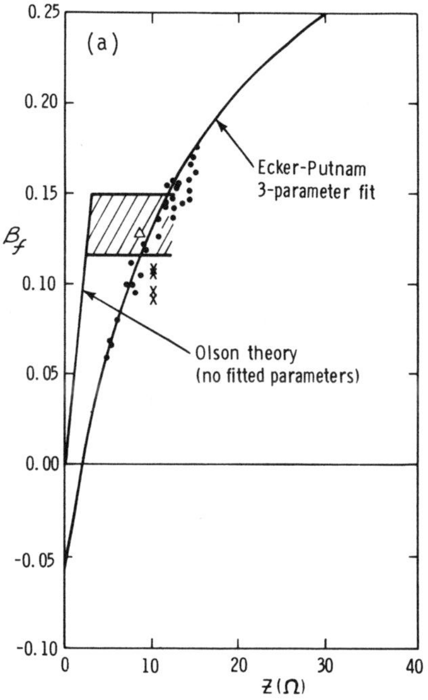

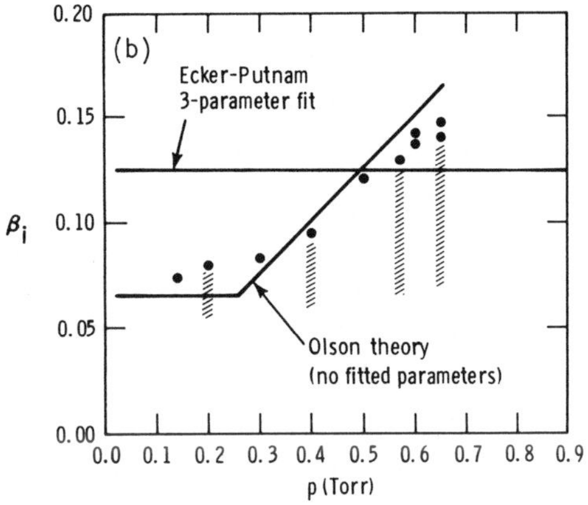

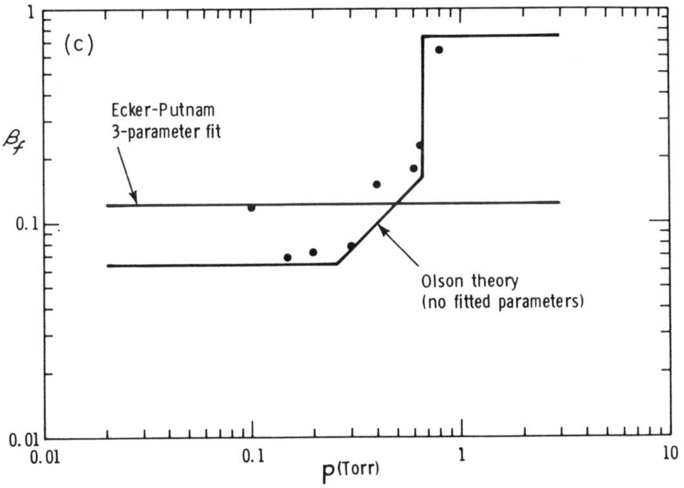

Fig. 25a - c. Comparison of the OLSON theory /16-18/ and the ECKER-PUTNAM 3-parameter curve fit /171/. (a) Beam front velocity ($\beta_f c$) vs beam impedance: data of ECKER and PUTNAM /171/. (b) Ion velocity ($\beta_i c$) vs pressure: data of ECKER et al. /167/. (c) Beam front velocity ($\beta_f c$) vs pressure: data of ECKER et al. /167/

propagate with some nonzero velocity. For large $\mathcal{Z}$ it is apparent that an ion acceleration cutoff occurs, although the "3-parameter fit" has no cutoff. In the Olson theory, two cutoff mechanisms are relevant here: (i) the runaway cutoff, which has already demonstrated good agreement with the data in Figs. 19a - d; and (ii) ion slipout /17/. Accordingly there are two cutoff scenarios, both contained in the Olson theory: (i) $r_b \approx r_c$ and the runaway cutoff is dominant, but ion slipout is also possible; and (ii) $r_b \ll r_c$ so the Olson power balance line has a much less stee slope, and ion slipout creates the cutoff. Note that in the Olson theory, as the run away cutoff is exceeded ($\mathcal{J} \geq \mathcal{J}_R$), the beam front velocity increases to the power balance limit. However, for the Ecker-Putnam 3-parameter fit, the runaway cutoff cannot be operative since the beam front β_f is already at the power balance limit. Thus in Fig. 25a, the Olson theory can account for the cutoff, whereas the 3-parameter fit cannot. In Figs. 25b, c we show the 3-parameter fit predictions for $\beta_i(p)$ and $\beta_f(p)$: in both cases, the Ecker-Putnam results show essentially no correlation with the data. Also, Fig. 25c shows how far the Ecker-Putnam β_f is below the Olson power balance β_f: this is a consequence of Ecker and Putnam's assumptions that $r_b << r_c$, $\kappa >> 1$, and that the full injected current I_e propagates downstream even for $\nu_e/\gamma_e >> 1$. It is also clear from data from other laboratories that the Olson power balance β_f is typically well above any β_i's [see Fig. (18)]. We note that TSYTOVICH has also now shown that the power balance β_f is $>> \beta_i$ in the ion acceleration regime /136/. From Figs. 25a - c and the above discussion, we must therefore conclude that the Ecker-Putnam 3-parameter fit is actually not able to account for the PI data, or data from other laboratories.

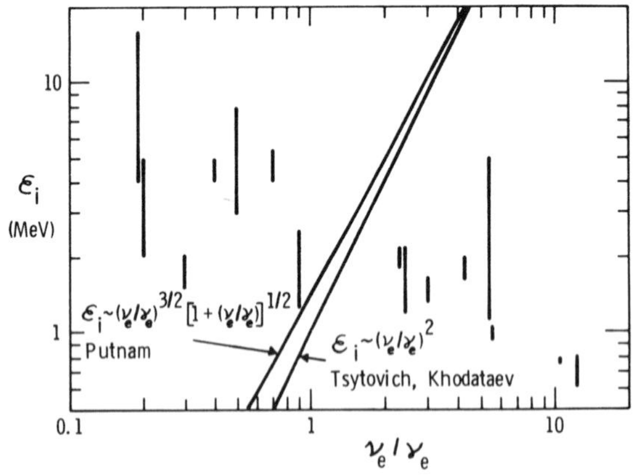

Fig. 26. Ion energy ($\mathcal{E}_e$) vs ν_e/γ_e for the data in Table 4. The pinch theory scalings of PUTNAM /203-206/, and of TSYTOVICH and KHODATAEV /133-136/, do not agree with the data

In Fig. 26, we consider the observed ion energies in the low pressure regime [fro Table 4] as a function of ν_e/γ_e. As we have shown earlier, the pinch theories of TSYTOVICH and KHODATAEV /133 - 136/ and PUTNAM /203 - 206/ are clearly not applicabl to the data [see Fig. 23]. However, if we ignore all of the problems and restriction discussed in Sect. 4.1.2, and simply plot the ion energy scaling [using (65) and (94)] as predicted by these theories, we obtain the results shown. Note that neither theoretical prediction correlates with the data. In the Olson theory, $\mathcal{E}_i$ scales as $Z\alpha\mathcal{E}_e$, as already shown in Fig. 16. In Fig. 26, $\mathcal{E}_i$'s are highest for low ν_e/γ_e (because some high γ_e beams were used) and lowest for high ν_e/γ_e (where low γ_e beams were used). Thus the data here show no advantage in using high ν_e/γ_e in regard to obtaining high ion energies. We conclude that the pinch theories cannot account for the data.

In summary, we have investigated and compared the various theories with the data in detail. We have shown that there is now extensive experimental and theoretical evidence that the acceleration process is as described by the Olson theory /16 - 18, 210 - 213/.

4.1.5 Limitations of the Observed Acceleration Process

Since the existing acceleration process takes place only over distances of the order of 10 cm, it is clearly of interest to determine if the acceleration length can be extended. The basic eleven parameters that determine the process have all been varied in the experiments, as summarized in Table 4 and in Sect. 4.1.4. Thus the acceleration process has already been extensively investigated in regard to variation of fixed parameters.

To achieve acceleration over substantially larger distances, two problems must be solved:

(i) The back of the potential well at the beam front must be kept steep. [In the final moving well equilibrium in the Olson theory (see Fig. 10), the well black length is much larger than the initial stopped beam well back length; the resulting E_z is therefore relatively small.]

(ii) The well must accelerate at a controlled rate so that the ions do not slip out the well back.

To solve (ii) it is reasonable to consider axial variation of parameters. The ionization front velocity $\mathscr{L}/\tau$ (60) suggests for fixed $\mathscr{L}$ that an axial increase in p should increase the front velocity as the beam propagates (since $\tau \sim p^{-1}$). Many authors have suggested this pressure gradient approach. Similarly, since $\mathscr{L} \sim \bar{R}$ (in the Olson theory in the high pressure regime), an axial increase in R (for fixed p) might increase the front velocity as the beam propagates. The moving well equilibrium in the Olson model [see Fig. (10)] has been investigated for propagation through an axial adiabatic change in p, and an axial adiabatic change in R. In both cases, the self-consistent moving well equilibrium was found to change its *length*, but not its *speed*, as a result of the gradient /212/.

Several experiments have been performed with gas pressure gradients, in an effort to obtain higher ion energies. KUSWA et al /176, 177/ studied both increasing and decreasing pressure gradients; in all cases, the ion energies obtained were actually less than those in the absence of the gradient. TKACH et al. /237/ studied beam front velocities in the presence of a pressure gradient, and found that an axial pressure increase of a factor of 10 produced a beam front velocity increase of only a factor of ~ 1.5. KOLOMENSKII et al. /186 - 190/ used a sectioned drift tube so that different pressures could be used in each of two sections; after substantial investigation and variation of pressures, essentially no enhancement in ion energies above those obtained for a single pressure chamber was found. MILLER and STRAW /182/, in a related experiment, used a drift tube at constant pressure but with a movable conducting center screen (ground plane); the presence of the screen always produced lower ion energies. Thus the existing pressure gradient experiments have not produced any substantial enhancement in ion energies.

Based on the above theoretical and experimental investigations, it does not appear possible to use passive means (such as pressure gradients) to achieve control of the beam front velocity. Also in these studies, no attempt was made to insure a steep well back. However, active control means have been proposed to both create a steep potential well back and control the well velocity; this is the ionization front accelerator (IFA) concept which we will discuss in Sects. 5 and 6.

4.2 Collective Ion Acceleration with Modified Drifting Beam Geometries

Naturally occurring collective ion acceleration has also been observed in several modified drifting beam geometries. These include (i) IREB injection into vacuum, (ii) IREB injection into a dielectric drift tube, and (iii) IREB injection into complicated geometries (annular beam in a cusp B_z field; foilless diode beam into a drift tube with a wall discontinuity, pressure gradient, and applied B_z). In each of these configurations, the basic collective acceleration process for IREB injection into neutral gas /16 - 18/ has been modified in a special way. In (i) and (ii), both the ion source and the means of obtaining charge neutralization are modified, whereas in (iii) the beam stopping effect is enhanced. We will discuss each of these categories briefly.

For IREB injection into vacuum, the experimental configuration is the same as in Fig. 6, but the pressure is reduced to zero. The collective acceleration process that results is a simple extension of the Olson theory /176, 177, 229/. The beam initially stops at the anode (for $I_e \gtrsim I_\ell$), and would remain stopped were it not for the fact that the beam will heat the anode foil and eventually create an anode plasma. This plasma acts as an ion source, and the initial deep well will accelerate ions from the source up to $\mathcal{E}_i \sim Z\alpha\mathcal{E}_e$. As ions are drawn out into the well, they will eventually create a background ion density sufficiently high to permit the beam front to begin propagation. A "quasi-propagation stage" will result in which charge neutrality behind the front is provided by a drawn-out dynamic ion background. Experiments by KUSWA et al. /176, 177, 229/ have demonstrated acceleration of anode ions with energies up to $\mathcal{E}_i/Z \approx 4$ MeV using an IREB with $\mathcal{E}_e = 1.8$ MeV. Estimates of foil heating and plasma production rates have been given by OLSON et al. /229/, where it was also shown that 2-D numerical simulations for the experimental parameters produced ion energies (protons) up to ~ 4 MeV. In the simulations, these ions resulted from assuming an ample ion source; it was then observed that the resultant ion current corresponded to the space charge limited ion current for acceleration into the potential well. At very low pressures, this acceleration process begins to compete with the neutral gas acceleration process. The transition between these two processes was actually observed by KUSWA /176/ and by SWAIN et al. /177/: for $0 < p \lesssim 0.05$ Torr (H_2), all shots looked like "vacuum shots", whereas for 0.05 Torr $\lesssim p \lesssim$ 0.15 Torr, all shots looked like "neutral gas shots" (here the upper pressure limit corresponds to the runaway cutoff). Thus for p above a pressure of ~ 0.05 Torr, an ion supply was created faster by the neutral gas ionization mechanism than by the anode foil mechanism. It has been proposed to use the concept of IREB injection into vacuum as an ion source for collective accelerators by OLSON et al. /229/, and by NATION et al. /238 - 240/. KIM and UHM /241/ have also investigated the process, and MILLER et al. /242/ have reported comparative studies of IREB injection into vacuum and neutral gas, both with and without an applied B_z.

For IREB injection into dielectric drift tubes, vacuum conditions are usually employed, and the drift tube is typically made of glass (quartz, pyrex) or plastic (lucite, acrylic). The boundary conditions are therefore changed from the usual metallic drift tube (perfectly conducting wall) case. When the dielectric surface is irradiated with an electron beam, plasma production and surface breakdown may result (although the precise details of these processes have not yet been investigated). Thus besides the possibility of ion production by the anode plasma, it is also now possible to have ion production from the dielectric wall plasmas. RANDER et al. /165/ performed experiments with an IREB ($\mathcal{E}_e$ = 850 keV, I_e = 100 kA) injected into a pyrex or lucite tube (R = 2.54 cm), with p = 0.08 Torr H_2. They observed high beam front velocities (β_f ~ 0.3), low net currents (~ 15 kA), and no accelerated ions. Presumably fast charge neutralization (from the wall plasmas, and from gas ionization) led to high beam front velocities and lack of ion trapping. MILLER and STRAW /182/ performed experiments with an IREB (r_c = 0.15 cm) injected into an acrylic plastic tube of radius R = 6.5 cm, which in turn was inside a metallic drift tube of radius 35.5 cm. The neutral gas pressure was varied (p = 0.15 - 0.40 Torr D_2). Ion acceleration was observed, and the acrylic tube (with or without the outer metallic boundary) behaved essentially as a metallic drift tube of about the same radius. It was postulated that rapid plasma formation on the acrylic tube walls was responsible for these results. GREENWALD et al. /243 - 244/ injected an IREB ($\mathcal{E}_e$ = 100 keV, $\nu_e/\gamma_e \approx 1$) into various dielectric drift tubes (with p = 0), and observed efficient current transport and accelerated ions with $\mathcal{E}_i/\mathcal{E}_e \approx 2 - 10$. Charge neutralization was said to be provided by ions removed from the dielectric surface, and the ion velocity correlated with the beam front velocity. PASOUR et al. /245, 246/ subsequently used a larger beam ($\mathcal{E}_e$ = 1.5 MeV, I_e = 65 kA, $\nu_e/\gamma_e \approx 1.4$, $t_b \approx 60$ nsec) to investigate scaling to higher energies. With a small diameter cathode and no anode foil, they detected proton energies up to 14 MeV. OLSON et al. /247/ performed experiments on IREB transport in various dielectric drift tubes (in vacuum and at various pressures). It was found that the drift tube material (dielectric, metal, or combination thereof) severely affected beam transport; in general, if sufficient dielectric wall material was present, transport increased at low pressures (≤ 0.02 Torr air) but decreased at higher pressures (~ 7 Torr), as compared to the metallic wall case. In summary, it appears that dielectric walls do strongly influence IREB propagation, and that they can be used to initiate a naturally occurring collective ion acceleration process.

Collective ion acceleration has also been observed in more complicated geometries. ROBERSON et al. /191, 192/ injected an annular IREB ($\mathscr{E}_e$ = 1.3 MeV, I_e = 50 kA) into a cusp B_z field in the presence of neutral gas (p = 0.3 Torr H_2). The relative number of accelerated ions ($\mathscr{E}_e$ < 4.5 MeV) was found to peak as the magnetic field approached the critical magnetic field for cusp transmission of the IREB. Near the critical field, beam stopping should occur, which may account for the peak in the ion output. RHEE et al. /248/ also injected a hollow IREB ($\mathscr{E}_e$ = 2.5 MeV) into a cusp B_z, and reported protons with energies up to $\mathscr{E}_i \approx$ 2 MeV. MAKO et al. /192, 249, 250/ injected a foilless diode beam ($\mathscr{E}_e$ = 300 keV, I_e = 20 kA) into a drift tube with (simultaneously) a sudden wall discontinuity (ΔR), a pressure gradient (∇p), and an applied B_z; accelerated ions ($\mathscr{E}_i \approx 6\,\mathscr{E}_e$) were reported only when the beam was scattered by carbon fibers. Electrostatic effects were thought to cause the acceleration /251, 252/. YOUNG and FRIEDMAN /253, 254/ injected an annular IREB ($\mathscr{E}_e$ = 750 keV, I_e = 250 kA, t_b = 90 nsec) through a metallic foil anode into an evacuated dielectric drift tube. With a deuterated polyethylene wire on axis, they observed up to 4×10^{13} deuterons with a mean energy of about 3 MeV directed along the axis of the wire. The acceleration was said to be due probably to an electrostatic acceleration mechanism. From these investigations, it is apparent that collectively accelerated ions may result in complicated geometries, although it is not clear that any substantial enhancements in ion output (over the basic metallic drift tube/neutral gas case) are acquired with the addition of these complications.

4.3 Collective Ion Acceleration in IREB Diodes

Collectively accelerated ions have also been observed in IREB diodes, for the basic diode configurations shown in Fig. 27. Ions are observed to be accelerated in the same direction that the electrons flow, i.e., the ions are accelerated *against* the applied potential U. The ion energies attained may greatly exceed the applied diode potential, $\mathscr{E}_i/Z \gg eU$. PLYUTTO /13/ first observed this phenomenon in 1960 with a plasma-filled diode device (a spark source with an extractor electrode). Subsequent work by PLYUTTO et al. has led to further observations of accelerated ions in plasma-filled diodes /33, 255 - 260/, and investigations of related diode phenomena /261-266/

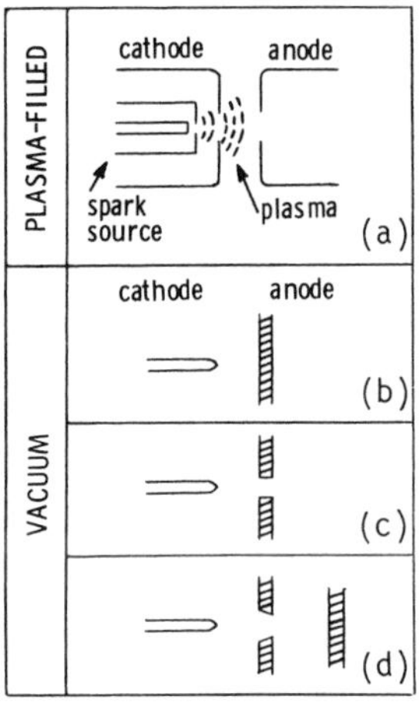

Fig. 27a - d. IREB diode configurations in which collectively accelerated ions have been observed

KOROP and PLYUTTO /267/ first observed collective acceleration phenomena in a vacuum diode in 1971. Subsequent work at eight different laboratories /175, 176, 268 - 300/ has resulted in extensive experimental investigation of these phenomena. A summary of the basic parameters associated with the collective acceleration of protons in diodes is given in Table 8. It should be noted that a great variety of electrode shapes, electrode materials, diode dimensions, and ion species are involved that cannot be included in a brief summary such as Table 8; for detailed information, the specific references must indeed be explored individually.

Although an extensive amount of experimental work has been done to confirm and characterize collective ion acceleration in diodes, relatively little theoretical work has been done to explain the process(es) involved. In all cases, time-dependent plasmas (cathode plasma, anode plasma, and/or injected plasma) must be considered, as well as the presence of the applied diode potential directly in the acceleration region. Here we will comment on the various experiments with plasma-filled diodes, and with vacuum diodes. We will then comment briefly on the status of possible theoretical explanations. It should be noted that since the diode length (and therefore the acceleration length) is always restricted to a few cm, it appears that these diode configurations in themselves can never be extended to attain high ion energies (~ GeV protons). However, these diode configurations may ultimately find applications as pulsed high current ion sources.

Plasma-Filled Diodes

PLYUTTO et al. /13, 33, 255 - 260/ at the Sukhumi Institute, Sukhumi, USSR, have investigated a variety of plasma-filled diodes. This work began as a study of spark sources /301/ with an extractor electrode /13/. In 1967, PLYUTTO et al. /255/ used the diode configuration of Fig. 27a with an applied gap voltage U of 0.2 - 0.3 MV, and obtained a broad spectrum of proton energies, with peak energy up to 4 - 5 MeV. In these and subsequent experiments /33, 256 - 260/, a plasma source (a spark source

Table 8. Collective acceleration of protons in IREB diodes

	Data Set	DIODE U(MV)	I_e(kA)	t_b(nsec) or ω	Configuration	A-K gap (cm)
plasma-filled	1	0.2-0.3			A	
	2	0-0.1	1-2	(1-2 MHz)	A	1-10
	3	~ 0.03	1-2		A	1-5
	4	0.003-0.03		(6.3 MHz)	A	1.5
	5	0.2-1		(6.3 MHz)	A	2-7
	6	0.0015			A	
vacuum diodes	7	0.2-0.3	5	~ 400	C	1-2
	8	0.06-0.12	0.1		~ C	1
	9	2.5	25	40	C	0.6
	10	5	50	120	B	1.8
	11	2.2	23	40	D	0.3, 1
	12	2	50-100	70	B	0.6
	13	0.1	10	150	C	
	14	2.3	5	50	B, C, D	
	15	2.5	30	30	D	
	16	1.5-2	60-100	30-60	D + lens	
	17	0.6	35	~ 30	D	0.25
	18	0.6-1	35	~ 30	D + lens	0.25
	19	0.6-1	35	~ 30	D + 2 lens	0.25
	20	1-2	20-40	30	D	0.6-0.7
	21	1-2	20-40	30	D + 2 lens	0.6-0.7
	22	0.3			D	
	23	4.5	50	35	~ D	1
	24	3	90	40	D	

Table 8 (continued)

Data Set	PROTONS $\mathscr{E}_i$ (MeV)	N	τ(nsec)	I_i(A)	Authors	References
1	to 4-5	10^{11}-10^{12}			Plyutto	255
2	to 2.4	10^{11}-10^{12}			Plyutto et al.	256
3	to 0.9	(~ 10^{12})	100	2	Mkheidze et al.	257
4	~ 0.4				Ryzhkov et al.	258
5	1-6				Plyutto et al.	259
6	0.01-0.08				Plyutto	260
7	2-3				Korop, Plyutto	267
8	to 0.4				Korop, Plyutto	268
9	~ 1	5 x 10^{12}			Kerns et al.	273,275
10	0.08-3				Johnson, Kerns	274
11	≤ 5.8				Hoeberling et al.	277,278
12	≤ 2.5	10^{10}-10^{12}			Bradley, Kuswa	175,176,279
13	0.1	> 10^{10}			Kuswa	175,176
14	20% > 1	10^{13}			Freeman et al.	285
15	up to 15	~ 10^{14}			Luce et al.	286
16	6-13, tail to 25	~3x10^{14}			Luce, Sahling	290, 291
17	4-6	> 10^{13}			Zorn et al.	292-294
18	tail to 8	> 10^{13}	5	180 A-12 kA	Zorn et al.	292-294
19	8, tail to 16	> 10^{13}			Zorn et al.	292-294
20	~ 7-11			~ 1 kA	Boyer et al.	239,296
21	to 13, max > 16			~ 1 kA	Boyer et al.	239,296
22	> 1.5	> 10^{12}			Williams et al.	297
23	5, tail to 10	10^{13}			Doggett,Bennett	298
24	~ 15, tail to 30		5	~ kA	Adamski et al.	299,300

or a Mg arc source) was initiated first, and the plasma created was permitted to drift into the acceleration gap before the gap voltage was applied. By varying the arc current and the delay time (before the gap voltage was applied) a wide variety of plasma densities and distributions in the gap were attainable. Both unipolar and AC voltages were applied to the gap. A characteristic feature of the resultant acceleration process is that it occurs only when the gap current exceeds some critical value. The minimum ion energies do not depend on the charge multiplicity, whereas the average ion energies (which are many times lower than the peak ion energies) do seem to scale with Z. In more recent work /259/, with applied gap potentials up to 1 MV, proton energies up to 6 MeV were observed. As seen from Table 8, the number of ions is of order 10^{12}, which occurs in a rather long pulse (~ 100 nsec). The data demonstrate that besides accelerated ions [$\mathcal{E}_i/(eU)$ up to ~ 30], accelerated electrons [$\mathcal{E}_e/(eU)$ up to ~ 3] also occur.

Vacuum Diodes

Collectively accelerated ions have also been observed in vacuum diodes at many laboratories /175, 176, 267 - 300/. Vacuum diodes develop cathode and anode plasmas, so that in time, they too become "plasma-filled". A variety of diode configurations has been used - the principal ones are shown in Figs. 27b - d. The cathode usually has a small radius, or is pointed; the anode is usually thick, may contain a central insert of a different material, and may have a hole on axis. Many materials (metals, dielectrics), coatings (especially those bearing deuterium), sizes, and shapes have been used for the cathode and the anode. Frequent use has been made of the "Bennett cathode" /302/ which is a narrow dielectric rod, and of the "Luce diode" /286, 290/, which has a rear target assembly as shown in Fig. 27d. Many of the experiments in the U.S. (especially during 1972 - 1974) were performed to determine if thermonuclear processes were taking place. Hence a frequent diagnostic in these experiments was total neutron yield (and its isotropy, or lack thereof). The overall conclusion of these investigations was that collectively accelerated ions were responsible for the neutron production through beam-target type interactions, and that thermonuclear effects, if any, were at least not the dominant effect. However, throughout these investigations, a sizeable amount of data relating to collective ion acceleration in vacuum diodes was obtained. To supplement Table 8, we will now briefly comment on some of these experiments.

KOROP et al. /33, 267 - 269/ at the Sukhumi Institute first observed accelerated protons in a vacuum diode, with ion energies up to 2 - 3 MeV for an applied gap voltage of ~ 0.3 MV. The similarities of the process to that which occurs in plasma-filled diodes were noted.

KERNS et al. /270 - 273/ at the Air Force Weapons Laboratory first observed high neutron yields in a vacuum diode using a Bennett cathode and a CD_2 target. Later JOHNSON et al. /274 - 276/ reported radially accelerated protons with energies up to 3 MeV using a 5 MV, 50 kA IREB. BRADLEY and KUSWA /175, 176, 279 - 282/ at Sandia

Laboratories first demonstrated that the neutron flux observed in vacuum diodes was produced by collectively accelerated ions. Observations of radially accelerated ions and anisotropies in the neutron production were reported. LUCE et al. /283 - 291/ at the Lawrence Livermore Laboratory developed and optimized the "Luce diode". Large numbers of energetic ions were inferred from neutron yields, and ion energies (especially a high energy tail) were determined by nuclear activation analysis. With a ~ 2 MeV, 60 - 100 kA IREB, proton energies in the range 6 - 13 MeV were reported, with a high energy tail extending out to > 25 MeV /291/.

Although the interest in neutron production subsided in 1974, there remained a continuing interest in collective acceleration with vacuum diodes. Several groups continued to investigate these phenomena, particularly in the Luce diode geometry, and sometimes with the addition of "dielectric lenses" /290/. ZORN et al. /292-294/ at the University of Maryland used a 0.6 - 1 MeV, 35 kA IREB with "floating lenses" in the Luce geometry to obtain protons with a high energy tail out to > 16 MeV. In subsequent work, BOYER et al. /236, 295, 296/ increased the electron energy to 1 - 2 MeV, and reported slightly higher proton energies. The peak ion current was originally inferred to be ~ 12 kA from the presence of a dip in the electron current monitored by a Faraday cup /293/; a direct measurement of the ion current yielded ~ 180 A /293/. In subsequent work, it was noted that proton beams "of 16 MeV energy, peak currents of 10 kA, and pulse lengths of 3 nsec" have been obtained /295/; however, this would make $N\mathscr{E}_i \gtrsim (\mathscr{E}_e I_e t_b)/e$ (which cannot occur). In reality, most of the ions have energy < 10 MeV, there is a high energy tail (> 16 MeV), and the peak ion current is probably in the range 0.2 - 1 kA /296/. WILLIAMS et al. /297/ at Cornell reported > 10^{12} protons with energies > 1.5 MeV using a Luce diode geometry with a 0.3 MeV IREB. DOGGETT and BENNETT /298/ used dielectric meshes or filaments for the anode and obtained 5 - 10 MeV protons with a 4.5 MeV, 50 kA IREB. ADAMSKI et al. /299 - 300/ at Boeing Aerospace Corporation reported mean proton energies of 15 MeV, a high energy tail to > 30 MeV, and proton currents in the kA range, using a 3 MeV, 90 kA IREB.

HOEBERLING et al. /277, 278/ at the Air Force Weapons Laboratory have recently investigated ion acceleration in the Luce geometry with emphasis on the vacuum drift region past the anode structure. It was found that the acceleration phenomenon is very similar to that which occurs for IREB injection into neutral gas, or for IREB injection into vacuum. A threshold for acceleration was observed at $I_e \approx I_\ell$, and the ion energies reported had $\mathscr{E}_i \lesssim 3Z\mathscr{E}_e$. It was noted that these results are consistent with an electrostatic acceleration mechanism /16 - 18, 178 - 185/.

Theories

For the general case of collective ion acceleration in IREB diodes, there is at present no definitive explanation for all of the results summarized in Table 8. We note that the basic diode problem differs from the drifting beam problem (IREB injection into neutral gas) in at least the following ways:

(i) In a diode, there is always an applied E_z (for drifting beams there is no applied E_z);

(ii) In a diode, moving plasma, and desorbed gases that become ionized, may provide charge neutrality (for drifting beams, gas ionization supplies charge neutrality);

(iii) In the diodes used, a narrow pointed cathode is frequently employed (for drifting beams, large uniform cathodes are typically used);

(iv) In the diodes used, the anode is frequently a thick, dielectric structure insulated from ground (with drifting beams, a thin grounded metallic foil anode is typically used).

With these differences, it is not unrealistic to expect diode acceleration to differ from drifting beam acceleration, although it is likely that a space charge mechanism (similar to that for the drifting beam case) is the dominant mechanism. Almost every possible acceleration mechanism discussed in Sect. 3 has been suggested to explain collective acceleration in IREB diodes /13, 31, 33, 255 - 300/. These include space charge mechanisms (beginning with an ambipolar diffusion space charge wave /13/), inverse coherent Cerenkov radiation, pinching effects, the "Linhart" effect /303/, the e-i two stream instability /219 - 221, 304/, turbulence effects /33, 305/, and plasma waves. We will not analyze these concepts in detail here, but we do note that many of them can be dismissed on the grounds that the fields produced are too small, that the initial conditions assumed are not met in the experiments, that 2-D instability thresholds are not exceeded, and/or that the diode E_z field was not taken into account. Thus although interesting numbers of energetic ions have been observed in a variety of diode configurations, we are still in need of a definitive theoretical explanation for the acceleration process(es) involved for all of the cases summarized in Table 8.

Summary

We conclude this section with some brief general comments about all naturally occurring collective ion acceleration processes. For IREB injection into neutral gases, the acceleration process now apparently has a valid theoretical explanation. With simple extensions, collective acceleration for IREB injection into vacuum or dielectric drift tubes may also be explained. The acceleration processes in IREB diodes involve more complications, and presently a comprehensive theoretical explanation is not available. However, accelerating fields of ~ 100 MV/cm have been observed in essentially all of the above categories, but with acceleration lengths of only ~ 10 cm. Existing attempts to increase the acceleration length (∇p, ∇R, plastic lenses, dielectric guides, wires, etc.) have not resulted in any large, scalable increases in ion energies. Although further efforts will undoubtedly be made to extend the acceleration length by such means, it is becoming apparent that significantly different configurations will be required to produce a high energy collective accelerator. In Sects. 5 and 6, we will discuss the status of progress toward this goal.

5. Collective Ion Accelerator Concepts Using Linear Electron Beams

Many collective accelerator concepts have been proposed that use linear electron beams. Some of these concepts are based on extensions of naturally occurring collective acceleration processes (as were discussed in Sect. 4). Here we present a "catalogue" of these concepts. From these, three scalable collective accelerators (IFA, ARA, CGA) have emerged that will be discussed in more detail in Sect. 6. In this section, we will consider both accelerators that use collective fields for acceleration and focusing, and accelerators that use collective fields only for focusing.

5.1 Accelerators that Use Collective Fields for Acceleration and Focusing

As summarized in Table 9, these accelerators may be categorized by the mechanism used to create the accelerating field - net space charge, envelope motion, drag effects, linear waves, or nonlinear waves. The largest categories presently are net space charge accelerators and linear wave accelerators.

5.1.1 Net Space Charge Accelerators

The ionization front accelerator (IFA) of OLSON /35 - 38/ is based on an extension of the acceleration process that occurs for IREB injection into neutral gas. In the IFA, a steep potential well is created at the front of the IREB and the well velocity is accurately controlled to permit trapping and acceleration of ions to high energies. In the IREB/neutral gas process, the potential well motion is determined by IREB-induced gas ionization effects; in the IFA, the potential well motion is controlled by laser ionization of a suitable background working gas. In the IREB/neutral gas process, a relatively long (low E_z) potential well moving at about constant velocity results (see Sect. 4.1.2 and Fig. 10); in the IFA, a very steep back (high E_z) potential well moving at an accurately programmed velocity results. The full space charge field ($E_z \sim 100$ MV/m for $\mathcal{E}_e = 1$ MeV, $I_e = 30$ kA, $r_b = 1$ cm) of an IREB may therefore be utilized for ion acceleration over long distances (many meters). The well velocity in the IFA may start at zero, so that ions can be accelerated directly from rest; this feature eliminates the need for a high energy ($\gtrsim 10$ MeV) ion source as is required for certain linear wave accelerators. Also the IFA may be programmed for any desired acceleration rate; this feature is especially useful for the slow phase velocity (but high E_z) acceleration required for partially stripped heavy ions. The IFA should produce a dense, short, energetic ion pulse with a peak power that

Table 9. Collective ion accelerator concepts that use collective fields from linear electron beams for both acceleration and focusing

Category	Concept	Authors	References
net space charge	ionization front accelerator (IFA)	Olson	31, 34 - 38, 45
	ionization front	Kucherov	306 - 308
	ionization front	Hoeberling	309
	virtual cathode control $R(z)$, $\gamma(z)$	Miller	184, 310 - 313
	virtual cathode control, helix	Boyer et al.	236
	transverse sweep	Alfven, Wernholm	4
	transverse sweep	Johnson	314
	scanning and rotation	Kolomensky, Logachev	24, 315 - 319
	transverse accelerator	Olson	35, 37, 320
	moving crossover	Agafanov et al.	321 - 323
	expanding plasma	Goldenbaum	325
	caviton sweep	Wong	327
	HIPAC	Janes et al.	329, 330
	IREB/torus	Rostoker	331
	spherical well	Verdeyen et al.	332, 333
envelope motion	moving magnetic mirror	Kovriznik	20
	moving pinch	Veksler; Benford, Benford	202, 208
	localized pinch	Putnam	203 - 206
	focusing instability	Tsytovich, Khodataev	133 - 136
	synchronized pinch	Irani, Rostoker	222

Table 9 (continued)

Category	Concept	Authors	References
drag accelerators	inverse coherent Cerenkov	Veksler; Doggett, Bennett	7, 8
	inverse coherent Bohr-Bethe	Lawson; Olson	23, 97
	inverse Cerenkov/torus	Irani; Rostoker	124
	e-i instability drag	Winterberg	335
linear waves	e-e instability	Mills	336
	autoresonant accelerator (ARA)	Sloan, Drummond	39, 40
	converging guide accelerator (CGA)	Sprangle et al.	41
	space charge waves	Yadavelli	340
	temporal γ control	Faehl, Godfrey	343
	focusing instability accelerator (FIA)	Tsytovich, Khodataev	136
	hybrid instability accelerator	Tsytovich, Khodataev	136
nonlinear waves	beat wave accelerator $R(z)$	Velikov et al.	348
	beat wave accelerator $r_b(z)$	Friedman	350, 351
	uncontrolled accelerator	Tsytovich, Khodataev	136

exceeds that of the IREB used to accelerate it. The basic IFA mechanism, its principal properties, and the status of feasibility experiments will be discussed in Sect. 6.1.

In related work, it should be noted that KUCHEROV /306 - 308/ has examined the acceleration fields possible for a sharp ionization front. Also HOEBERLING /309/ has considered a variation of the IFA concept that uses laser multiphoton ionization

An acceleration concept that employs a moving virtual cathode has been proposed by MILLER /184, 310 - 131/. This concept arose by noting that the condition for beam stopping and deep well formation, $I \geq I_\ell$ /16 - 18, 178 - 185/ depends on I, γ, R, r_b, and f_e. Using (6), this condition may be written as

$$I(z,t) \geq \beta(z,t)[\gamma(z,t) - 1](mc^3/e)\{1 + 2 \ln[R(z,t)/r_b(z,t)]\}^{-1}[1 - f_e(t)]^{-1}. \quad (111)$$

In the IFA, $f_e(z,t)$ is controlled (and $I_e \geq I_\ell$ is desirable, but not required). MILLER noted that the position of IREB stopping and well formation (for $I_e \geq I_\ell$) could in principle be controlled in space and time by varying any combination of I(z,t), $\gamma(z,t)$, R(z,t), $r_b(z,t)$ or $f_e(z,t)$. In particular, he considered the case of an unneutralized ($f_e = 0$) beam propagating along a strong B_z inside a metallic drift tube for the case of I(t) and R(z), with γ_e and r_b fixed. For a finite current risetime beam, a functional form for R(z) was calculated to produce a well acceleration toward the anode. Computer simulations /311, 313/ and experiments /312, 313/ demonstrated that a moving virtual cathode does indeed form. However, the well back turns out to be very long (which yields low E_z), and electrons reflected off of the virtua cathode lead to a two-stream instability with the incoming electrons, which eventually disrupts the beam. This problem was subsequently avoided in simulations /311/ by using a hollow beam and a relatively weak B_z, so electrons could escape radially rather than reflect back along the axis. In a 100 cm acceleration experiment with a 2 MeV, 20 kA IREB, an 8 kG B_z, and a neutral gas background in a very narrow pressure range, some accelerated deuterons with energy above 6.6 MeV were observed /312, 313/. It was noted from the simulations that the E_z field of the well toward the anode side was larger than that on the far side. It was therefore suggested that this field might be used to accelerate negative ions, or electrons, toward the anode /310, 311/. (However, it should be noted that the IREB space charge field is radially defocusing for negative particles, so that a large B_z would be required for radial containment of the accelerated particles.) Further investigations are planned.

A different virtual cathode control scheme has been proposed by BOYER et al. /236, in which $\gamma(z,t)$ in (111) is controlled by use of a helical slow wave structure. In this scheme, a helical structure is inserted into a drift tube, but is electrically isolated from it. The parameters are chosen so that $I_e < I_\ell$ if the helix is grounded

However, early in the pulse, the isolated helix will charge up, and electrons inside the helix will have the relativistic factor $\gamma(z,t)$

$$\gamma(z,t) = \gamma_e - e[\phi(z,t)]/(mc^2), \qquad (112)$$

where γ_e corresponds to the injected IREB. For $\gamma \ll \gamma_e$, the beam will stop ($I_e > I_\ell$). Thus if the input end of the helix is shorted to ground with a spark gap switch, a wave front will travel down the helix with a velocity determined by the pitch of the helix. The virtual cathode will move and continue to form near the wave front where $I > I_\ell$. By slowly increasing the helix pitch, the wave can be made to accelerate. Preliminary experiments /236/ have shown no increase in ion energies (above those from the source which was a Luce diode), but the parameters were apparently such that $I_e > I_\ell$ occurred even when the helix was grounded. Further experiments with more appropriate parameters are planned.

A totally different accelerator approach is to use the *radial* field E_r of an unneutralized (or partially neutralized) IREB to accelerate ions by displacing the beam transversely in time. ALFVEN and WERNHOLM /4/ suggested this concept in 1952 for a low current electron beam. To achieve high E fields, it was proposed to focus the beam; in the "crossover" (focus) region, high fields could then be obtained. JOHNSON /314/ in 1968 proposed a similar transverse accelerator that also used a low current, but highly focused, electron beam. KOLOMENSKY and LOGACHEV /24, 315 - 319/ in 1971 proposed several concepts for "scanning" and "rotation" of IREB's to accelerate ions. Scanning refers to schemes in which the beam is displaced transversely in time; ions are trapped in the beam and may also slip along the beam axis. Rotation refers to schemes in which an electron beam is bent into a closed configuration which is then rotated. One example (the "gyrotron") concerns a beam bent into (e.g.) a ring, and rotated about an axis so that ions injected on the axis would be trapped and accelerated outward by the centrifugal force. Although extremely complicated, such a scheme could in principle provide a quasi-continuous ion output. OLSON /35, 37, 320/ in 1973 proposed a transverse collective accelerator, in which an IREB is transported through a charge-neutralizing region where the deflection occurs, and then enters a vacuum drift tube (from the side, through a metallic foil window). Ion acceleration would occur inside the drift tube as the beam was swept along the tube. To date, no transverse accelerator schemes have actually been investigated in IREB experiments.

AGAFONOV et al. /321 - 323/ have proposed to accelerate ions in the moving "crossover" (focus) of an IREB focused by a gas lens. In this scheme, an IREB is injected through a gas-filled gap into a vacuum region. As the gas becomes ionized in the gap, the force neutral condition ($f_e = \gamma_e^{-2}$) will first be met; for subsequent times, the beam will contract ($\gamma_e^{-2} < f_e < 1$) in the gap. Thus downstream in the vacuum region, the focus will be swept *toward* the gas lens, and it was proposed to accelerate ions in the potential well associated with this focus. In examples /321, 322/, it was estimated that moderate proton energies (tens of MeV) may be achieved with a 100 kA, $\gamma_e = 5$ IREB. We note here, however, that space charge limiting effects were not considered in the analysis, which is strictly valid only for $I_e \ll I_\ell$. The scheme may still be considered, but only for $I_e < I_\ell$. (For $I_e \gtrsim I_\ell$, as in the example, the beam should actually stop in the gas lens.) Further, if I_e is close to I_ℓ, a virtual cathode will be created in the vacuum region as soon as the beam contracts. The principal effect would then be motion of a virtual cathode toward the gas lens, which could, however, still be used to accelerate ions.

For completeness, we also note that net space charge acceleration schemes exist that use plasma effects, rather than IREB's, to accelerate ions. WIDNER et al. /324/ in 1971 numerically examined the problem of plasma expansion into a vacuum; they concluded that ion energies up to about 10 times the electron thermal energy were possible. GOLDENBAUM /325/ used this concept in 1975 to propose an accelerator scheme in which a low impedance Blumlein supplies power to a stabilized Z pinch plasma; the plasma is heated and the plasma electrons expand into a vacuum region, dragging ions with them. CROW et al. /326/ reexamined the plasma expansion problem in 1975, and concluded that ion energies much higher than 10 times the electron thermal energy should be possible. A different plasma accelerator scheme has been proposed by WONG /327/ based on the concept of ion acceleration by "cavitons". If a plasma with a density gradient is irradiated with electromagnetic waves of frequency ω_0, resonant absorption will occur in a layer where $\omega_0 = \omega_{pe}(z)$. Cavitons are density cavities in the plasma caused by the ponderomotive force $[-\nabla\langle E^2\rangle/(8\pi)]$ which results from the resonant absorption. Ion energies up to 7 times the electron thermal energy have been reported in experiments in which a plasma was irradiated with microwaves /328/. It was then proposed /327/ that a continuing ion acceleration might be effected by sweeping the microwave frequency down in time, so that a quasi-continuous supply of cavitons would be created down the density gradient. It should be noted that the power source for accelerating ions in the plasma expansion scheme is a suddenly created hot plasma, whereas microwave power is essentially used to accelerate ions in the caviton scheme.

Ion acceleration with net space charge "clouds" has also been investigated. In the HIPAC (heavy ion plasma accelerator) scheme proposed by JANES et al. /329, 330/ in 1965, a toroidal electron cloud is created and contained by a magnetic field. Ions are accelerated and stripped as they oscillate radially back and forth in the net potential well of the electron cloud. A related toroidal device, but with a very high circulating current and a deeper potential well, was proposed by ROSTOKER /331/ in 1973. Spherical electron space charge clouds have been investigated by VERDEYEN et al. /332, 333/ and LARANT'EV /334/; here a spherical potential well is created, and ions are accelerated to the center of the sphere. For parabolic well shapes, ions originating at different radii will all reach the center of the sphere at the same time. This results in a very high power, short pulse, ion bunch that may be useful for pellet fusion. In all of these space charge cloud schemes, it should be noted that the potential well is stationary, and that the ion energies are therefore limited to the potential well depth.

5.1.2 Envelope Motion Accelerators

Envelope motion accelerators fall into two categories - those in which the envelope motion is externally induced, and those in which the envelope motion is created automatically by the presence of an ion bunch or an ion background. In both categories, the beam is either unneutralized ($f_e = 0$) or partially neutralized ($f_e = \gamma_e^{-2}$ usually), so that an envelope constriction results in an electrostatic field E_z due to the potential well created. As the constriction moves, an inductive field E_z is also created, but this is typically neglected. In the first category, KOVRIZNIK, according to RABINOVICH /20/, proposed to make a traveling constriction in an IREB propagating along a B_z field by externally producing a moving magnetic mirror field (a local enhancement in B_z). A moving potential well would be created at the moving constriction, if a means for creating a strong moving magnetic mirror field could be devised (none were proposed). In regard to the second category, several envelope motion acceleration *mechanisms* were discussed in Sect. 4 (see Table 7). These included (i) the pinch concept of VEKSLER /202/, (ii) the localized pinch of PUTNAM /203 - 206/, (iii) the pinch overshoot of BENFORD and BENFORD /208/, (iv) the focusing instability of TSYTOVICH and KHODATAEV /133 - 136/, and (v) the synchronized pinch of IRANI and ROSTOKER /222/. Concepts (ii) - (v) were proposed to explain ion acceleration during IREB injection into neutral gas; however, as explained in Sect. 4.3, these concepts are unable to account for the observed acceleration. Nonetheless, in principle, each of (i) - (v) could still be considered as an accelerator concept. [Concept (iv) *has* been proposed as an accelerator concept /136/, and we will discuss it further under wave accelerators.] Here it must be noted though, that since $f_e = \gamma_e^{-2}$ is required for all of these concepts, the current is restricted by space charge limiting current effects to be $I_e < I_\ell \beta_e^{-2}$, and therefore ν_e/γ_e must be sufficiently less than unity to satisfy (68). These restrictions are required for the beam to propagate at all with these schemes, if the IREB is created in the conventional manner. GODFREY et al. /227, 228/ have numerically examined the accelerator characteristics of a moving envelope constriction. Under ideal conditions, they found that the acceleration is phase unstable - the ion bunch disperses and ion energies never exceed 1.5 times the well depth. It was concluded that such a moving envelope constriction is unsuitable for high energy ion acceleration. No moving envelope constriction experiments have been reported.

5.1.3 Drag Accelerators

Drag acceleration mechanisms include those such as the inverse coherent Cerenkov radiation mechanism proposed by VEKSLER /7, 8/, and subsequently investigated by WACHTEL and EASTLUND /209/, DOGGET and BENNETT /93 - 96/, and OLSON /91, 97/. As discussed in Sect. 3.3, the accelerating fields possible with the inverse Cerenkov mechanism, or the inverse Bohr-Bethe mechanism, were shown to be relatively small (see Fig. 5), so that these mechanisms are relatively uninteresting for linear collective accelerators. However, IRANI and ROSTOKER /124/ have considered the cyclic case of an IREB in a torus. For this case, they found that focusing fields can hold an ion bunch together if the distance around the torus is an integral multiple of the dominant plasma wavelength. The accelerating fields remain relatively small, but this is not necessarily a disadvantage for the cyclic case. WINTERBERG /335/ proposed a drag accelerator concept based on the e-i two stream instability, where it was hoped that the ions would acquire the same γ value as the electrons in a time equal to one e-fold growth for this instability. In an example with a presently unattainable high-γ IREB [$\gamma_e = 1000$ ($\mathcal{E}_e \approx 500$ MeV), $I_e = 200$ kA, $r_b = 0.1$ cm], it was estimated that protons would reach $\gamma \approx 1000$ ($\mathcal{E}_i \approx 1000$ GeV) in a distance of 10 cm. However, there are several drawbacks with this concept. (1) For the parameters considered, the threshold of this instability [eq. (40)] is not met, so the instability would not be initiated (see Sect. 3.4 for a discussion of the e-i two stream instability). (2) If the instability could be initiated, the fastest growing waves would have a fixed phase velocity [eq. (39)], and the ion velocity could at best attain only this value (and then the ion γ would be much less than the electron γ). (3) The accelerating fields available, even for a saturated, nonlinear, stage of the instability, are at most of order $(I_e/\beta c)/r_b$; for the example used, this give $E_z \sim 6 \times 10^7$ V/cm, whereas the attainment of 1000 GeV protons in 10 cm would require $E_z \geq 10^{11}$ V/cm. We must therefore conclude that, at present, *linear* drag accelerator concepts are relatively uninteresting.

5.1.4 Linear Wave Accelerators

Linear wave acceleration mechanisms were discussed in Sect. 3.4, where the pioneering contributions of FAINBERG et al. /11, 125 - 127/ to this area were noted. Here we define linear wave accelerators as those whose fundamental wave properties can be described by a linear dispersion relation. Note though, e.g., that this definition does not necessarily prohibit the attainment of wave field saturation by nonlinear effects. Specific linear wave accelerator schemes are as follows.

MILLS /336/ proposed a linear wave accelerator concept in 1965 in which an IREB streams through a plasma background. The dispersion relation for this scheme demonstrated that phase velocity control could, in principle, be achieved by axial variation of the plasma density; however, the required density variations were so small a to be impractical. Nonetheless, this is apparently the first specific linear wave accelerator scheme in which the waves would automatically grow in amplitude, and in which phase velocity control would be accomplished by axial variation of a parameter.

The auto-resonant accelerator (ARA) proposed by SLOAN and DRUMMOND /39, 40/ in 1973 is based on phase velocity control of the slow cyclotron wave mode of an unneutralized ($f_e = 0$) IREB, propagating along a strong axial magnetic field B_z. The approximate linear dispersion relation for this mode yields a phase velocity

$$v_\phi = [\omega_0/(\omega_0 + \omega_{ce})]\, v_z \,, \tag{113}$$

where ω_0 is the constant (initially excited) frequency of the mode, $\omega_{ce} = eB(\gamma mc)^{-1}$ is the relativistic cyclotron frequency, and v_z is the electron flow velocity. The cyclotron wave mode is a negative energy mode which means that as energy is dissipated from the IREB/wave system, the wave fields will actually grow at the expense of the IREB particle kinetic energies /337, 338/. This feature means that as ions are accelerated in the wave (thereby removing energy from the IREB/wave system), the wave fields will grow. However, it also means that initially special means must be used to select and grow a single mode, because all negative energy modes are capable of growing. Phase velocity control in the ARA is achieved by spatially varying B_z, i.e., from (113), $\omega_{ce}(z)$ can be programmed to give the desired $v_\phi(z)$. We will discuss the ARA concept, its principal features, and the status of feasibility experiments in Sect. 6.2.

The converging guide accelerator (CGA) of SPRANGLE et al. /41/ is based on phase velocity control of the slow space charge wave on an unneutralized ($f_e = 0$) IREB, propagating along a strong axial magnetic field B_z. The linear dispersion relation for this mode yields the approximate phase velocity

$$v_\phi = \{\omega_0/[\omega_0 + (k_z/k)(\omega_{pe})]\}\, v_z \,, \tag{114}$$

where ω_0 is the constant (initially excited) frequency of the wave, $\omega_{pe} = [4\pi n_b e^2/(\gamma^3 m)]^{1/2}$ is the (longitudinal) relativistic plasma frequency, and v_z is the electron flow velocity. The slow space charge wave is also a negative energy wave. In the CGA, use is made of the fact that the local value of the electron relativistic factor on axis is

$$\gamma = \gamma_e - eI_e(\beta_z c)^{-1}[1 + 2\ln(R/r_b)](mc^2)^{-1} \,. \tag{115}$$

Thus R(z) may be programmed to vary $\gamma(z)$ and $[k_z(z)]/[k(z)]$ in (114), to give the desired $v_\phi(z)$. With a converging guide tube, $v_\phi(z)$ increases. However, to achieve low phase velocities at the beginning of the accelerator ($v_\phi \ll c$), it is necessary to have I_e very close to I_ℓ (so the beam is on the verge of stopping), as noted by BRIGGS /339/. Thus an ion injector is apparently required with this scheme, so that the initial phase velocity can be relatively high (~ 0.2 c or higher). The CGA concept, its principal features, and the status of feasibility experiments will be discussed in Sect. 6.3.

YADAVILLI /340/ has also considered slow space charge wave accelerator schemes, but for the nonrelativistic case ($\beta_e \ll 1$, $\gamma_e \approx 1$), for use as ion injectors. Here the phase velocity is

$$v_\phi = \{\omega_0/[\omega_0 + (\mathscr{R}\omega_{pe})]\} v_z , \tag{116}$$

where now $\omega_{pe} = [4\pi n_b e^2/m]^{1/2}$, and transverse boundary effects are included in the plasma reduction factor $\mathscr{R}$, which is a function of $(\omega_0 R/v_z)$ and (r_b/R) /337, 341/. It was suggested to control the phase velocity by using a strong B_z and programming R(z); or by using $B_z = 0$ and utilizing the $r_b(z)$ caused by beam spreading (for $I_e \ll I_\ell$). Two other schemes that use highly rotating electron beams were also proposed, in which phase velocity control is achieved using $B_z(z)$ or R(z). However, SHANAHAN /342/ has recently reported that rotating beams in the presence of a $B_z(z)$ are subject to an instability which is driven by an axial shear in the rotation velocity. Nonetheless, some of these schemes may find application as low energy ($\beta_i \ll 1$), relatively low current (since $I_i < I_e < I_\ell$ and I_ℓ is small because $\beta_e \ll 1$) ion sources.

FAEHL and GODFREY /343/ have proposed a variation of the ARA concept, in which temporal modulation of γ_e is used to vary v_ϕ. In this concept, R and B_z are axially constant, whereas $\gamma_e(t)$ is programmed to increase in time to make the wave accelerate. Due to the temporal character of this concept, the ion output must necessarily be pulsed. Also, the accelerating field still decreases along the accelerator, and the new complication of accurately programming $\gamma_e(t)$ must be solved.

The focusing instability accelerator (FIA) of TSYTOVICH and KHODATAEV /136/ is based on the focusing (pinch) instability of a force-neutral IREB /133 - 136/. In Sect. 4.1, it was shown that this instability was unable to account for the collective acceleration process observed during IREB injection into neutral gas. Here we briefly reconsider this instability as a possible linear wave accelerator concept. For this instability mechanism to exist, conditions (67) - (68) must hold, e.g.,

$$I_e/I_\ell < \beta_e^{-2} \tag{67}$$

if the IREB is created in the conventional manner. Note that the electron relativistic factor on axis inside the acceleration region (γ) is

$$(\gamma - 1) = (\gamma_e - 1)[1 - (I_e/I_\ell)] . \tag{117}$$

As $I_e \to I_\ell$, $\gamma \to 1$, i.e., the electron kinetic energy is depleted in overcoming the net space charge potential depression. To maintain a reasonable electron kinetic energy density, it is thus necessary to have $I_e \ll I_\ell$; this, in turn, means the accelerating fields will be relatively small. To initiate this instability requires that the threshold (100) be exceeded,

$$I_e/I_\ell \geq \gamma_e(\gamma_e + 1)\ \xi^{-2}\ , \tag{100'}$$

where $1 < \xi \lesssim 2$ typically. The requirements (67) and (100') are compatible only for low energy beams ($\gamma_e \lesssim 2$). Also note that as discussed in Sect. 4.1.2, if the focusing instability begins, the beam will become kink unstable and deflect to the wall [see (101), (102)]; this effect may preclude realization of the FIA concept altogether. Lastly we note that the phase velocity of the FIA wave, (93), may be written as

$$v_\phi = 2^{-2/3}\ \lambda^{-1/3}\ [I_e/(mc^3/e)]^{1/3}\ [Zm/M]^{1/3}\ c\ . \tag{93'}$$

In regard to possible axial variation of parameters to program v_ϕ, it should be noted that the phase velocity (93') has *no* dependence on B_z or γ_e, and only a weak dependence on R through λ. Even apart from all of these problems, it is not clear that the required delicate equilibrium of a propagating force-neutral ($f_e = \gamma_e^{-2}$) IREB could be achieved in the first place. We must therefore conclude, that as a linear wave accelerator, the FIA does not appear to be a viable concept.

Other linear wave accelerator concepts exist that are based on electron-ion instability excitation of waves in the ARA, CGA, FIA, or combinations thereof. First, it should be mentioned that the ARA and the CGA do not *require* any ion background for wave initiation and wave growth (see Sect. 6); thus, e.g., the ARA is *not* basically an e-i instability as claimed by TSYTOVICH and KHODATAEV /136/. However, if an ion background is present, wave excitation and growth may occur through e-i instabilities. INDYKUL et al. /344, 345/ considered such a case of resonant wave excitation of the ARA by an e-i instability. DRUMMOND et al. /346/ considered several cases of e-i instabilities as possible means of wave excitation for the ARA, but then actually abandoned this method for use in ARA feasibility experiments /347/. TSYTOVICH and KHODATAEV /136/ later considered a combination of the ARA and FIA. For $f_e = \gamma_e^{-2}$ and $B_z \neq 0$, they obtained the dispersion relation

$$\omega_{pi}^2 k_z^2 r_b^2\ \lambda^{-1}(\omega - k_z u)^{-2} = 1 - \{\omega_{pe}^2[(\omega - k_z v_z)^2 + \beta_e^2\omega_{pe}^2 - \omega_{ce}^2]^{-1}\}, \tag{118}$$

where $\omega_{pe}^2 = 4\pi n_b e^2\gamma_e^{-1}m^{-1}$, $\omega_{ce} = eB_z(\gamma mc)^{-1}$, u is the axial streaming velocity of the ion species, and v_z is the axial streaming velocity of the electron species. For $\omega_{ce} = 0$, (118) corresponds to the FIA case (87); for $k_z^2 \ll k^2$ and $\beta^2\omega_{pe}^2 \ll \omega_{ce}^2$, (118) corresponds to the e-i ARA case /344, 345/. The dispersion relation (118) may be analyzed for unstable waves in various regions of (ω,k) space; in particular, this was done for the spatial growth ($\mathrm{Im}\{k\} \neq 0$) case /136/. If the values of ω and $\mathrm{Re}\{k\}$ are restricted in (118) so that $\omega/(\mathrm{Re}\{k\}) = u$, then the growing wave that results is said to be due to a "resonance" instability. On the other hand, if the same dispersion relation (118) is analyzed for the fastest growing waves [anywhere in (ω,k) space so $\omega/(\mathrm{Re}\{k\}) = u$ is not required to hold], then the growing waves that result are said to be due to an "aperiodic" instability since the growth rate is very fast ($|\mathrm{Im}\{k\}| \approx |\mathrm{Re}\{k\}|$). Thus both a resonance instability and an aperiodic instability occur for the e-i ARA case, for the FIA case, and for the hybrid case. For any given case, unless special means are used, both instabilities will occur but the aperiodic instability will dominate. The aperiodic instability case leads to a nonlinear state which we will discuss below under nonlinear wave accelerators. The resonance instability for the hybrid case has

$$v_\phi = \{\omega_0/[(\omega_{pe}^2/\gamma_e^2) + \omega_{ce}^2]^{1/2}\}\, v_z \,. \tag{119}$$

Note that it would be substantially more difficult to axially program this phase velocity [than either (113) or (114)] because γ_e, n, and B_z may all change simultaneously. Thus for *linear* wave accelerators, we see no inherent advantage in complicating the phase velocity control beyond that implied for the ARA (113) or the CGA (114).

5.1.5 Nonlinear Wave Accelerators

VELIKOV et al. /348/ have proposed a nonlinear wave accelerator based on the interaction of two waves. In this concept, an IREB is propagated along a strong B_z in vacuum inside a conducting drift tube. A stationary wave is created by rippling the drift tube wall axially, and a travelling wave is created by modulating the injected IREB current. The beat waves produced may be used for collective ion acceleration. To demonstrate this concept, note that the potential on axis may be written as [see (5)]

$$\phi(z,t) = [I_e(z,t)](\beta_z c)^{-1}\{1 + 2\ \ln[R(z)/r_b]\} \,. \tag{120}$$

Then for a rippled wall, and a modulated IREB, we may choose

$$R(z) = R[1 + \xi_0 \sin(k_1 z)] \tag{121}$$

$$I_e(z,t) = I_e[1 + \xi_2 \sin(k_2 z - \omega t)] \,, \tag{122}$$

where $R(1 - |\xi_0|) > r_b$ and $|\xi_2| < 1$. Using (121) and (122) in (120) gives for $|\xi_0| \ll 1$

$$\phi(z,t) \approx \phi_0[1 + \xi_1 \sin(k_1 z)][1 + \xi_2 \sin(k_2 z - \omega t)], \tag{123}$$

where $\phi_0 \equiv I_e(\beta_z c)^{-1}[1 + 2 \ln(R/r_b)]$ and $\xi_1 = 2 \xi_0 [1 + 2 \ln(R/r_b)]^{-1}$. Noting that

$$[\sin(k_1 z)][\sin(k_2 z - \omega t)] = 1/2\{\cos[(k_1 - k_2) z + \omega t] - \cos[(k_1 + k_2) - \omega t]\}, \tag{124}$$

(123) may be written as

$$\phi(z,t) \approx \phi_0\{1 + \xi_1 \sin(k_1 z) + \xi_2 \sin(k_2 z - \omega t) + (\xi_1\xi_2/2) \cos[(k_1 - k_2)z + \omega t] - (\xi_1\xi_2/2) \cos[(k_1 + k_2) z - \omega t]\} . \tag{125}$$

Then $E_z = -\partial\phi(z,t)/\partial z$ is

$$E_z(z,t) = -\phi_0\{\xi_1 k_1 \cos(k_1 z) + \xi_2 k_2 \cos(k_2 z - \omega t) - [(\xi_1\xi_2/2)(k_1 - k_2)] \cdot \sin[(k_1 - k_2)z + \omega t] + [(\xi_1\xi_2/2)(k_1 + k_2)] \sin[(k_1 + k_2) z - \omega t]\} . \tag{126}$$

Note that four waves are produced. For $\omega = 2\pi/\tau$, $k_1 = 2\pi/L$, and $k_2 = 2\pi/(v_z\tau)$, the phase velocities for the two nonlinear (beat) waves are

$$v_\phi = [L/(L \pm v_z\tau)] \, v_z \tag{127}$$

which may be programmed by axially varying L. Small phase velocities ($v_\phi \ll c$) require $v_z\tau \gg L$, in which case $k_2 \ll k_1$ and the amplitude of the wave is

$$E_z^{max} \approx \phi_0\xi_1\xi_2\pi/L . \tag{128}$$

For the above analysis to be applicable, $\gamma(z)$ must be essentially constant. This requires $I_e \ll I_\ell$ where I_ℓ is evaluated using $R^{max} = R(1 + \xi_0)$.

VELIKOV et al. /348/ reported a small beam experiment using this concept. A 3 keV, 0.5 A e-beam ($I_e \ll I_\ell$) was modulated (2 - 4 MHz) and ions (Fe^+, Ar^+, or N_2^+) were injected with 8.5 keV energy. At the output (five spatial periods downstream), the ion energy increased 5 keV. LEBEDEV and PAZIN /349/ have analyzed the rippled wall generation of the stationary wave in more detail; they concluded that the highest fields could be produced with the lowest currents if a thin, annular beam is used. In related work, we note that FRIEDMAN /350, 351/ has proposed a variation of this accelerator concept which uses a rippled $r_b(z)$ [created by using a fixed guide tube radius R and a rippled magnetic field $B_z(z)$] rather than a rippled R(z) [as used in (121)].

In regard to applying this accelerator concept to IREB's, it should be noted that the requirement $I_e \ll I_\ell$ means $e\phi_0 << \mathscr{E}_e$, so (128) is

$$E_z^{max} << (\mathscr{E}_e/e)\ \xi_1\xi_2\pi/L\ . \qquad (129)$$

Therefore, to achieve high accelerating fields requires high $\mathscr{E}_e$, large modulations ξ_1 and ξ_2, and initially a small L. To increase v_ϕ requires L to increase; this in turn means that the accelerating field (129) must *decrease* along to accelerator. Note that this scheme is more complicated than the ARA or the CGA in that the injected IREB must be strongly modulated at high frequency (~ GHz), and that the desired wave is automatically accompanied by three undesired waves (126).

A different nonlinear wave accelerator concept has been suggested by TSYTOVICH and KHODATAEV /136/, who proposed to use a nonlinear wave that is supposed to automatically bunch, focus, and synchronously accelerate ions with high efficiency; no axial variation of parameters or other means of control are to be used. This concept is associated with the FIA. For the linear stage of the FIA [see (87) - (102)], we have already shown that v_ϕ is not amenable to axial control since it has no dependence on B_z or γ_e [see (93')]. In the nonlinear stage, one has even less of a chance to control v_ϕ, so it is not surprising that TSYTOVICH and KHODATAEV /136/ simply *hoped* that the nonlinear stage might synchronously accelerate the ions. Although this is a pleasant concept, there is presently no convincing physical basis for it. In fact, GODFREY et al. /226, 227/ have recently concluded, on the basis of numerical simulations, that the "self-synchronized" motion of an ion bunch cannot be used to form the basis of an accelerator.

5.2 Accelerators that Use Collective Fields only for Focusing

Accelerators have also been proposed that use collective fields only for radial focusing, while using conventional methods to produce axial accelerating fields. BUDKER /10/ in 1956 proposed use of an intense beam ($\gamma_e^{-2} < f_e < 1$) in a cyclic configuration for this purpose. The beam was to dissipate its transverse energy through synchrotron radiation and thereby shrink down to a fine filament. An equilibrium would be reached when scattering effects caused by the ions (which should increase the radius) were balanced by radiation losses (which should decrease the radius). The enormous electric and magnetic fields that would be produced were to form the basis for a compact very-strong-focusing, high-energy accelerator. However, as noted by LAWSON /23/, the desired beam state could not be reached in practice due to instability problems, and the fact that the relaxation time to reach the small radius equilibrium is extremely long.

ROSTOKER /331/ in 1972 proposed creation of a high ν/γ IREB in vacuum, in a toroidal magnetic field, inside a torus. Here it was proposed to inject a low ν/γ beam into a rising toroidal magnetic field at the edge of a torus; as the field rises, the beam would sweep into the torus. A toroidal electric field would also be employed to further accelerate the beam. The final high ν/γ beam equilibria in vacuum would supply large radial focusing fields for ions, while a conventional applied E_z would be used to accelerate the ions.

ROSTOKER et al. /352 - 355/ subsequently proposed a different scheme in 1976, which is more closely related to the HIPAC. In this scheme, the electron cloud does not circulate, but is confined in a "bumpy" torus consisting of a series of mild magnetic mirrors. If a small induced toroidal electric field is applied, the electrons will remain trapped (and not circulate), while ions (specifically heavy ions) would be continually accelerated around the torus. Thus the confined electrons would radially focus the ions, while conventional induced fields would accelerate them. Continuing investigations of this concept are planned.

6. Scalable Collective Ion Accelerators Using Linear Electron Beams

Many collective accelerator concepts were discussed in Sect. 5 that use collective fields for both acceleration and focusing. From amongst these, three scalable accelerator schemes have emerged, for which feasibility experiments are currently in progress. Here we consider these accelerators (IFA, ARA, CGA) in more detail.

6.1 Ionization Front Accelerator (IFA)

6.1.1 Description

In the IFA /35 - 38/, as shown in Fig. 28, external means are used to accurately control the ionization of a special background working gas. No magnetic field is required, and the conducting drift tube is taken to have $R \approx r_b$. The gas pressure is chosen low enough so that IREB-induced ionization processes cannot substantially ionize the gas before the external ionizer does. The injected IREB will propagate freely in the ionized region behind the front, but at the front a potential well with a steep well back will be created. If the ionization front is moved, the potential well will follow it synchronously. Here the IREB supplies the fields and the power to accelerate and focus the ions; the external ionizer supplies only a very small amount of energy to control the IREB motion. Note that absolute control of the potential well motion is possible with this scheme, and that this scheme does *not* correspond to pressure control of the CRAVATH-LOEB velocity $\mathscr{L}/\tau$ [see (60)] as erroneously interpreted by some authors /136, 356/. The IFA will produce a dense, short, energetic ion bunch with a peak power higher (up to orders of magnitude higher)

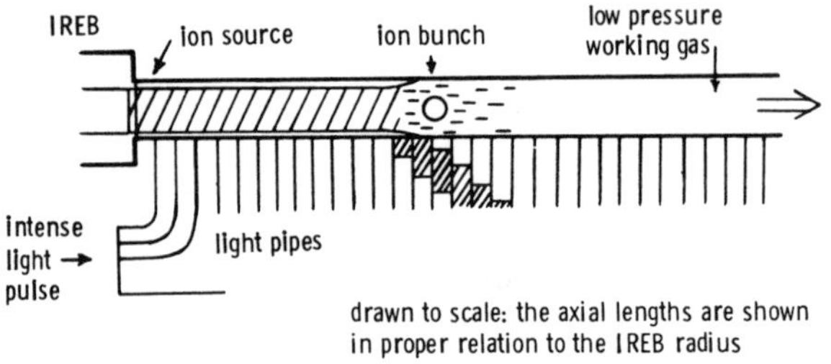

Fig. 28. Ionization front accelerator (IFA)

than the IREB used to accelerate it. It should be noted that moderate fluctuations in time of the beam parameters (γ_e, I_e, r_b) are not critical in IFA; these effects may cause variations in the potential well amplitude, but will not affect the potential well motion, which is determined entirely by the external ionizer.

For the *charge-neutralized* IFA /35 - 38/, we would have $f_e = 1$ behind the front and $f_e = 0$ ahead of the front. For this case, $\nu_e/\gamma_e < 1$ is required for the IREB to freely propagate in the region behind the beam front, and a steep well back will be created for either $I_e \gtrsim I_\ell$, or for $I_e < I_\ell$. For the *current-neutralized* IFA /35, 36/, we would have $f_e = 1$, $f_m \approx 1$ behind the front and $f_e = 0$, $f_m = 0$ ahead of the front. For this case, $\nu_e/\gamma_e > 1$ is allowed. The current-neutralized IFA is interesting for scaling toward relativistic ion energies, and for large ion numbers per pulse. However, all of the basic IFA features may be demonstrated in the charge-neutralized case, for which feasibility experiments are currently in progress.

In applying the IFA concept, several choices have been considered for the background gas, the ionization method, the external ionizer, and the sweep method /35-38/. The currently optimized choices include Cs as the working gas, 2-step photoionization as the ionization method, lasers as the external ionizers, and a light pipe array as the sweep method. The accelerated ion species, which may be created in a number of different ways /229/, is typically different than the working gas species. Ions from the working gas acquire only a very small Z/A value, and are therefore not trapped and accelerated in the moving well. Two-step photoionization was selected because the intense light source powers required are substantially lower than those for single-step photoionization. Lasers were selected as the intense light sources in the feasibility experiments, although other sources (e.g., IREB-powered SRL sources) may eventually replace the lasers altogether. To create the required programmed, light sweep, a sharp risetime light pulse is to be injected into a light pipe array as shown in Fig. 28; the light pipe lengths are systematically increased along the drift tube so that the pulses will arrive at the drift tube in a programmed sequence, thus affecting a sweep. Although synchronization of the intense light sources with the IREB is required, it should be noted that the required intense light sweep is then actually produced *passively*.

We summarize here briefly the key operating parameters of a charge-neutralized IFA. For $\nu_e/\gamma_e < 1$, the average axial electron velocity in the region behind the front ($f_e = 1$) is /35 - 38/

$$\beta_z \approx \beta_e[1 + (\nu_e/\gamma_e)]^{-1/2} \tag{130}$$

so that for a sharp ionization front (rise length $\lesssim r_b$), the axial accelerating field is

$$E_0 \approx I_e[1 + (\nu_e/\gamma_e)]^{1/2} (\beta_e c r_b)^{-1} . \tag{131}$$

The desired well motion, during which (131) is to be maintained, is relativistically

$$z = (c^2/a^*)\{[1 + (a^* t/c)^2]^{1/2} - 1\} , \tag{132}$$

where $a^* \lesssim Ze\, E_0/M$. For nonrelativistic velocities, (132) reduces to $z = a^* t^2/2$, and the acceleration time T is

$$T = 2L(\beta_i c)^{-1} , \tag{133}$$

where $\beta_i c$ is the final ion velocity, and L is the acceleration length. The ion bunch parameters are

$$\mathcal{E}_i = Ze\, E_0 L \tag{134}$$

$$\Delta\mathcal{E}_i/\mathcal{E}_i < (Z/A)(e\phi_0)(\mathcal{E}_i/A)^{-1} \tag{135}$$

$$N \approx r_b^3\, n_b\, Z^{-1} \tag{136}$$

$$T_i \approx r_b(\beta_i c)^{-1} , \tag{137}$$

where $e\phi_0$ is the potential well depth, and T_i is the ion bunch pulse length.

Numerical simulations have been performed by POUKEY and OLSON /34, 45, 357, 358/ which demonstrate that the IFA concept should indeed work. For the feasibility experiment parameters (discussed below), simulations have shown that the IREB head follows the moving ionization front synchronously, and makes there a potential well whose depth is of the order of the electron beam energy. Various initial spatial distributions of the ions were considered and it was shown that a large fraction of these ions become trapped and are accelerated with the front.

6.1.2 IFA Feasibility Experiments

A series of experiments have been undertaken by OLSON et al. /247, 358, 359/ at Sandia Laboratories to demonstrate feasibility of the IFA. The parameters chosen for these experiments are summarized in Table 10. The first light source is a dye laser (8521 Å) for Cs excitation; the second light source is a frequency-doubled ruby laser for photoionization of the Cs from the excited state. The acceleration section should produce 10 MeV protons in a distance of 10 cm ($E_0 \approx 100$ MV/m). Note that the expected ion output (10^{12}, 10 MeV with a 120 J, 0.6 MeV IREB) is more than three orders of magnitude more efficient than the closest IREB/gas data ($\sim 3 \times 10^{10}$, 10 MeV protons with a 10 kJ, 1 MeV IREB - see data set 12 in Table 4 /167, 173/). Also, even in this experiment, the ion power should exceed the IREB power. The main value of the experiment, however, is to demonstrate the IFA principle of acceleration - controlled potential well motion - and open the route to higher energy IFA systems.

The required apparatus were all assembled and made operational in 1977 - these include the high $\mathcal{Z}$ IREB, the dye laser, the fast shutter (to produce an optical rise time of < 1 nsec for the dye laser output), the Cs drift chamber, and several light pipe arrays (with various sweep rates). IFA feasibility is to be demonstrated by achieving the following goals:

I. Demonstrate Cs is a feasible working gas

II. Demonstrate laser-controlled IREB beam-front motion

III. Demonstrate IFA collective ion acceleration

Goal I was achieved /358/ in a series of IREB drift studies in Cs from which the effective ionization time for IREB-induced ionization of Cs was deduced. These results showed that a neutral Cs density of up to 10^{15} cm^{-3} could be used before IREB-induced ionization would begin to interfere with the IFA process. Since the IREB density is typically $\sim 10^{12}$ cm^{-3}, this means the Cs has to be ionized only $\sim$ 0.1% to make the IFA functional.

Goal II involves operation of the entire system. To achieve this goal, five prerequisite problems had to be solved - the first four have been solved, and the fifth is near solution /247/; (i) A "transparent/conducting" drift tube was developed that allows the swept dye laser pulse to enter the drift tube from the sides, but still has perfectly conducting boundary conditions (see Sect. 4.2). This special drift tube is required for IFA operation. (ii) A uniform, sharp-front, dye laser sweep had to be demonstrated. This was accomplished by observing the output from the dye laser/shutter/light pipe array system with a streak camera. However, to make the 8521 Å light visible to the P-11 photocathode of the streak camera, a frequency-doubling crystal was inserted just before the light pipe array. The streak camera pictures demonstrated that a sharp-front, accurately programmed, swept light front had been achieved. (iii) To observe the beam front motion in the controlled beam front experiments, it was necessary to develop a new time-resolved beam front diagnostic. This consists of a scintillator lining the inside of the drift tube walls;

Table 10. Parameters for IFA feasibility experiments

IREB	$\mathscr{E}_e$	600 keV
	I_e	20 kA
	t_b	> 10 nsec
	r_b	0.5 cm
	$\mathscr{P}_e$	1.2×10^{10} W
	$\mathscr{E}_b$	> 120 J
DRIFT	ℓ	10 cm
	E_o	10^6 V/cm
	p	0.03 Torr
	$\mathscr{E}_{ionize}$	3×10^{-5} J
LIGHT	$\mathscr{P}_1$	5×10^6 W
	τ_1	1 nsec
	$\mathscr{E}_1$	0.005 J
	$\mathscr{P}_2$	1.5×10^8 W
	τ_2	10 nsec
	$\mathscr{E}_2$	1.5 J
PROTONS	$\mathscr{E}_i$	10 MeV
	N	10^{12}
	$\mathscr{P}_i$	1.5×10^{10} W
	T_i	0.1 nsec
	$\mathscr{E}$	1.5 J

IREB electrons strike the scintillator, and the emitted light is recorded with a streak camera. This technique was perfected, so that it is now possible to see, with subnanosecond resolution, e.g., stopped IREB's (for injection into vacuum) and fully propagating IREB's (at optimum transport pressures). (iv) Because the experimental chamber must be heated to ~ 250 ^{0}C to achieve the desired Cs operating pressure, a number of complications arise concerning materials problems. These problems were adequately solved and it is now possible to operate at the required temperatures rather routinely. (v) To achieve the required synchronization of the lasers and the IREB it was necessary to minimize the jitter and delay times as much as possible. Present jitter measurements yield standard deviation values of ~ 6 nsec for the IREB (start of blumlein charging to pulse out), < 4 nsec for the lasers (trigger to pulse out), and ~ 1 nsec for the fast shutter (trigger to shutter pulse out). The first system shots began in September 1977, in which the fast shutter was not used (so the dye laser pulse had a rather long risetime) and laser pretriggering problems arose (which were later solved). Nonetheless, in some of these first system shots, the synchronization was favorable, and the streak pictures did roughly indicate that the beam front was controlled by the lasers. However, further tests, with the fast shutter, will be made before this conclusion can be finalized.

Goal III involves adding ion diagnostics to the complete IFA system, and investigating ion trapping and acceleration for various sweep rates. A magnetic spectrometer with cellulose nitrate detectors was recently added to the system, and full IFA collective ion acceleration investigations are now being initiated.

6.1.3 IFA Scaling

Here we consider the scaling of the IFA in regard to maximum ion energy, peak power, efficiency, control accuracy, and general accelerator characteristics (length, acceleration time, etc.).

We first consider the maximum ion energy possible for a *charge-neutralized* IFA. Several effects can limit the maximum beam-front velocity, which in turn, determines the maximum ion energy. Possible limiting effects include the injected electron velocity limit

$$\gamma_i = \gamma_e \, , \tag{138}$$

the magnetic stopping limit

$$\nu_e/\gamma_e = 1 \, , \tag{139}$$

and the power balance limit for magnetic field creation, (79), which may be written as

$$\beta_i = 4(\gamma_e - 1)(\beta_e\gamma_e)^{-1}(\nu_e/\gamma_e)^{-1}[1 + 4 \ln (R/r_b)]^{-1} . \tag{140}$$

A self-consistent power balance limit may be written from (78) as /213/

$$\mathscr{E}_e I_e/e = (I_e^2/c)[1 + 4 \ln (R/r_b)](\beta_f/4) + (\mathscr{E}_e I_e/e)\, \xi(\beta_f/\beta_z) + (\gamma_f - 1)\, mc^2 I_e/e, \tag{141}$$

where

$$\left.\begin{aligned} \xi &= 3^{-1}(\nu_e/\gamma_e)[1 + (\nu_e/\gamma_e)]^{1/2}\, \gamma_e(\gamma_e - 1)^{-1} \\ &\qquad \text{for } (\nu_e/\gamma_e)[1 + (\nu_e/\gamma_e)]^{1/2}\, \gamma_e(\gamma_e - 1)^{-1} < 1 \\ \xi &= 3^{-1} \qquad \text{for } (\nu_e/\gamma_e)[1 + (\nu_e/\gamma_e)]^{1/2}\, \gamma_e(\gamma_e - 1)^{-1} > 1 . \end{aligned}\right\} \tag{142}$$

Here β_e, γ_e refer to the injected electrons; β_i, γ_i refer to the ions; and β_f, γ_f refer to the beam front ($\beta_i = \beta_f$, $\gamma_i = \gamma_f$). The ξ term in (141) represents the secondary electron power loss, and the factor 3^{-1} in ξ arises from averaging the possible secondary electron escape energies over the beam cross section /213/. Result (141) may be solved for γ_f by iteration. A kinematic limit arises due to the trajectory oscillation of the beam electrons in the beam's self-magnetic field, (130); this may be written as

$$\gamma_f = \gamma_e[1 + (\nu_e/\gamma_e)]^{1/2}\, [1 + \gamma_e^2(\nu_e/\gamma_e)]^{-1/2} . \tag{143}$$

Lastly, a final limit arises by using the power balance limit γ from (141) in (143), An acceleration time limit may be derived by combining (131), (132) and (134); using $a^* = ZeE_0/M$, this gives the exact relativistic result

$$\mathscr{E}_i/A = M_p c^2 \left\{\left[1 + \left\{K(\nu_e/\gamma_e)[1 + (\nu_e/\gamma_e)]^{1/2}\right\}^2\right]^{1/2} - 1\right\} \tag{144}$$

for

$$(\nu_e/\gamma_e)[1 + (\nu_e/\gamma_e)]^{1/2}\, \gamma_e(\gamma_e - 1)^{-1} < 1, \tag{145}$$

where

$$K = T\, mc\gamma_e(r_b\, M_p)^{-1}\, (Z/A), \tag{146}$$

M_p is the proton rest mass, Z is the ion charge, and A is the ion atomic number. When (145) does not hold, the expression for E_0 differs from (131) and result (144) changes accordingly /213/.

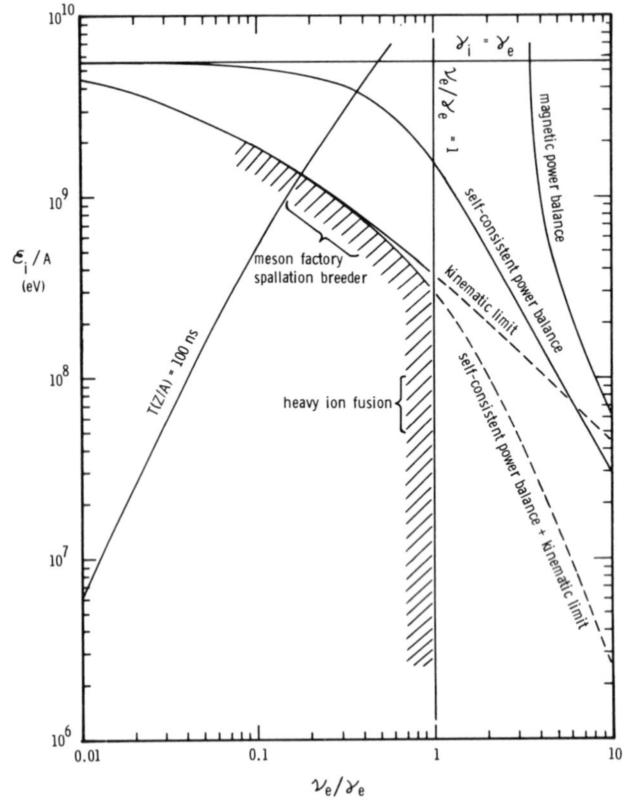

Fig. 29. Ion energy per nucleon ($\mathcal{E}_i$/A) vs ν_e/γ_e for the IFA, for $\gamma_e = 7$ ($\mathcal{E}_e \approx 3$ MeV). The various loci and limits are discussed in the text. The proposed operating regime is shown hatched /35-38/

The seven limits included in (138) - (144) are plotted in Fig. 29 for the charge-neutralized IFA case of $\gamma_e = 7$ ($\mathcal{E}_e \approx 3$ MeV), $R = r_b$, and $K = 11.4$ (which correspond to TZ/A = 100 nsec, and $r_b = 1$ cm). Here the kinematic results are drawn dashed for $\nu_e/\gamma_e > 1$ to emphasize that they are strictly applicable only for $\nu_e/\gamma_e < 1$. Note from Fig. 29, that in the IFA operating regime, power balance effects are essentially negligible. The key limit is the kinematic limit; note that no power loss is created by this limit, and that almost all of the beam power is still available for accelerating ions at this limit. The kinematic limit was discussed in the original IFA papers /35 - 38/, where it was noted that $\nu_e/\gamma_e < 1$ is desirable to achieve a fast beam front velocity, whereas $I_e/I_\ell \sim 1$ is also desirable to achieve as large an E_z as possible. Thus the *charge-neutralized* IFA operating regime ($\nu_e/\gamma_e \sim 0.1 - 0.3$ for $\mathcal{E}_i/A \sim 1$ GeV, and $\nu_e/\gamma_e \sim 1$ for $\mathcal{E}_i/A \ll 1$ GeV) was proposed /35 - 38/, as is shown in Fig. 29. The time limit line is not really a limit but a design parameter, in that it defines the beam pulse length required to achieve a desired ion energy. As shown, a 100 nsec IREB is more than sufficient to achieve 1 GeV protons (IREB's with slightly shorter pulses were actually used in the original examples of 1 GeV proton IFA's /35 - 38/). Also shown in Fig. 29 are the operating regimes for some potential IFA applications.

In related work, we note that TSYTOVICH and KHODATAEV /136/ attempted to define similar limits for a charge-neutralized IFA. However, their power balance equation [the equivalent of (141)] is both conceptually and dimensionally wrong; further, in plotting this limit, they failed to mention what values they selected for γ_e, R/r_b, and α (a parameter that is zero for hollow beams and unity for uniform beams). Lastly, they then plotted several IREB/gas data points for which γ_e, R/r_b, and α varied drastically, and proceeded to compare them with their single limit based on their single (unnamed) choices for γ_e, R/r_b, and α. Their results, therefore, are quantitatively meaningless. PUTNAM /355/ attempted to determine an IFA ion energy limit, but erroneously assumed $I_e > I_\ell$ was required for IFA operation. In fact, in most of the above calculations, and especially for high ion energies, $I_e < I_\ell$ is definitely used.

Although the first goal of any collective accelerator scheme should be to span the ion energy regime of $\mathcal{E}_i/A = 0$ to $\mathcal{E}_i/A \approx 1$ GeV, i.e., $\beta_i = 0$ to $\beta_i = 0.87$, (as we have shown is possible for a charge-neutralized IFA), it is also of interest to examine the possibility of obtaining very relativistic ion energies ($\mathcal{E}_i/A >> 1$ GeV, i.e., $0.87 < \beta_i < 1$). The limit $\gamma_i = \gamma_e$ may be approached for a charge-neutralized IFA by going to higher γ_e with I_e fixed (i.e., lower ν_e/γ_e), at the expense of overall efficiency. Also, use of a B_z and/or a hollow IREB would reduce the kinematic limit effect, and permit γ_i to approach γ_e. For a current-neutralized IFA, $\nu_e/\gamma_e > 1$ is possible and the limit $\gamma_i = \gamma_e$ should be readily approachable; however, higher external ionizer powers would be required to achieve the necessary current neutralizing background. To achieve $\gamma_i > \gamma_e$, it is necessary to have the unneutralized (or, at most, partially neutralized) IREB propagate first, and then sweep the ionization front along the already propagating beam. In principle, this could be accomplished by initially propagating a partially neutralized IREB (with $\gamma_e^{-2} < f_e << 1$ along the channel initially), or by propagating an unneutralized IREB along a strong B_z ($f_e=0$ initially). In both cases, the background gas pressure would have to be reduced to prevent IREB-induced ionization before the moving front arrived. In any event, the attainment of $\mathcal{E}_i/A = 1$ GeV definitely appears possible with the IFA, and higher energies may apparently be attained with a number of minor IFA variations.

The peak ion power attained in the IFA is $\mathcal{P}_i = N\mathcal{E}_i/T_i$, whereas the peak electron power used is $\mathcal{P}_e = \mathcal{E}_e I_e/e$. The ratio of these powers is, assuming N is given by (136) and T_i is given by (137),

$$\mathcal{P}_i/\mathcal{P}_e = \pi^{-1}[(\mathcal{E}_i/A)/\mathcal{E}_e](\beta_i/\beta_z)(A/Z). \tag{147}$$

For a 1 GeV proton IFA with $\mathcal{E}_e$ = 3 MeV, this gives $\mathcal{P}_i/\mathcal{P}_e \approx 100$. For a 25 GeV U^{60+} ($\mathcal{E}_i/A \approx 100$ MeV) IFA with $\mathcal{E}_e$ = 1 MeV, this gives $\mathcal{P}_i/\mathcal{P}_e \approx 80$. The feature $\mathcal{P}_i/\mathcal{P}_e \gg 1$ is unique to the IFA and may be especially useful for some applications (e.g., heavy ion fusion).

The efficiency of the charge-neutralized IFA may be determined by considering the instantaneous power used to accelerate the ions,

$$\mathcal{P}_{accel} = d(N\mathcal{E}_i)/dt = NZeE_0[\beta_i(t)]c. \tag{148}$$

Assuming E_0 is given by (131) and N is given by (136), the ratio $\mathcal{P}_{accel}/\mathcal{P}_e$ is, for $R \approx r_b$ and $I_e \leq I_\ell$,

$$\mathcal{P}_{accel}/\mathcal{P}_e = \pi^{-1}(I_e/I_\ell)[\beta_i(t)]/\beta_z. \tag{149}$$

This shows that the most efficient IFA operation occurs for $\beta_i(t) \approx \beta_z$ and $I_e \approx I_\ell$. The ion/electron energy efficiency η is

$$\eta = \left(\int_0^T \mathcal{P}_{accel}\, dt\right) \Bigg/ \left(\int_0^{t_b} \mathcal{P}_e\, dt\right). \tag{150}$$

For optimum efficiency we choose $t_b = T - L/(\beta_z c)$ so that the IREB expires at the *end* of the acceleration region just as the ion bunch exits; then we find

$$\eta = \pi^{-1}(I_e/I_\ell)[(T\beta_z c/L) - 1]^{-1}, \tag{151}$$

where for a given L, the exact relativistic value for T may be found from (132). For nonrelativistic ion energies, (133) may be used to give

$$\eta \approx \pi^{-1}(I_e/I_\ell)[(2\,\beta_z/\beta_i) - 1]^{-1}. \tag{152}$$

For a 1 GeV proton IFA with $\mathcal{E}_e$ = 3 MeV and $\nu_e/\gamma_e = 0.27$ (as in Fig. 29), this gives $\eta \approx 10\%$. For a 300 MeV proton IFA with $\mathcal{E}_e$ = 3 MeV and $\nu_e/\gamma_e \sim 1$ (as in Fig. 29), this gives $\eta \approx 32\%$. For a 25 GeV U^{60+} ($\mathcal{E}_i/A \approx 100$ MeV) IFA with $\mathcal{E}_e$ = 1 MeV, this gives $\eta \approx 16\%$. The overall IFA system efficiency should be close to η since the high-$\mathcal{Z}$ IREB machine can be made to be very efficient ($\geq 50\%$), and because the ionization energy required to ionize the working gas is typically ~ 6 orders of magnitude smaller than the IREB energy (see examples in /38/). This means that even with an inefficient ionization process, and inefficient intense light sources, it is still possible to have the ionization source energy small in comparison to the IREB energy

Lastly, it should be noted that $\eta \sim N$, and that the N value used, (136), corresponds to the number of ions trapped and accelerated in the naturally occurring IREB gas acceleration process (see Sect. 4). If N is higher than that given by (136), then η could even be higher than that given by (151).

In related work, we note that PUTNAM /356/ arbitrarily assumed N to be ~ 6 times smaller than that given by (136) and also assumed an IREB efficiency of only 50%; he accordingly found the IFA system efficiency to be 1 - 2%. We see no reason to assume such low estimates when N, as given by (136), has already been observed in a naturally occurring acceleration process, and when highly efficient high-$\mathcal{Z}$ IREB machines can readily be built. Thus IFA efficiencies of order 10% and higher appear quite justifiable. In fact, if efforts are made to recoup the lost IREB energy, then substantially higher efficiencies may ultimately be achievable.

The control accuracy of the swept light front in the IFA is determined simply by the accuracy of the lengths of the light pipes. The desired arrival time t at a distance z along the accelerator axis is given by (132). To insure a smooth, continuous, sweep of a steep ionization front, the axial width Δz of one light pipe at the accelerator is chosen to be $\Delta z \ll r_b$. To keep the swept light front motion very close to the desired motion (132), it is necessary that any given light pipe delay have a time accuracy Δt

$$\Delta t(z) << r_b[\beta_f(z)]^{-1}\, c^{-1} \; . \tag{153}$$

The increment in light pipe length $\Delta\ell$ which corresponds to a time delay increment Δt is $\Delta\ell = c\Delta t/n$, where n is the index of refraction of the light pipe core. Thus (153) may be written as

$$\Delta\ell << r_b \; (n\beta_f)^{-1} \; . \tag{154}$$

For an extreme case ($\beta_f \approx 1$, $r_b = 1$ cm, $n = 1.6$) corresponding to relativistic ions, (154) requires $\Delta\ell << 0.63$ cm which can be easily met. It should be noted that a 1 mm delay length of light pipe with n = 1.6 gives a 5.3 psec delay. Thus extremely accurate and precise control is readily possible in the IFA. (In related work, we note that PUTNAM /356/ attempted to define the control accuracy required for the IFA, but considered the actually irrelevant case of τ_N control of the Cravath-Loeb velocity $v = \mathcal{L}/\tau$ [see (60)]; that case corresponds to pressure gradient control of τ_N, and *not* the IFA case.)

The axial accelerating field, the acceleration length, and the acceleration time have already been given in (131) - (134). It should be noted that for a given IREB, the maximum axial electric field (131) is realizable with the IFA. In addition, since this field can be maintained during acceleration, the IFA offers the shortest acceleration length and the shortest acceleration time possible for *any* collective accelerator scheme employing the given beam. Further the ion energy is determined entirely by the programmed light pipes; this means the ion energy will not vary from shot to shot, but will be precisely as programmed; fluctuations in the IREB current and energy may cause potential well fluctuations, but cannot influence the programmed potential well motion. Also, since no B_z is required, the ions may exit the accelerator without ever acquiring any angular momentum from crossing diverging magnetic field lines. Precise ion energy control and lack of ion angular momentum are features unique to the IFA; these features should greatly facilitate ion beam transport, control, and focusing past the accelerator.

6.2 Autoresonant Accelerator (ARA)

6.2.1 Description

In the ARA /39, 40/, as shown in Fig. 30, an IREB is injected into vacuum along a strong axial magnetic field B_z. A negative-energy, slow cyclotron wave is excited and grown on the beam. Ions are then added to this low phase velocity wave. In the main acceleration section, the magnetic field $B_z(z)$ decreases so that the wave phase velocity

$$\begin{aligned} v_\phi(z) &= \{\omega_0/[\omega_0 + \omega_{ce}(z)]\}\, v_z(z) \\ \omega_{ce}(z) &= eB_z(z)[\gamma(z)\ mc]^{-1} \end{aligned} \tag{155}$$

increases, thus accelerating the trapped ions. The ARA is thus a linear wave accelerator (see Sect. 5). The ARA is also a quasi-continuous accelerator so that in principle, if long IREB pulses can be produced, long ion pulses may be obtained. A unique feature associated with the negative energy character of the wave, is that as ions extract energy from the IREB/wave system, the wave will actually grow (at the expense of the IREB energy). For quasi-continuous operation, and also to justify the linear wave mechanism, it follows that $\mathscr{P}_i \ll \mathscr{P}_e$ for the ARA, i.e., the ion power must necessarily be less than the IREB power. Also since the phase velocity (155) depends on both $B_z(z)$ and $\gamma(z)$, it is necessary that $B_z(z)$ be accurately controlled, and that $\gamma(z)$ [actually $\gamma(r,z,t)$] be as constant in space and time as possible.

In applying the ARA concept, several choices have been considered /40, 346, 347, 360 - 362/ for the wave excitation, the wave growth, and the ion loading. An RF exciter has been selected to initially excite the desired wave, which will then be grown in the growth section downstream /347/. Growth mechanisms considered include liner mechanisms (resistive, dielectric), e-i instabilities, and slow wave structures. Suppression of resistive and hydrodynamical kink (m = 1) modes were found to be particularly troublesome, while trying to simultaneously maximize growth of the desired m = 0 cyclotron mode. The favored mechanisms to date include helical slow

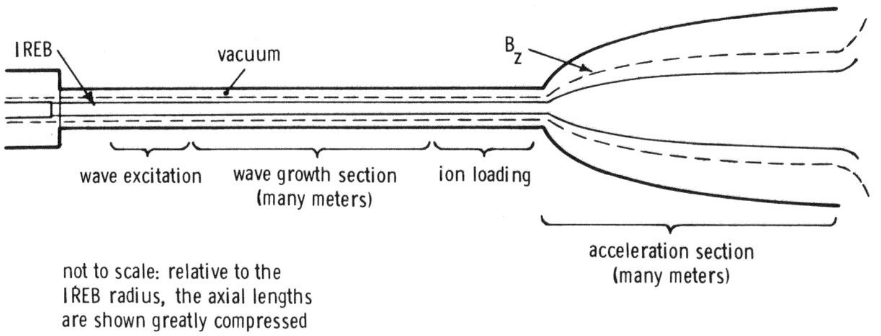

Fig. 30. Autoresonant accelerator (ARA)

wave structures, and resistive liners with azimuthal shorting rings. Ion loading may be accomplished actively with an ion injector, or passively by IREB ionization of neutral gas from a puff valve. For the later case, the initial wave phase velocity must be chosen small enough to permit the moving potential well to trap the ions. In the main acceleration section, $B_z(z)$ and $R(z)$ are flared gradually enough to accelerate the wave without losing the ions. The drift tube radius is chosen so that $R(z) \approx r_b^{max}(z)$.

The basic requirements for the initial equilibrium of the IREB propagating in vacuum are /39, 40/

$$\omega_{pe}^2 < \gamma\omega_{ce}c/r_b \qquad \text{or} \qquad \nu/\gamma < 4^{-1}\,(\beta_z/\beta)(r_b e B_z)(mc^2)^{-1} \tag{156}$$

$$\omega_{pe}^2 < \gamma^2\omega_{ce}^2/2 \qquad \text{or} \qquad \nu/\gamma < 8^{-1}(\beta_z/\beta)(r_b e B_z)^2(mc^2)^{-2} \tag{157}$$

$$I_e < I_\ell \qquad \text{or} \qquad \nu/\gamma < (\beta_z/\beta)[(\gamma - 1)/\gamma][1 + 2\ \ln(R/r_b^{min})]^{-1}\ , \tag{158}$$

where $\omega_{pe}^2 = 4\pi n_b e^2(\gamma m)^{-1} = 4(\nu/\gamma)(\beta/\beta_z)\ c^2/r_b^2$. Requirements (156) and (157) insure that the radially confining $\underline{v} \times \underline{B}$ force (including the beam's self-magnetic field) is larger than the radial electric field force, and the centrifugal force associated with the $\underline{E} \times \underline{B}$ equilibrium rotation of the beam. Requirement (158) is necessary for the beam to propagate, and should be strongly satisfied to justify use of the linear wave mechanism. An additional requirement for the stability of the equilibrium reduces to (156) for $R = r_b$, which is the case under consideration. Note that (156) and (157) can always be satisfied in principle, by applying a large enough B_z. On the other hand, (158) requires $\nu_e/\gamma_e \ll 1$ regardless of the value of B_z. MILLER et al. /363 - 365/ have performed a series of experiments in which I_e/I_ℓ and B_z were varied; their results demonstrated the above thresholds for both equilibrium and stability.

The linear wave dispersion relation for the slow cyclotron wave is /40/

$$\omega = k_z v_z - \omega_{ce}(1 + \omega_{pe}^2 k^{-2} c^{-2}) . \qquad (159)$$

Since

$$\omega_{pe}^2 k^{-2} c^{-2} \approx (0.69)(\nu_e/\gamma_e)(\beta_e/\beta_z) \qquad (160)$$

for $k_{\perp} r_b = 2.4$, we find the ω_{pe}^2 term in (159) may be neglected for $(\nu_e/\gamma_e) \ll 1$, in which case the phase velocity $v_\phi = \omega/k_z$ is given by (155). Note that this phase velocity depends on the local values of B_z, γ, and v_z. Thus, e.g., the effects of radial shear in γ and v_z, and the effects of axial variation of γ and v_z must be considered in programming the phase velocity of the desired single, large-amplitude, cyclotron wave.

The axial electric field in the ARA is produced by a periodic modulation of the beam envelope /366/. Using (5), the net well depth $\Delta\phi = \phi^{max} - \phi^{min}$ is

$$\Delta\phi/(\gamma_e mc^2/e) = (\nu_e/\gamma_e)(\beta_e/\beta_z)\ 2\ \ln(r_b^{max}/r_b^{min}). \qquad (161)$$

To achieve a substantial $\Delta\phi$ therefore requires a substantial variation in r_b. For example, GODFREY /366/ notes that $\Delta\phi/(\gamma_e mc^2/e) \approx \nu_e/\gamma_e$ for $r_b^{max}/r_b^{min} = 1.65$. The exact value of $\Delta\phi$ attainable depends, of course, on nonlinear saturation effects. The peak axial electric field associated with (161) is

$$E_z = k_z(\Delta\phi)(1 - v_\phi v_z c^{-2}), \qquad (162)$$

where we have included the inductive correction (see Sect. 3.2) which becomes important for high phase velocities. For fixed $\Delta\phi$, the maximum electric field should occur at the beginning of the accelerator, $E_z(0) \approx (\Delta\phi)[k_z(0)]$. As $v_\phi(z)$ increases, $k_z(z)$ decreases, and therefore $E_z(z)$ decreases also. For example, for acceleration of protons from rest to 1 GeV, $[v_\phi(z)]/c$ must increase from a small value ($\lesssim 0.05$) to a large value (~ 0.9), and the corresponding $E_z(z)$ must *decrease* by the same factor ($\sim$ 10 - 20). Thus for a given IREB, the acceleration time and the acceleration length for the ARA must be substantially longer than, e.g., for the IFA.

Numerical simulations relating to the ARA have been performed by GODFREY et al. /366 - 370/. For IREB injection into vacuum through an anode foil, they found that stationary cyclotron waves were excited due to beam pinching effects near the anode (where $E_r \approx 0$ but $B_\theta \neq 0$). These waves are undesirable and must be avoided, which may require a complicated IREB injection geometry. Simulations were also performed, assuming the desired cyclotron wave was imposed on an IREB, and the wave phase velocity was programmed to increase from $v_\phi/c = 0.05$ to $v_\phi/c = 0.16$ (i.e., a ten-fold increase in ion energy). Test ions were loaded on the wave, but acquired only an eight-fold increase in energy over a distance of 50 cm. This was apparently due to the fact that radial shear in γ caused phase mixing of the cyclotron waves near the end of the acceleration region. It was noted that this coherence loss must be overcome if substantial acceleration lengths are to be achieved. Also, from these investigations, it was tentatively suggested that for optimum ARA operation, ν_e/γ_e should be in the range $0.25 \lesssim \nu_e/\gamma_e \lesssim 0.33$. Numerical simulations of various aspects of the ARA (especially wave growth and saturation for the wave growth section) have also been performed by DRUMMOND et al. /345, 346, 371, 372/.

6.2.2 ARA Feasibility Experiments

A series of experiments has been planned by DRUMMOND et al. /346/ at Austin Research Associated to demonstrate feasibility of the ARA. The parameters selected for these experiments are summarized in Table 11. To minimize anode foil damage in these experiments, the beam is to be created with a large radius (3 cm), and then magnetically compressed to a small radius (1 cm). An active RF exciter is to be used to

Table 11. Parameters for ARA feasibility experiments

IREB	$\mathscr{E}_e = 3$ MeV $I_e = 30$ kA $t_b = 200$ nsec $r_b = 3$ cm $\mathscr{P}_e = 9 \times 10^{10}$ W $\mathscr{E}_{IREB} = 18{,}000$ J $HF_{ripple} < 0.5\%$
COMPRESSION	$B_{in} = 2.4$ kG $\ell = 100$ cm $r_{out} = 1$ cm $B_{out} = 24$ kG
RF EXCITE	$f_o = 240$ MHz

WAVE GROWTH	$\ell = 500$ cm
ACCELERATION SECTION	$\ell = 400$ cm $B_{in} \equiv 24$ kG $B_{out} = 2$ kG $r_{in} = 1$ cm $r_{out} = 3.5$ cm
PROTONS	$\mathscr{E}_i = 30$ MeV $I_i = 30$ A $\mathscr{P}_i = 9 \times 10^8$ W $\tau \approx 100$ nsec $N \approx 2 \times 10^{13}$ $\mathscr{E}_i \approx 90$ J

initiate the wave at frequency ω_0. Wave growth will then presumably be achieved with a sheath helix liner, and will take place in a 5 meter growth section. The actual acceleration section is to produce 30 MeV protons in a distance of 4 meters (average E_z = 7.5 MeV/m). Note that quasi-continuous operation should be demonstrated ($\tau \sim$ 100 nsec). Also note that the expected ion power is two orders of magnitude *lower* than the IREB power.

In order to maintain $\gamma_e(t)$ sufficiently constant, it is necessary for the IREB to have an extremely low level of high frequency ripple ($\lesssim 0.5\%$ above 10 MHz). If this requirement is not met, sudden phase shifts will shift the wave and the ions will become untrapped and be lost.

An IREB machine has been acquired with the desired electrical parameters, and is undergoing testing. The first experiments will involve IREB beam transport through the compression section, and studies of wave excitation and growth in the growth section. These experiments should presumably begin in the near future.

6.2.3 ARA Scaling

The maximum ion energy possible in the ARA is determined by the maximum phase velocity possible; this in turn depends partially on the axial electron velocity

$$v_z(r,z) = \{[v(r,z)]^2 - [v_r(r,z)]^2 - [v_\theta(r,z)]^2\}^{1/2} , \tag{163}$$

where we have assumed azimuthal symmetry, and v_r, v_θ, and v_z are the r, θ, and z components of the electron velocity $\underline{v}(r,z)$. All terms on the right hand side of (163) contribute to making $v_z(r,z) < v_e$, where v_e is the injected electron velocity. In general, $v(r,z)$ is the smallest for $r = 0$, while $v_r(r,z)$ and $v_\theta(r,z)$ are the largest for $r \approx r_b$. On axis, the electron velocity $v(z) \equiv v(r = 0, z)$ is determined from $\gamma(z) = \{1 - [v(z)/c]^2\}^{-1/2}$, where

$$\gamma(z) = 1 + (\gamma_e - 1)[1 - (I_e/I_\ell)][1 - \varepsilon(z)]. \tag{164}$$

Here, the function $\varepsilon(z)$ is defined as the ratio of the power used to accelerate ions up to the axial location z, to the power available at z if no ions were accelerated. Applying (164) at the end of the accelerator gives

$$\gamma(L) = 1 + (\gamma_e - 1)\{1 - (\nu_e/\gamma_e)(\beta_e/\beta_z)(\gamma_e)(\gamma_e - 1)^{-1} \cdot [1 + 2\ln(R/r_b^{min})]\}[1 - \varepsilon(L)]. \tag{165}$$

Henceforth the subscript (e) refers to the injected IREB parameters, the notation (0) refers to quantities evaluated at the beginning of the accelerator ($z = 0$), and the notation (L) refers to quantities evaluated at the end of the accelerator. The peak radial electron velocity corresponding to an electron trajectory at the nominal radius r_b is

$$v_r(r,z) = v_z(r,z)[k_z(z)]\ \Delta r(r/r_b), \tag{166}$$

where Δr represents the magnitude of the beam envelope oscillation. Since $k_z(z) = \omega_0/[v_\phi(z)]$, we may use (155) to obtain

$$v_r(r,z) = v_z(r,z)[v_\phi(0)][v_\phi(z)]^{-1}\ [\beta_e(0) - \beta_\phi(0)]^{-1}\ [r_b e B_z(0)] \cdot \cdot[\gamma(0)mc^2]^{-1}\ (\Delta r/r_b)(r/r_b). \tag{167}$$

For $v_\theta \approx 0$, (163) and (167) give

$$v_z(L) = v(L)\cos\left[\mathrm{TAN}^{-1}\{[v_\phi(0)][v_\phi(L)]^{-1}\ [\beta_e(0) - \beta_\phi(0)]^{-1}\ [r_b e B_z(0)] \cdot[\gamma(0)mc^2]^{-1}\ (\Delta r/r_b)(r/r_b)\}\right]. \tag{168}$$

The azimuthal velocity $v_\theta(r,z)$ is determined from the rigid rotor equilibrium of the IREB /373, 374/ which gives a rotation frequency ω_θ

$$\omega_\theta = (\omega_{ce}/2)\left\{1 - [1 - (2\ \omega_{pe}^2)(\gamma^2\ \omega_{ce}^2)^{-1}]^{1/2}\right\} \tag{169}$$

which is real, provided (157) holds. If (157) is strongly satisfied,

$$2\ \omega_{pe}^2 << \gamma^2\ \omega_{ce}^2 \quad \text{or} \quad \nu/\gamma << (23.2)^{-1}[r_b(\text{cm})\ B_z(\text{kG})]^2\ (\beta_z/\beta), \tag{170}$$

we find

$$v_\theta \approx 2(\nu/\gamma)[\gamma(r_b e B_z)(mc^2)^{-1}\ (\beta_z/\beta)]^{-1}\ (r/r_b)c \tag{171}$$

which is typically very small. On the other hand, for the extreme case of Brillouin flow ($2\ \omega_{pe}^2 = \gamma^2\ \omega_{ce}^2$), we find

$$v_\theta = (1.41)(\nu/\gamma)^{1/2}\ (\beta/\beta_z)^{1/2}\ (r/r_b)\gamma^{-1}c \tag{172}$$

which can be sizeable. Thus (170) must hold to insure that v_θ is negligible in (163). We conclude from these results that $v(r,z)$ is significantly limited by (164), that $v_r(r,z)$ may be significant depending on the parameters in (168), and that $v_\theta(r,z)$ may be neglected provided (170) holds. The phase velocity (155), which is the final limitation to the ion energy, gives

$$v_\phi(L) = v_z(L)\Big\{1 + [B_z(L)][B_z(0)]^{-1}[\gamma_e(0)][\gamma_e(L)]^{-1}\,[v_z(0) - v_\phi(0)] \cdot [v_\phi(0)]^{-1}\Big\}^{-1}. \tag{173}$$

Together, results (163) - (173) determine the maximum ion energy possible in the ARA.

An acceleration time limit may also be derived. This limit applies to all wave accelerators, including the ARA, for the case of constant potential well depth with

$$v_\phi(z) = v_\phi(0)[\lambda(z)/\lambda(0)] \tag{174}$$

$$E_z(z) = E_z(0)[\lambda(0)/\lambda(z)]. \tag{175}$$

Assuming an ion of mass M and charge Z undergoes the acceleration $a(z) = ZeE_z(z)/M$, we find for the nonrelativistic case

$$\lambda(z) = \lambda(0)\{1 + [3ZeE_z(0)]M^{-1}\,[v_\phi(0)]^{-2}\,z\}^{1/3}. \tag{176}$$

The ions will acquire the energy

$$\mathcal{E}_i(L) = Ze\langle E_z\rangle L + \mathcal{E}_i(0) \tag{177}$$

over a length L, where $\mathcal{E}_i(0) = M[v_\phi(0)]^2/2$, and the average accelerating field is

$$\langle E_z\rangle = M[v_\phi(0)]^2\,(2eL)^{-1}\Big[\{1 + (3Ze)[E_z(0)]M^{-1}[v_\phi(0)]^{-2}L\}^{2/3} - 1\Big]\,Z^{-1}. \tag{178}$$

Alternatively, we find

$$L = [\mathcal{E}_i(L) - \mathcal{E}_i(0)](Ze\langle E_z\rangle)^{-1} \tag{179}$$

$$T = L[v_\phi(0)]^{-1}\langle E_z\rangle[E_z(0)]^{-1}, \tag{180}$$

where T is the acceleration time. For the usual case of $Ze[E_z(0)]L \gg M[v_\phi(0)]^2$, (178) reduces to

$$\langle E_z\rangle\,[E_z(0)]^{-1} \approx (3^{2/3}/2)\Big[M[v_\phi(0)]^2\,\{Ze[E_z(0)]L\}^{-1}\Big]^{1/3} \tag{181}$$

which demonstrates the large reduction in accelerating field due to (175). Results (177) and (180) combine to give

$$[\mathcal{E}_i(L)]/A = (Z/A)\,e[E_z(0)]\,T[v_\phi(0)] + [\mathcal{E}_i(0)]/A \tag{182}$$

which is an exact, nonrelativistic result, where $E_z(0) = [\Delta\phi(0)]\, k_z(0)$, $\Delta\phi$ is given by (161), $M = AM_p$, and M_p is the proton rest mass.

For the relativistic case, the ion motion is described by

$$\partial z/\partial t = v_\phi(0)[\lambda(z)/\lambda(0)] \tag{183}$$

$$\partial^2 z/\partial t^2 = [ZeE_z(0)]M^{-1}\{1 - [v_\phi(0)/c]^2[\lambda(z)/\lambda(0)]^2\}^{3/2}\,[\lambda(0)/\lambda(z)] \tag{184}$$

which replace (174) and (175). Here we have again assumed $\Delta\phi$ is constant and given by (161). Also, we have ignored the inductive correction in (162) which should be included, and which would make the acceleration substantially slower than that which we calculate here. Solution of (183) and (184) yields

$$\begin{aligned}&[\lambda(z)/\lambda(0)]\{1 - [\beta_\phi(0)]^2[\lambda(z)/\lambda(0)]^2\}^{-1/2} - [\beta_\phi(0)]^{-1}\,\mathrm{SIN}^{-1}\{[\beta_\phi(0)]\,\lambda(z)/\lambda(0)\}\\ &- \{1 - [\beta_\phi(0)]^2\}^{-1/2} + [\beta_\phi(0)]^{-1}\,\mathrm{SIN}^{-1}[\beta_\phi(0)] = Ze[E_z(0)](Mc^2)^{-1}\, z\end{aligned} \tag{185}$$

which is the relativistic analogue of (176). The acceleration length L and acceleration time T are then found to be

$$\begin{aligned}L = {}& Mc^2\{[\beta_\phi(0)]\, Ze[E_z(0)]\}^{-1}\,\{\beta_\phi(L)\gamma_\phi(L) - \mathrm{SIN}^{-1}[\beta_\phi(L)]\\ &-\beta_\phi(0)\gamma_\phi(0) + \mathrm{SIN}^{-1}[\beta_\phi(0)]\}\end{aligned} \tag{186}$$

$$T = Mc^2\{[v_\phi(0)]\, Ze[E_z(0)]\}^{-1}\,[\gamma_\phi(L) - \gamma_\phi(0)], \tag{187}$$

and the final ion energy is

$$[\mathscr{E}_i(L)]/A = (Z/A)\, e[E_z(0)]\, T[v_\phi(0)] + [\mathscr{E}_i(0)]/A \tag{188}$$

which is an exact, relativistic result [which matches the nonrelativistic result (182)].

The amplitude of the accelerating field for the ARA is given by $E_z(z) = [\Delta\phi(z)]\cdot[k_z(z)]$, if we ignore the inductive correction for high phase velocities in (162). A maximum value for this field may be estimated as follows. The potential amplitude $\Delta\phi$ is given by (161), *provided* $k_z r_b \ll 1$, since (161) was obtained by implicitly assuming r_b varies slowly in z. For $k_z r_b \gg 1$, $\Delta\phi$ is much *less* than that given by (161). This may be seen by imagining a cylindrical Gaussian pillbox concentric with the beam at the axial location of a constriction (r_b^{min}). Then if the box radius is r_b^{min} and its axial thickness is $\Delta\ell$, we find

$$E_r 2\pi r_b \Delta\ell + 2[\pi r_b^2 (\Delta\phi\ k_z)\ k_z\ (\Delta\ell/2)] = 4\pi^2 r_b^2 n_b e\Delta\ell\ , \tag{189}$$

where the second term was obtained by assuming the axial field E_z is constant over both faces of the box. Cancelling the $\Delta\ell$'s in (189) gives

$$E_r = 2I_e\ (\beta_z c r_b)^{-1}\ \{1 - (\Delta\phi/\phi_o)(k_z r_b/2)^2\}, \tag{190}$$

where $\phi_o = I_e(\beta_z c)^{-1}$, and $\Delta\phi/\phi_o = 2\ \ln\ (r_b^{max}/r_b^{min})$ from (161). Now we note that result (1) was used to obtain (5), which in turn was used to obtain (161). But (190) reduces to (1) only if the bracketed expression {} is negligible. For $\Delta\phi/\phi_o = 1$ (a desirable case), this requires

$$k_z r_b << 2. \tag{191}$$

It follows that $\Delta\phi\ k_z$ is small for $k_z r_b \gg 1$ or for $k_z r_b \ll 1$. A reasonable estimate for the maximum field is therefore $E_z \approx \Delta\phi/r_b$ (i.e., $k_z r_b \approx 1$).

Lastly, we note that at z = 0, we must specify $v_\phi(0)$, ω_o, $k_z(0)$, and $B_z(0)$. But these four quantities are restricted by the relations $v_\phi(0) = \omega_o/k_z(0)$ and $k_z(0) = [\omega_o + \omega_{ce}(0)]/v_z(0)$. Thus only two of the four quantities may be chosen. The prudent choices are to specify (1) $v_\phi(0)$ so ion trapping is possible, and (2) $k_z(0)$ so as to maximize $E_z(0)$ [i.e., choose $k_z(0) = r_b^{-1}$]. By specifying $v_\phi(0)$ and $k_z(0)$, ω_o and $B_z(0)$ are then automatically determined to be

$$\omega_o = v_\phi(0)\ k_z(0) \tag{192}$$

$$B_z(0) = \gamma_e(0)\ (mc^2/e)\ k_z(0)\ [v_z(0) - v_\phi(0)]/c\ . \tag{193}$$

Along the accelerator, $B_z(z)$ will decrease, but (156) - (158) must always be satisfied.

The above ARA scalings are demonstrated in Fig. 31 for the case of an IREB with $\gamma_e = 7$ ($\mathscr{E}_e \approx 3$ MeV), $R(z) \approx r_b^{max}(z)$, $r_b^{max}/r_b^{min} = 1.65$ so $\Delta\phi/(\gamma_e mc^2/e) \approx \nu_e/\gamma_e$, $k_z(0) = [r_b(0)]^{-1}$, and $\varepsilon = 10\%$. The initial phase velocity is taken to be $M[v_\phi(0)]^2/2 = 2^{-1} \cdot e(\Delta\phi)$ to permit trapping and acceleration of initially stationary ions. The input parameters are taken to be $r_b(0) = [k_z(0)]^{-1} = 1$ cm, so $B_z(0) = 12$ kG from (193). The output parameters are taken to be $B_z(L) = 0.5$ kG and $r_b(L) = 4.9$ cm, since $B_z(0) \cdot [r_b(0)]^2 = B_z(L)[r_b(L)]^2$. With these parameters, $E_z(0) = 1.07$ MV/cm for $\nu_e/\gamma_e = 0.3$. Also, for these parameters, $v_\theta(z)$ and $v_r(z)$ [see (166) - (172)] make negligible contributions to $v_z(z)$ as given by (163). Thus as shown, the main limitation to $v_z(L)$ comes from (165); here it should be noted that if ε were > 10%, $v_z(L)$ would be even lower than indicated. The most serious restriction is the phase velocity limit (173

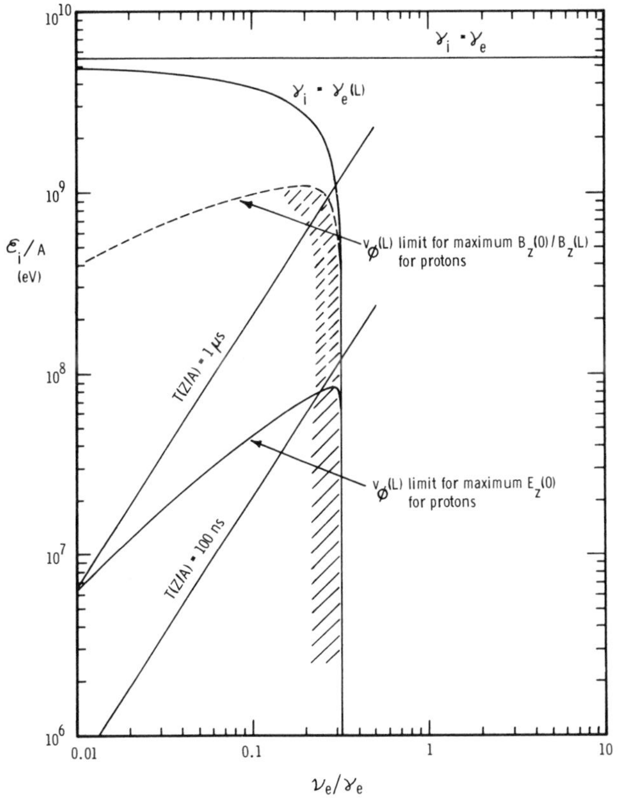

Fig. 31. Ion energy per nucleon ($\mathscr{E}_i/A$) vs ν_e/γ_e for the ARA, for γ_e=7 ($\mathscr{E}_e \approx 3$ MeV). The various loci and limits are discussed in the text

Note that this limit decreases for low ν_e/γ_e since $v_\phi(0)$ decreases as ν_e/γ_e decreases. The time limit (188) is also shown for TZ/A = 100 nsec and TZ/A = 1 μsec. Note that relatively long acceleration times are required even for modest ion energies, because $E_z(z)$ decreases according to (184).

The phase velocity limit (173) can be increased if we drop the initial choice of maximizing $E_z(0)$, and simply maximize $B_z(0)/B_z(L)$ without any concern for the resultant E_z field. If we choose $B_z(0)$ = 200 kG (essentially at the technologically feasible limit) and keep $B_z(L)$ = 0.5 kG [but now $r_b(L)$ = 20 cm], then the phase velocity locus would be as shown by the dashed line in Fig. 31. Now, however, $k_z(0)\cdot r_b(0) = 16.6$ so that $E_z(0) \ll (\Delta\phi)[k_z(0)]$ for $\Delta\phi$ given by (161), and the accelerating fields are small. If we could ignore this restriction *(which we cannot)* and simply assume the maximum value $E_z(0) = \Delta\phi/r_b$ in (186), we would still find long IREB pulses (> 1 μsec) are required to achieve high energies; for an optimum case (ν_e/γ_e = 0.25), 1 GeV protons would require T = 1.2 μsec and L = 253 meters. Even for this hypothetical case, comparing Fig. 31 with Fig. 29, we see that for 1 GeV protons, L and T would *still* be more than an order of magnitude larger for the ARA than for the IFA.

The acceleration length and the acceleration time for the ARA may be made more competitive by, e.g., using higher γ_e, using an ion injector, or staging. Using higher γ_e permits higher E_z (in *both* the ARA and the IFA), and therefore smaller L and T. Use of an ion injector in the ARA permits higher $v_\phi(0)$, which in turn reduces L and T. Lastly, the concept of staging has been suggested; this means two or more ARA's would be used in series so that the phase velocity range $[v_\phi(L)]/[v_\phi(0)]$ of each stage could be kept to a more reasonable size. However, problems of phase matching, and ion transport between the large output radius of one stage and the small input radius of the succeeding stage, appear to be drawbacks for this concept. A realistic staging scenario remains to be given.

The peak ion power $\mathscr{P}_i$, and the relative efficiency η for the ARA are related by

$$\eta = \mathscr{P}_i/\mathscr{P}_e \ll \{1 - (\nu_e/\gamma_e)(\beta_e/\beta_z)\,\gamma_e(\gamma_e - 1)^{-1}\,[1 + 2\,\ln(R/r_b^{min})]\}(1-\varepsilon)\,, \quad (194)$$

where the injected electron power is $\mathscr{P}_e = I_e\,\mathscr{E}_e/e$. The large bracketed term in (194) represents the fractional electron beam power available *after* the beam has overcome its own space charge depression. Efficiencies of order 10% and larger have been suggested for the ARA. For total system efficiency, it should be noted that the magnetic field energy required in just the acceleration section is

$$\mathscr{E}_B = 8^{-1}\,r_b^2[B_z(0)]^2[v_\phi(0)][v_e(0) - v_\phi(0)]^{-1}[v_e(0)T - L] \quad (195)$$

which is typically of the same order as the IREB energy. Also, a potential complication concerning the efficiency may arise, since the flared $B_z(z)$ and $R(z)$ must be programmed for known $\gamma_e(z)$, which depends on η. In addition, substantial variation in $\eta(t)$ may result in substantial variations in $\gamma_e(t)$ which will cause $\mathscr{E}_i(L,T)$ to vary.

The desired axial dependence of the magnetic field in the ARA may be readily calculated. For the nonrelativistic case, we find, using (174), (176), and (173)

$$\begin{aligned} B_z(z) = {} & B_z(0)[\gamma_e(z)/\gamma_e(0)]\;v_\phi(0)[v_z(0) - v_\phi(0)]^{-1} \\ & \cdot\{[v_z(z)/v_\phi(0)]\,[1 + Az]^{-1/3} - 1\}, \end{aligned} \quad (196)$$

where $A = 3ZeE_z(0)\,M^{-1}[v_\phi(0)]^{-2}$. For the "test ion" case [for which $\eta = \varepsilon = 0$ so that $\gamma_e(z) = \gamma_e(0)$, we find

$$\partial B_z(z)/\partial z = -B_z(0)[v_z(z)]\,[v_z(0) - v_\phi(0)]^{-1}\,(A/3)(1 + Az)^{-4/3}. \quad (197)$$

Table 12. Example of $B_z(z)$ and $\partial B_z(z)/\partial z$ for the ARA

Z(cm)	$\beta_\phi(z)$	$\mathcal{E}_i$(MeV)	B_z(gauss)	$-\partial B_z/\partial z$ (gauss/cm)	$(-\partial B_z/\partial z)/B_z$ (cm^{-1})
0	0.031	0.45	12,000	12,408	1.03
1	0.049	1.1	7,407	1,954	0.26
10	0.097	4.5	3,540	127	0.036
100	0.21	21.0	1,442	6.2	0.0043
300	0.30	45.0	875	1.4	0.0016
500	0.36	65.3	674	0.72	0.00107

Results (196) and (197) are evaluated in Table 12 for protons for the test ion case corresponding to $\nu_e/\gamma_e = 0.25$ in Fig. 31 [$\gamma_e = 7$, $r_b(0) = 1$ cm, $B_z(0) = 12$ kG, $r_b^{max}/r_b^{min} = 1.65$, $R = r_b^{max}$, $v_\phi(0)/c = 0.031$, and $E_z(0) = 0.89$ MV/cm]. Two characteristic features are readily evident. First, near z = 0, $B_z(z)$ changes too rapidly to permit the required adiabatic change in $k_z(z)$. It has been suggested /346/ to spread the initial acceleration out over a longer distance to avoid this problem; however, this means lowering the effective accelerating field for this region. The second feature is that further along the accelerator, the required $\partial B_z(z)/\partial z$ becomes very small; e.g., at z = 5 meters, $B_z(z)$ is 674 gauss and is supposed to decrease at the rate of 0.72 gauss/cm (i.e., less than a 0.11% change per cm). The required $B_z(z)$ control becomes even more delicate as $\beta_\phi(z)$ approaches unity.

The control accuracy required for $\gamma_e(t)$ and $B_z(z)$ has been estimated with a 1-D numerical model for the ARA feasibility experiment parameters /346/. For these parameters (see Table 11), it was found that the ions would maintain phase stability and reach the programmed energy of 30 MeV provided the high frequency ripple in the injected electron beam was $\Delta\gamma_e(z)/\gamma_e(0) \lesssim 0.5\%$ (> 10 MHz), and provided that any long wavelength perturbations in $B_z(z)$ deviated from the desired $B_z(z)$ by less than 5%. Both of these requirements become even more severe for higher ion energies. For example, since the wave phase $\varphi(z) \sim \int_0^z k_z(z)\,dz$ and $k_z(z) \sim [\gamma_e(z)]^{-1}$ [at least for $\omega_0 \ll \omega_{ce}(z)$], it follows that $|\Delta\varphi/\varphi| \sim |\Delta\gamma/\gamma|$. Since $\Delta\varphi/\varphi \ll \lambda(L)/L$ is required to maintain phase stability, this means

$$\Delta\gamma_e(t)/\gamma_e(0) << [\lambda(L)]/L \tag{198}$$

is required to prevent ion loss. The high frequency ripple requirement (198) clearly becomes proportionately more severe for higher ion energies (larger L).

In reference to ion parameters, it should be noted that if γ_e varies during the IREB pulse, then the final ion energy $\mathcal{E}_i(L)$ will vary also. Further, the ions are born in a high magnetic field [$B_z(0)$] and must therefore necessarily cross field lines to exit the accelerator. This means that the ions will acquire a finite angula momentum. Lastly, the accelerator exit radius is typically much larger than the inpu radius. All of these features [non-constant $\mathscr{E}_i(t)$, finite ion angular momentum, and large radius exit] suggest that transport and focusing of these ion beams may be quite difficult.

6.3 Converging Guide Accelerator (CGA)

6.3.1 Description

The CGA /41, 375 - 377/, as shown in Fig. 32, is similar to the ARA in that it employs an IREB propagating along a strong axial magnetic field in vacuum. Here, however, a negative-energy, longitudinal space charge wave is used, and phase velocity control is achieved by varying the radius of the guide tube R(z) while keeping the beam radius r_b constant. By converging the guide, $v_\phi(z)$ can be made to increase. Like the ARA, the CGA is a quasi-continuous accelerator so that although long ion pulses may ultimately be achieved, the ion beam power is limited to $\mathscr{P}_i \ll \mathscr{P}_e$.

The dispersion relation for the desired slow space charge wave is /40, 41/

$$\omega = k_z v_z - \omega_b \gamma^{-3/2} \left[(k_z^2 c^2 - \omega^2)/(k^2 c^2 - \omega^2)\right]^{1/2} , \tag{199}$$

where $\omega_b = (4\pi n e^2/m)^{1/2}$. The phase velocity $v_\phi = \omega/k_z$ may therefore be written as

$$v_\phi(z) = \{\omega_0/[\omega_0 + \alpha(z)\, \omega_{pe}(z)]\}\, v_z(z) \tag{200}$$

$$\omega_{pe}(z) = \{4\pi\, n_b(z)\, e^2\, [\gamma(z)]^{-3}\, m^{-1}\}^{1/2} , \tag{201}$$

where ω_0 is the constant frequency, $\alpha(z) = \{[k_z(z)]^2 c^2 - \omega_0^2\}^{1/2} \cdot \{[k(z)]^2 c^2 - \omega_0^2\}^{-1/2}$ and $\alpha(z) \approx [k_z(z)]/[(k(z)]$ for $v_\phi(z) \ll c$. Dividing (199) by k_z and noting that $\omega_{pe} = (2/\gamma)(\nu/\gamma)^{1/2} (\beta/\beta_z)^{1/2}\, c/r_b$, we also find for $v_\phi(z) \ll c$,

$$v_\phi = v_z\{1 - (\gamma^{-1})[(\nu/\gamma)^{1/2}\, (\beta/\beta_z)^{1/2}\, (c/v_z)(k r_b/2)^{-1}]\} , \tag{202}$$

where all quantities (except r_b) depend on z. It is useful to note

$$k(z) = \{[k_\perp(z)]^2 + [k_z(z)]^2\}^{1/2} \tag{203}$$

$$k_z(z) = k_z(0)[\lambda(0)/\lambda(z)] \tag{204}$$

$$k_\perp(z) \approx (2.4/r_b)\{1 + 2\, \ln[R(z)/r_b]\}^{-1} \tag{205}$$

$$\gamma(z) = \gamma_e - e I_0 [\beta_z(z)\, c]^{-1}\, [mc^2]^{-1}\, \{1 + 2\, \ln[R(z)/r_b]\} . \tag{206}$$

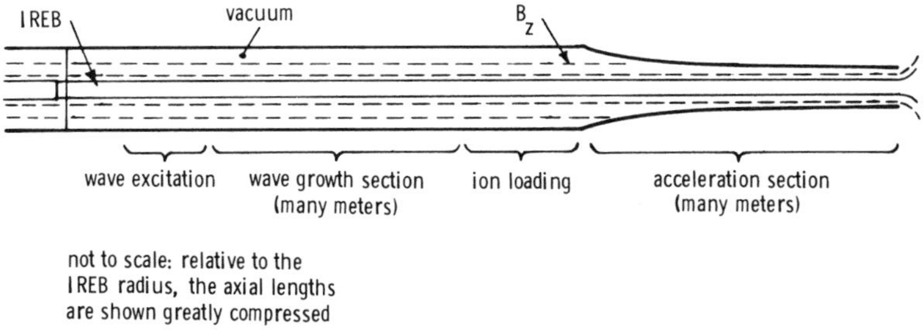

Fig. 32. Converging guide accelerator (CGA)

For the basic CGA concept, R(z) is varied to control $\gamma(z)$ by the potential depression in (206); the resultant $\gamma(z)$ varies $\omega_{pe}(z)$ in (200) to control $v_\phi(z)$. Of course, the true axial dependence of $v_\phi(z)$ in (200) involves $\gamma(z)$, $n_b(z)$, $k_z(z)$, and k(z).

A basic problem with the CGA concerns its difficulty in achieving low phase velocities. This is most easily seen from (202). If relativistic electron velocities are to be maintained in the accelerator ($\beta_z \sim 1$), it follows that the bracketed expression in (202) is at most of order unity, which means that the *minimum* phase velocity possible is

$$v_\phi^{min} \approx v_z \, (1 - \gamma^{-1}). \tag{207}$$

For $\gamma \gg 1$, $v_\phi^{min} \approx v_z$ which makes this mode essentially unusable for accelerating ions from rest or low initial velocities. Realizing this, SLOAN and DRUMMOND /39, 40/ dismissed this mode in favor of the cyclotron mode when they proposed the ARA. SPRANGLE et al. /41/ reconsidered this mode for the nonrelativistic case ($\gamma \sim 1$), for which v_ϕ^{min} can be made moderately small. However, the attainment of low phase velocities then requires I_e to approach I_ℓ so the beam is on the verge of stopping. This is a very delicate situation if energetic IREB's are to be employed in the CGA. For a thin, hollow beam, BRIGGS /339/ has shown that $v_\phi \to 0$ as $I_e \to I_\ell$, but that the slope $|\partial v_\phi / \partial (I_e/I_\ell)| \to \infty$. Similarly, GODFREY /370, 378, 379/ has considered more general beam profiles and shown that the slope does indeed approach infinity for virtually all cases of interest. These results mean that it would be very difficult to achieve low phase velocities in the CGA for interesting IREB parameters. It has therefore been suggested by NATION et al. /297, 377/ to use an ion injector with the CGA so that the initial phase velocity can be relatively high (perhaps $[v_\phi(0)]/c \sim 0.2$).

Many basic properties of the CGA are similar to those of the ARA. For example, the IREB equilibrium and stability requirements for the CGA are given by (156)-(158). Also, the accelerating field is given by (162), but now the effective potential well depth $\Delta\phi$ is determined by nonlinear saturation effects on the longitudinal space charge wave.

6.3.2 CGA Feasibility Experiments

A series of experiments has been undertaken by NATION et al. /297, 375 - 377/ at Cornell University to demonstrate feasibility of the CGA. Research to date has concentrated on (1) wave excitation and growth, and (2) ion injector investigations. In the wave excitation studies, a disc loaded slow wave structure was used to passively excite slow space charge waves in the 0.5 - 2 MHz regime, using an IREB with $\mathscr{E}_e$ = 300 keV, I_e = 2 kA, and t_b = 300 nsec /375, 376/. In more recent work /377/, with slightly different parameters ($\mathscr{E}_e$ = 250 - 950 keV, $I_e \approx$ 1 kA, $t_b \approx$ 100 nsec, and $B_z \geq$ 10 kG), a 1.25 GHz wave was obtained with growth rates as high as 0.3 db/cm These studies produced wave fields in the range 20 - 30 kV/cm. Wave phase velocities reported were v_ϕ/c = 0.70 ± 0.07 (for $\mathscr{E}_e$ = 700 keV and I_e/I_ℓ = 0.2) and v_ϕ/c = 0.30 ± 0.10 (for $\mathscr{E}_e$ = 270 keV and I_e/I_ℓ = 0.5). Substantial beam modulation was achieved with $|\phi^{max}|/(\mathscr{E}_e/e)$ = 0.13, and 0.34, respectively for the two cases. Thus the initia excitation and growth of a space charge mode has been demonstrated with a passive structure (albeit for relatively high phase velocities). Future experiments will investigate scaling of these effects to higher beam currents (and to lower phase velocities).

The Luce diode has been selected as an ion injector for the CGA /297, 377/. The injector must produce ions with $\beta_i \gtrsim 0.2$, and of sufficient beam quality to permit capture and acceleration by the space charge wave. Also, apparently, the injected ion beam will have to be injected concentrically with the main IREB. The status of collective acceleration research with the Luce diode, and other vacuum diodes, has been summarized in Sect. 4.3 (see Table 8).

6.3.3 CGA Scaling

For scaling purposes, results for the CGA are very similar to results (163) - (188) for the ARA, except that now $r_b(z)$ is constant [so $v_r(r,z)$ may be neglected in (163) and the substitution $\omega_{ce}(z) \to [k_z(z)/k(z)]\ \omega_{pe}(z)$, i.e.,

$$eB(z)\ [\gamma(z)\ mc]^{-1} \to [k_z(z)]\ \{[k_z(z)]^2 + [k_\perp(z)]^2\}^{-1/2} \cdot \cdot\{4\pi\ n_b(z)\ e^2\ [\gamma(z)]^{-3}\ m^{-1}\}^{1/2} \tag{208}$$

must be made. Results for the ion energy, acceleration length, and acceleration time (173) - (188) are directly applicable except that now $\Delta\phi$ refers to the potential variation for the slow space charge wave. Since an ion injector is to be used, this means the required range of phase velocities $v_\phi(L)/v_\phi(0)$ in the CGA accelerator section is relatively small. Thus the added complication of an ion injector in the CGA brings with it the benefit that $\langle E_z \rangle$ will be closer to $E_z(0)$ than in the ARA [see (181)]. On the other hand, an ion injector/ARA system would have essentially the same ion acceleration characteristics as an ion injector/CGA system.

The CGA efficiency, like the ARA efficiency, is bounded by (194). SPRANGLE et al. /41/ have estimated the CGA efficiency using E_z as given by (162) with $\Delta\phi = \alpha\mathscr{E}_e/e$. For a specific case, they calculated $\eta = \mathscr{P}_i/\mathscr{P}_e = 0.5\%$ for $\alpha = 1$. For $\alpha = 10$ they calculated $\eta = 50\%$; however $\alpha = 10$ means $\Delta\phi \approx 10\mathscr{E}_e/e$, which is a non-physical situation (the beam would stop). In the experiments discussed above, $\alpha = 0.13 - 0.34$ was measured at the end of a wave growth section. Also the linear wave analysis and phase velocity control [(199) - (206)] proposed for the CGA are strictly valid only for $\alpha \ll 1$. As a working estimate for the CGA (and the ARA), it therefore appears reasonable to assume $\eta \lesssim 10\%$ until wave saturation and ion acceleration experiments demonstrate otherwise.

The desired axial dependence of R(z) for the CGA may be calculated from (200) - (206) using (176) or (185). For the nonrelativistic case, we find using (176),

$$v_\phi(0)[1 + Az]^{1/3} = v_z(z)\left\{1 - [I_e/(mc^3/e)]^{1/2}\left[\beta_z(z)\,\gamma_e - [I_e/(mc^3/e)]\cdot\{1 + 2\ln[R(z)/r_b]\}\right]^{-3/2}\left[(2.4)^2\{1 + 2\ln[R(z)/r_b]\}^{-2} + [r_b k_z(0)]^2\cdot[1 + Az]^{-2/3}\right]^{-1/2}\right\}^{2}, \tag{209}$$

where $v_z(z)$ may be found from (206), and $A = 3Ze\,E_z(0)\,M^{-1}[v_\phi(0)]^{-2}$. Result (209) defines R(z). To estimate the basic scaling effect, we consider the simplified case of $v_z(z) \approx c$ and $k(z)r_b/2 \approx 1$; then (209) yields

$$R(z) = r_b \exp\left\{2^{-1}\left[\gamma_e[I_e/(mc^3/e)]^{-1} - 1 - [I_e/(mc^3/e)]^{-2/3}\cdot\left\{1 - [\beta_\phi(0)](1 + Az)^{1/3}\right\}^{-2/3}\right]\right\} \tag{210}$$

$$\partial R(z)/\partial z = -\,[R(z)](1/9)[I_e/(mc^3/e)]^{-2/3}\cdot\left\{1 - [\beta_\phi(0)]\,(1 + Az)^{1/3}\right\}^{-5/3}[\beta_\phi(0)]\,(1 + Az)^{-2/3}\,A. \tag{211}$$

For comparison with the ARA example (see Table 12), we consider the case of proton acceleration with $\gamma_e = 7$ ($\mathscr{E}_e \approx 3$ MeV), $\nu_e/\gamma_e = 0.25$ ($I_e \approx 30$ kA), $E_z(0) = 0.89$ MV/cm, $\beta_\phi(0) = 0.2$ [i.e., injected $\mathscr{E}_i(0) = 19.3$ MeV], and $r_b = 1$ cm. For these parameters, results (210) and (211) are evaluated in Table 13. A characteristic feature of the CGA is readily evident from this example: this is that the required control accuracy on $R(z)$ and $\partial R(z)/\partial z$ is extremely delicate. For example, $R(z)$ is supposed to decrease only 45 *microns* per cm at $z = 0$, and even less (~ 13 microns per cm) at $z = 300$ cm. Furthermore, this control requires $r_b(z)$ to be constant to better than this variation [i.e., $|r_b(z) - r_b(0)| \ll 13$ microns]. It is indeed difficult to imagine how this required control accuracy could be realized experimentally.

Table 13. Example of $R(z)$ and $\partial R(z)/\partial z$ for the CGA

z(cm)	$\beta_\phi(z)$	$\mathcal{E}_i$(MeV)	R(z) [cm]	$-\partial R(z)/\partial z$	$[-\partial R(z)/\partial z]/R(z)$ [cm^{-1}]
0	0.20	19.3	2.97	0.0045	0.0015
100	0.40	84.6	2.73	0.0017	0.00062
200	0.49	139	2.58	0.0014	0.00054
300	0.56	192	2.45	0.0013	0.00053

A further complication arises concerning the radial shear that causes $\gamma(r,z)$ to vary across the beam cross section. Such radial variation in γ can easily be as large as the required axial variation in γ. Thus, radial shear effects must clearly be incorporated into a convincing CGA design.

The ion pulse characteristics of the CGA should be somewhat similar to those of the ARA. If γ_e varies during the pulse, then so will $\mathcal{E}_i(L)$. Also, if the ions are born in a strong B_z, then they must necessarily cross field lines to exit the accelerator; the ions will accordingly acquire a finite angular momentum. However, this effect should be less important in the CGA than in the ARA, because a moderate, constant B_z is used in the CGA, whereas an initially large $B_z(0)$ must be used in the ARA. In any event, nonconstant $\mathcal{E}_i(t)$ and finite angular momentum effects might ultimately complicate the transport and focusing of CGA ion beams.

7. Conclusions

Collective accelerators that use linear electron beams show promise of producing a new breed of compact, high-gradient, high energy ion accelerators. As should be apparent from the diversity of ideas discussed here, there are many possibilities for the realization of a working collective ion accelerator. A summary of the field has been given, including discussions of collective ion acceleration mechanisms, naturally occurring collective ion acceleration processes, collective ion accelerator concepts, and present *scalable* collective ion accelerator concepts for which feasibility experiments are in progress.

Collective ion acceleration mechanisms that use linear electron beams were investigated, and it was concluded that net space charge mechanisms and wave mechanisms appear to be the most promising. Mechanisms that employ induced fields, inverse coherent drag effects, stochastic processes, or impact processes, were shown to be substantially less interesting.

Naturally occurring collective ion acceleration processes were discussed, and it was noted that very large accelerating fields (100 MV/m) have been observed in a variety of experiments, but only over short distances (~ 10 cm). The acceleration process that accompanies IREB injection into neutral gas was examined in depth. With extensive comparisons between theory, experiments, and numerical simulations, it was demonstrated that there is a well-substantiated theoretical explanation for this process. Other naturally occurring collective ion acceleration processes were also examined (modified drift geometries, IREB diodes), and it was concluded that a convincing theoretical explanation for the various diode acceleration phenomena is still needed.

A discussion of thirty-seven collective ion accelerator concepts that use linear electron beams was given. The basic goal of these concepts is to use large accelerating fields (~ 100 MV/m) over large distances. It was concluded that net space charge accelerator concepts and wave accelerator concepts appear to be the most promising.

Three scalable collective ion accelerators (IFA, ARA, CGA) were investigated in detail. The principles of operation, the status of feasibility experiments, and the predictions of several scaling laws were discussed.

Potential applications for collective ion accelerators that use linear electron beams are numerous. They include such diverse uses as:

high energy nuclear physics
heavy ion physics
meson factories
intense neutron sources
radiography
cancer therapy (p, π^-, heavy ion, or n)
spallation breeders
heavy ion fusion

While we will not discuss these potential applications in detail, we note that the application of collective accelerators to heavy ion fusion /3, 34, 45, 380/ is particularly interesting since certain collective accelerators (e.g., the IFA) are particularly well qualified to produce extremely high power bursts of high energy heavy ions. In addition, a unique IFA heavy ion reactor concept exists, which exhibits the standoff capabilities required for a practical reactor concept /35, 45, 380/.

It is evident that the field of collective ion accelerators that use linear electron beams has grown from a number of interesting experimental observations of naturally occurring collective ion acceleration processes, to the stage where many collective ion accelerator concepts have been proposed. Serious efforts have now been initiated to demonstrate specific collective ion accelerator concepts. It is anticipated that one or more of these collective ion accelerator concepts will be demonstrated experimentally in the very near future.

Acknowledgments

This work was supported in part by the United States Department of Energy and the United States Air Force Office of Scientific Research.

References

1 Proc. ERDA Information Meeting on Accelerator Breeding, Brookhaven National Laboratory, January 18-19, 1977 (National Technical Information Service, Springfield, Virginia, 1977)

2 C.A. Robinson, Jr.: Aviation Week & Space Technology, May 2, 1977, p. 16; R. Hotz: ibid., p. 11

3 ERDA Summer Study of Heavy Ions for Inertial Fusion, Berkeley, California, July 19-30, 1976 (National Technical Information Service, Springfield, Virginia, 1976)

4 H. Alfven and P. Wernholm: Arkiv Fysik 5, 175 (1952)

5 S.R.B.R. Harvie: AERE memorandum G/M 87 (1951)

6 W. Raudorf: Wireless Engineer 28, 215 (1951)

7 V.I. Veksler: Proc. CERN Symp. High Energy Accel., Geneva, June 11-23, 1956 (CERN, Geneva, 1956) Vol. 1, p. 80

8 V.I. Veksler: Atomnaya Energiya 2, 525 (1957) [English transl.: Sov. J. Atom. En. 2, 525 (1957)]

9 V.I. Veksler: Usp. Fiz. Nauk 66, 99 (1958) [English transl.: Sov. Phys.-Usp. 66, 54 (1958)]

10 G.I. Budker: Proc. CERN Symp. High Energy Accel., Geneva, June 11-23, 1956 (CERN, Geneva, 1956) Vol. 1, p. 68; G.I. Budker and A.A. Naumov, ibid., p. 76

11 Ya.B. Fainberg: Proc. CERN Symp. High Energy Accel., Geneva, June 11-23, 1956 (CERN, Geneva, 1956) Vol. 1, p. 84

12 V.I. Veksler, V.P. Sarantsev, A.G. Bonch-Osmolovskii, G.V. Dolbilov, G.A. Ivanov, I.N. Ivanov, M.L. Iovnovitch, I.V. Kozhukhov, A.B. Kuznetsov, V.G. Makhankov, E.A. Perelstein, V.P. Rashevsky, K.A. Reshetnikova, N.B. Rubin, S.B. Rubin P.I. Ryl'tsev, and O.I. Yarkovoy: Proc. VI Int. Conf. High Energy Accel. (CEAL, Cambridge, Massachusetts, 1967) p. 289

13 A.A. Plyutto: Zh. Eksper. I. Teor. Fiz 39, 1589 (1960) [English transl.: Sov. Phys.-JETP 12, 1106 (1961)]

14 S.E. Graybill and J.R Uglum: J. Appl. Phys. 41, 236 (1970)

15 S. Graybill, G. Ames, J. Rizzo, J. Uglum, W. McNeill, F.K. Childers, and S.V. Nablo: Report DASA 2315, Ion Physics Corporation (1968)

16 C.L. Olson: Bull. Am. Phys. Soc. 18, 1356 (1973)

17 C.L. Olson: Phys. Fluids 18, 585 (1975)

18 C.L. Olson: Phys. Fluids 18, 598 (1975)

19 J.W. Poukey and C.L. Olson: Phys. Rev. A11, 691 (1975); J.W. Poukey: private communication

20 M.S. Rabinovich: Preprint No. 36, Lebedev Institute, Moscow, USSR (1969)
21 A.M. Sessler: Comments on Nuclear and Particle Physics 3, 93 (1969)
22 G. Yonas: Report PIIR-36-71, Physics International Company (1971)
23 J.D. Lawson: Particle Accel. 3, 21 (1972)
24 A.A. Kolomensky: Particle Accel. 5, 73 (1973)
25 M.S. Rabinovich and V.N. Tsytovich: Particle Accel. 5, 99 (1973)
26 G. Yonas: Particle Accel. 5, 81 (1973)
27 G. Yonas: presented at Int. Summer School Appl. Physics, Erice, Sicily (1973)
28 M.S. Rabinovich and V.N. Tsytovich: Preprint No. 55, Lebedev Institute, Moscow, USSR (1973)
29 A.A. Kolomensky: Proc. IX Int. Conf. High Energy Accel., SLAC, Stanford, California, May 2-7, 1974 (National Technical Information Service, Springfield, Virginia, 1974) p. 254
30 G. Wallis, K. Sauer, D. Sunder, S.E. Rosinskii, A.A. Rukhadze, and V.G. Rukhlin: Usp. Fiz. Nauk 113, 435 (1974) [English transl.: Sov. Phys.-Usp. 17, 492 (1975)]
31 C.L. Olson: IEEE Trans. Nucl. Sci. NS-22 #3, 962 (1975)
32 D. Keefe: Proc. 1976 Proton Linear Accel. Conf., Ontario, Canada (Preprint LBL-5536, Lawrence Berkeley Laboratory, 1976)
33 A.A. Plyutto: Preprint SFTI-1, Sukhumi Institute of Physics and Technology, Sukhumi, USSR (1977)
34 C.L. Olson: Proc. II Symp. Collective Methods Accel., Dubna, USSR, Sept. 29 - Oct. 2, 1976 (JINR, Dubna, USSR, 1977) p. 101
35 C.L. Olson: Bull. Am. Phys. Soc. 18, 1369 (1973)
36 C.L. Olson: Report SLA-73-0865, Sandia Laboratories (January 1974)
37 C.L. Olson: Proc. IX Int. Conf. High Energy Accel., SLAC, Stanford, California, May 2-7, 1974 (National Technical Information Service, Springfield, Virginia, 1974) p. 272
38 C.L. Olson: Proc. Int. Top. Conf. E-Beam Res. and Tech., Albuquerque, New Mexico, Nov. 3-5, 1975 (National Technical Information Service, Springfield, Virginia, 1976) Vol. 2, p. 312
39 M.L. Sloan and W.E. Drummond: Phys. Rev. Lett. 31, 1234 (1973)
40 M.L. Sloan and W.E. Drummond: Proc. IX Int. Conf. High Energy Accel., SLAC, Stanford, California, May 2-7, 1974 (National Technical Information Service, Springfield, Virginia, 1974) p. 283
41 P. Sprangle, A.T. Drobot, and W.H. Manheimer: Phys. Rev. Lett. 36, 1180 (1976)
42 J.C. Martin, AWRE, Aldermaston, England: private communication
43 S.E. Graybill and S.V. Nablo: IEEE Trans. Nucl. Sci. NS-14 #3, 782 (1967)
44 W.T. Link: IEEE Trans. Nucl Sci. NS-14 #3, 777 (1967)
45 C.L. Olson: Fiz. Plazmy 3, 465 (1977) [English transl.: Sov. J. Plasma Phys. 3, 259 (1977)]
46 W.H. Bennett: Phys. Rev. 45, 890 (1934)
47 W.H. Bennett: Phys. Rev. 98, 1584 (1955)
48 H. Alfven: Phys. Rev. 55, 425 (1939)
49 J.D. Lawson: J. Electronic Control 3, 587 (1957)
50 J.D. Lawson: J. Electronic Control 5, 146 (1958)
51 J.D. Lawson: Particle Accel. 1, 41 (1970)
52 J.D. Lawson, P.M. Lapostolle, and R.L. Gluckstern: Particle Accel. 5, 61 (1973)
53 J.D. Lawson: Plasma Phys. 17, 567 (1975)
54 C.L. Olson and J.W. Poukey: Phys. Rev. A 9, 2631 (1974)
55 M. Nezlin: Usp. Fiz. Nauk 102, 105 (1970) [English transl.: Sov. Phys.-Usp. 13, 608 (1971)]
56 V.R. Bursian and V.I. Pavlov: J. Russ. Phys. Chem. Soc. 55, 71 (1923)
57 L.P. Smith and P.L. Hartman: J. Appl. Phys. 11, 220 (1940)
58 L.S. Bogdankevich, I.L. Zhelyazkov, and A.A. Rukhadze: Zh. Eksper. I. Teor. Fiz. 57, 315 (1969) [English transl.: Sov. Phys.-JETP 30, 174 (1970)]
59 L.S. Bogdankevich and A.A. Rukhadze: Usp. Fiz. Nauk 103, 609 (1971) [English transl.: Sov. Phys.-Usp. 14, 163 (1971)]
60 T.J. Fessenden: Report UCID-16527, Lawrence Livermore Laboratory (1974)
61 B.N. Breizman and D.D. Ryutov: Nucl. Fusion 14, 873 (1974)
62 D.D. Ryutov: Preprint 75-97, Institute of Nuclear Physics, Novosibirsk, USSR (1975)

63 V.E. Nechaev: Fiz. Plazmy 3, 112 (1977) [English transl.: Sov. J. Plasma Phys. 3, 64 (1977)]
64 J.R. Thompson and M.L. Sloan: Bull. Am. Phys. Soc. 22, 1133 (1977)
65 J.A. Nation and M. Read: Appl. Phys. Lett. 23, 426 (1973)
66 M.E. Read and J.A. Nation: J. Plasma Phys. 13, 127 (1975)
67 J.W. Poukey and J.R. Freeman: Phys. Fluids 17, 1917 (1974)
68 M. Reiser: Phys. Fluids 20, 477 (1977)
69 D.A. Hammer and N. Rostoker: Phys. Fluids 13, 1831 (1970)
70 G. Benford, D.L. Book, and R.N. Sudan: Phys. Fluids 13, 2621 (1970)
71 G. Benford and D.L. Book: in *Advances in Plasma Physics, Vol. 4,* ed. by A. Simon and W.B. Thompson (Interscience, New York, 1971) p. 125
72 G. Yonas and P. Spence: in *Record 10th Symp. Electron, Ion, and Laser Beam Technology,* ed. by L. Marton (San Francisco Press, San Francisco, 1969) p. 143
73 M.L. Andrews, H.E. Davitian, H.H. Fleischmann, D.A. Hammer, and J.A. Nation: Appl. Phys. Lett. 16, 98 (1970)
74 P. Auer: Phys. Fluids 17, 148 (1974)
75 S. Yoshikawa: Phys. Rev. Lett. 26, 295 (1971)
76 J.R. Kan and H. Lai: Phys. Fluids 15, 2041 (1972)
77 J.D. Lawson: Phys. Fluids 16, 1298 (1973)
78 J. Hieronymus: Ph.D. Thesis, Cornell University (1971)
79 A. Chodorow and A. Erteza: Phys. Fluids 19, 1779 (1976)
80 F. Friedlander, R. Hecktel, H.R. Jory, and C. Mosher: Report DASA-2173, Varian Associates (1973)
81 J.M. Creedon: J. Appl. Phys. 46, 2946 (1975)
82 J.R. Pierce: J. Appl. Phys. 15, 721 (1944)
83 J. Frey and C.K. Birdsall: J. Appl. Phys. 37, 2051 (1966)
84 J.R. Pierce: J. Appl. Phys. 19, 231 (1948)
85 O. Buneman: Phys. Rev. Lett. 1, 8 (1958)
86 O. Buneman: Phys. Rev. 115, 503 (1959)
87 M.V. Nezlin: Zh. Eksper. I. Teor. Fiz. 41, 1015 (1961) [English transl.: Sov. Phys.-JETP 14, 726 (1962)]
88 M.V. Nezlin and A.M. Solntsev: Zh. Eksper. I. Teor. Fiz. 53, 437 (1967) [English transl.: Sov. Phys.-JETP 26, 290 (1968)]
89 M.V. Nezlin, M.I. Taktakishvili, and A.S. Trubnikov: Zh. Eksper. I. Teor. Fiz. 60, 1012 (1971) [English transl.: Sov. Phys.-JETP 33, 548 (1971)]
90 M.D. Raizer, A.A. Rukhadze, and P.S. Strelkov: Zh. Techn. Fiz. 38, 776 (1968) [English transl.: Sov. Phys.-Techn. Phys. 13, 587 (1968)]
91 C.L. Olson: Particle Accel. 6, 107 (1975)
92 C.L. Olson: Phys. Fluids 16, 529 (1973)
93 W.O. Doggett and W.H. Bennett: Bull. Am. Phys. Soc. 14, 1048 (1969)
94 W.O. Doggett and W.H. Bennett: Bull. Am. Phys. Soc. 15, 641 (1970)
95 W.O. Doggett: Bull. Am. Phys. Soc. 15, 1347 (1970)
96 W.H. Bennett: Annals N.Y.A.S. 251, 213 (1975)
97 C.L. Olson: Bull. Am. Phys. Soc. 20, 1385 (1975)
98 N. Bohr: Phil. Mag. 30, 581 (1915)
99 E. Segre: *Nuclei and Particles* (W.A. Benjamin, Inc., Reading, Massachusetts, 1965) Chap. 2
100 J.D. Jackson: *Classical Electrodynamics* (John Wiley and Sons, New York, 1963) Chap. 13
101 H.A. Bethe: Ann. Phys. 5, 325 (1930)
102 P.A. Cerenkov: Phys. Rev. 52, 378 (1937)
103 I.M. Frank and I. Tamm: Dokl. Akad. Nauk SSSR 14, 109 (1937)
104 G.B. Collins and V.G. Reiling: Phys. Rev. 54, 499 (1938)
105 A.A. Kolomenskii: Dokl. Akad. Nauk SSSR 106, 982 (1956) [English transl.: Sov. Phys.-Dokl. 1, 133 (1956)]
106 J.F. McKenzie: Philos. Trans. Roy. Soc. Lond. A255, 71 (1963)
107 P.D. Coleman: in *Proc. Symp. Quasi-Optics XIV* (Polytechnic Press, Brooklyn, N.W., 1964) p. 199
108 W. Sollfrey and H.T. Yura: Phys. Rev. 139, A48 (1965)
109 P. Sasiela and J.P. Freidberg: Rad. Sci. 2, 703 (1967)
110 H. Kikuchi: in *Proc. Symp. Quasi-Optics XIV* (Polytechnic Press, Brooklyn, N.Y. 1964) p. 235

111 D. Pines and D. Bohm: Phys. Rev. 85, 338 (1952)
112 A. Ahiezer: Suppl. Nuovo Cimento 3, 581 (1956)
113 Y.L. Klimontovich: Zh. Eksper. I. Teor. Fiz. 36, 1405 (1959) [English transl.: Sov. Phys.-JETP 36, 999 (1959)]
114 O. Aono: J. Phys. Soc. Japan 17, 853 (1962)
115 S.T. Butler and M.J. Buckingham: Phys. Rev. 126, 1 (1962)
116 A.J.R. Prentice: Plasma Physics 9, 433 (1967)
117 P.C. Clemmow and J.P. Dougherty: *Electrodynamics of Particles and Plasmas* (Addison-Wesley, Reading, Massachusetts, 1969) Chap. 3
118 V.N. Tsytovich: *Nonlinear Effects in Plasma* (Plenum, New York, 1970) p. 151
119 I.F. Kharchenko. Ya.B. Fainberg, P.M. Nikolayev, E.A. Kornilov, E.A. Lutsenko, and N.S. Pedenko: Proc. 4th Int. Conf. Ion. Phen. Gases, Uppsala, Sweden, Aug. 17-21, 1959 (North Holland Pub. Co., Amsterdam, 1960) Vol. 2, p. IIIB 671
120 I.F. Kharchenko, Ya.B. Fainberg, R.M. Nikolaev, E.A. Kornilov, E.A. Lutsenko, and N.S. Pedenko: Zh. Eksper. I. Teor. Fiz. 38, 685 (1960) [English transl.: Sov. Phys.-JETP 11, 493 (1960)]
121 A.K. Berezin, Ya.B. Fainberg, G.P. Berezina, L.I. Bolotin, and V.G. Stupak: Plasma Phys. (J. Nucl. En. Part C) 4, 291 (1962)
122 H.M. Epstein, W.J. Gallagher, P.J. Mallozzi, and M.R. Vanderlind: Proc. 8th Int. Conf. Phen. Ionized Gases, Vienna, Austria, Aug. 27 - Sept. 2, 1967 (Springer-Verlag, Vienna, 1967) p. 372
123 A. Irani and N. Rostoker: Bull. Am. Phys. Soc. 20, 1382 (1975)
124 A.A. Irani and N. Rostoker: Report UCI #75-56, University of California, Irvine (1976)
125 Ya.B. Fainberg Plasma Phys.: (J. Nucl. En. Part C) 4, 203 (1962)
126 Ya.B. Fainberg: Usp. Fiz. Nauk 93, 617 (1967) [English transl.: Sov. Phys.-Usp. 10, 750 (1968)]
127 Ya.B. Fainberg: Particle Accel. 6, 95 (1975)
128 L.E. Thode and R.N. Sudan: Phys. Rev. Lett. 30, 732 (1973)
129 L.E. Thode and R.N. Sudan: Phys. Fluids 18, 1552 (1975)
130 R.V. Lovelace and R.N. Sudan: Phys. Rev. Lett. 27, 1256 (1975)
131 L.E. Thode and R.N. Sudan: Phys. Fluids 18, 1564 (1975)
132 L.E. Thode: Phys. Fluids 19, 831 (1976)
133 V.N. Tsytovich: Proc. 12th Int. Conf. Phen. Ionized Gases, Eindhoven, The Netherlands (1975) Vol. 2, p. 141
134 K.V. Khodataev and V.N. Tsytovich: Fiz. Plazmy 2, 301 (1976) [English transl.: Sov. J. Plasma Phys. 2, 164 (1976)]
135 K.V. Khodataev and V.N. Tsytovich: Proc. II Symp. Collective Methods Accel., Dubna, USSR, Sept. 29 - Oct. 2, 1976 (JINR, Dubna, USSR, 1977) p. 124
136 V.N. Tsytovich and K.V. Khodataev: Comments on Plasma Physics and Controlled Fusion 3, 71 (1977)
137 I.B. Bernstein and R.M. Kulsrud: Phys. Fluids 3, 937 (1960)
138 S.A. Bludman, K.M. Watson, and M.N. Rosenbluth: Phys. Fluids 3, 747 (1960)
139 M. Yoshikawa: Nucl. Fusion 1, 167 (1961) [In this reference, a factor γ^3 is accidently missing from the left hand side of Eq. (25), as can be seen from Eqs. (21) and (23)]
140 A.C. Scott, F.Y.F. Chu, and D.W. McLaughlin: Proc. IEEE 61 #10, 1443 (1973)
141 V.N. Tsytovich: JINR Preprint P9-5090, Dubna, USSR (1970)
142 J. Jancarik and V.N. Tsytovich: Nucl. Fusion 13, 807 (1973)
143 V.N. Tsytovich: Dokl. Akad. Nauk SSSR 142, 63 (1962) [English transl.: Sov. Phys.-Dokl. 7, 31 (1962)]
144 Ya.B. Fainberg: Atomnaya Energiya 6, 447 (1959) [English transl.: Sov. J. Atom. En. 6, 311 (1959)]
145 Ya.B. Fainberg, V.D. Shapiro, and V.I. Shevchenko: Zh. Eksper. I. Teor. Fiz. Pis. Red. 11, 410 (1970) [English transl.: Sov. Phys.-JETP Lett. 11, 277 (1970)]
146 V.B. Krasovitskii: Zh. Techn. Fiz. 37, 493 (1967) [English transl.: Sov. Phys.-Techn. Phys. 12, 354 (1967)]
147 V.B. Krasovitskii, V.I. Kurilko, and M.A. Strzhemechnyi: Atomnaya Energiya 24, 545 (1968) [English transl.: Sov. J. Atom. En. 24, 670 (1968)]
148 V.B. Krasovitskii: Zh. Eksper. I. Teor. Fiz. 59, 176 (1970) [English transl.: Sov. Phys. JETP 32, 98 (1971)]

149 V.N. Tsytovich: Usp. Fiz. Nauk 89, 89 (1966) [English transl.: Sov. Phys.-Usp 9, 370 (1966)]
150 S.A. Kaplan and V.N. Tsytovich: Phys. Reports (Sect. C of Phys. Lett.) 7, 1 (1973)
151 N.G. Koval'skii, B.I. Khripanov, and S.A. Chuvatin: Zh. Tekhn. Fiz. 41, 308 (1971) [English transl.: Sov. Phys.-Techn. Phys. 16, 232 (1971)]
152 G.P. Berezina, A.K. Berezin, and V.P. Zeidlets: Zh. Eksper. I. Teor. Fiz. Pis. Red. 14, 77 (1971) [English transl.: Sov. Phys.-JETP Lett. 14, 49 (1971)]
153 V.I. Veksler and V.N. Tsytovich: Proc. Int. Conf. High Energy Accel.-CERN 1959 Geneva, Switzerland, Sept. 14-19, 1959 (CERN, Geneva, Switzerland, 1959) p. 1
154 A.G. Bonch-Osmolovskii: Atomnaya Energiya 31, 127 (1971) [English transl.: Sov. J. Atom. En. 31, 835 (1971)]
155 S.E. Graybill and J.R. Uglum: Bull. Am. Phys. Soc. 14, 1048 (1969)
156 S.E. Graybill, J.R. Uglum, W.H. McNeill, J.E. Rizzo, R. Lowell, and G. Ames: Report DASA 2477, Ion Physics Corporation (1970)
157 S.E. Graybill: IEEE Trans. Nucl. Sci. NS-18 #3, 438 (1971)
158 S.E. Graybill: IEEE Trans. Nucl. Sci. NS-19 #2, 292 (1972)
159 S.E. Graybill: private communication (November 1973)
160 G. Yonas, P. Spence, B. Ecker, and J. Rander: Report PIFR-106-2, Physics International Company (1969)
161 J. Rander, B. Ecker, and G. Yonas: Report PIIR-9-70, Physics International Company (1969)
162 J. Rander, B. Ecker, G. Yonas, and D.J. Drickey: Bull. Am. Phys. Soc. 14, 1048 (1969)
163 J. Rander, B. Ecker, G. Yonas, and D.J. Drickey: Phys. Rev. Lett. 24, 283 (1970)
164 J. Rander: Phys. Rev. Lett. 25, 893 (1970)
165 J. Rander, J. Benford, B. Ecker, G. Loda, and G. Yonas: Report PIIR-2-71, Physics International Company (1971)
166 G. Yonas: IEEE Trans. Nucl. Sci. NS-19 #2, 297 (1972)
167 B. Ecker, S. Putnam, and D. Drickey: IEEE Trans. Nucl. Sci. NS-20 #3, 301 (1973)
168 D. Drickey, B. Ecker, and S. Putnam: Proc. IX Int. Conf. High Energy Accel., SLAC, Stanford, California, May 2-7, 1974 (National Technical Information Service, Springfield, Virginia, 1974) p. 236
169 B. Ecker and S. Putnam: Proc. II Symp. Collective Methods Accel., Dubna, USSR, Sept. 29 - Oct. 2, 1976 (JINR, Dubna, USSR, 1977) p. 152
170 B. Ecker and S. Putnam: Bull. Am. Phys. Soc. 21, 1059 (1976)
171 B. Ecker and S. Putnam: Report PIIR-8-76, Physics International Company (1976)
172 B. Ecker and S. Putnam: IEEE Trans. Nucl. Sci. NS-24 #3, 1665 (1977)
173 B. Ecker, R.B. Miller, and D.C. Straw: to be published
174 J.P. VanDevender: Report UCRL-50935, Lawrence Radiation Laboratory (1970)
175 G.W. Kuswa, L.P. Bradley, and G. Yonas: IEEE Trans. Nucl. Sci. NS-20 #3, 305 (1973)
176 G.W. Kuswa: Annals N.Y.A.S. 251, 514 (1975)
177 D. Swain, G. Kuswa, J. Poukey, and C. Olson: Proc. IX Int. Conf. High Energy Accel., SLAC, Stanford, California, May 2-7, 1974 (National Technical Information Service, Springfield, Virginia, 1974) p. 268
178 D.C. Straw and R.B. Miller: Appl. Phys. Lett. 25, 379 (1974)
179 R.B. Miller and D.C. Straw: IEEE Trans. Nucl. Sci. NS-22 #3, 1022 (1975)
180 R.B. Miller and D.C. Straw: J. Appl. Phys. 47, 1897 (1976)
181 R.B. Miller and D.C. Straw: Proc. Int. Top. Conf. E-Beam Res. and Tech., Albuquerque, New Mexico, November 3-5, 1975 (National Technical Information Service, Springfield, Virginia, 1976) Vol. 2, p. 368
182 R.B. Miller and D.C. Straw: Report AFWL-TR-75-236, Air Force Weapons Laborato (1976)
183 D.C. Straw and R.B. Miller: J. Appl. Phys. 47, 4681 (1976)
184 R.B. Miller: Report AFWL-TR-76-169, Air Force Weapons Laboratory (1977)
185 D.C. Straw and R.B. Miller: IEEE Trans. Nucl. Sci. NS-24 #3, 1645 (1977)
186 A.A. Kolomenskii, V.M. Likhachev, I.V. Sinil'shchikova, O.A. Smit, and V.N. Ivanov: Proc. IV All Union Conf. Charge Particle Accelerators, Moscow, USSR (1974)

187 A.A. Kolomenskii, V.M. Likhachev, I.V. Sinil'shchikova, O.A. Smit, and V.N. Ivanov: Zh. Eksper. I. Teor. Fiz. 68, 51 (1975) [English transl.: Sov. Phys.-JETP 41, 26 (1975)]
188 A.A. Kolomensky: Proc. Int. Top. Conf. E-Beam Res. and Tech., Albuquerque, New Mexico, Nov. 3-5, 1975 (National Technical Information Center, Springfield, Virginia, 1976) Vol. 2, p. 295
189 A.A. Kolomensky, V.M. Likhachyov, and B.N. Yablokov: IEEE Trans. Nucl. Sci. NS-22 #3 (1975)
190 V.N. Ivanov, A.A. Kolomensky, V.M. Likhachev, I.V. Sinil'shchikova, and O.A. Smit: Proc. II Symp. Collective Methods Accel., Dubna, USSR, Sept. 29 - Oct. 2, 1976 (JINR, Dubna, USSR, 1977),p. 114
191 C.W. Roberson, S. Eckhouse, A. Fisher, S. Robertson, and N. Rostoker: Phys. Rev. Lett. 36, 1457 (1976)
192 S. Eckhouse, A. Fisher, R. Mako, C.W. Roberson, S. Robertson, and N. Rostoker: Proc. II Symp. Collective Methods Accel., Dubna, USSR, Sept. 29 - Oct. 2, 1976 (JINR, Dubna, USSR, 1977) p. 146
193 V.M. Bystritskii, V.I. Podkatov, A.G. Sterligov, G.E. Remnev, and Yu.P. Usov: Pis. Zh. Tekhn. Fiz. 2, 80 (1976) [English transl.: Sov. Phys.-Techn. Phys. Lett. 2, 30 (1976)]
194 V.M. Bystritskii, V.I. Podkatov, Yu.P. Usov, and V.N. Shustova: presented at II Symp. Collective Methods Accel., JINR, Dubna, USSR, September 29 - October 2, 1976
195 S.E. Graybill and F.C. Young: Bull. Am. Phys. Soc. 21, 1058 (1976)
196 N. Rostoker: Report LPS 21, Cornell University (1969); Proc. VII Int. Conf. High Energy Accel., Yerevan, Armenian SSR (1969), Vol. II, p. 509
197 N. Rostoker: Bull. Am. Phys. Soc. 14, 1047 (1969)
198 J.R. Uglum, S.E. Graybill, and W.H. McNeill: Bull. Am. Phys. Soc. 14, 1047 (1969)
199 S.E. Graybill, W.H. McNeill, and J.R. Uglum: in *Record 11th Symp. Electron, Ion, and Laser Beam Technology*, ed. by R.F.M. Thornley (San Francisco Press, San Francisco, 1971) p. 577
200 J.W. Poukey and N. Rostoker: Plasma Phys. 13, 897 (1971)
201 S.E. Rosinskii, A.A. Rukhadze, and V.G. Rukhlin: Zh. Eksper. I. Teor. Fiz. Pis. Red. 14, 53 (1971) [English transl.: Sov. Phys.-JETP Lett. 14, 34 (1971)]
202 V.I. Veksler: Proc. CERN Symp. High Energy Accel., Geneva, June 11-23, 1956 (CERN, Geneva, 1956) Vol. 1, p. 99
203 S.D. Putnam: Bull. Am. Phys. Soc. 14, 1048 (1969)
204 S.D. Putnam: Phys. Rev. Lett. 25, 1129 (1970)
205 S. Putnam: IEEE Trans. Nucl. Sci. NS-18 #3, 496 (1971)
206 S. Putnam: Report PIFR-72-105, Physics International Company (1971)
207 I. Kapchinski and V. Vladimirski: Proc. Int. Conf. High Energy Accel.-CERN 1959, Geneva, Sept. 14-19, 1959 (CERN, Geneva, 1959) p. 274
208 J. Benford and G. Benford: New Scientist, November 30, 1972, p. 514
209 J.M. Wachtel and B.J. Eastlund: Bull. Am. Phys. Soc. 14, 1047 (1969)
210 C.L. Olson: Phys. Rev. A11, 288 (1975)
211 C.L. Olson: Annals N.Y.A.S. 251, 536 (1975)
212 C.L. Olson: Bull. Am. Phys. Soc. 19, 914 (1974)
213 C.L. Olson: to be published
214 K.F. Alexander, E. Hantzsche, and P. Siemroth: Report (in German), Zentralinstitut für Elektronenphysik, Berlin, DDR (December 1973)
215 K. Alexander, E. Hantzsche, and P. Siemroth: Zh. Eksper. I. Teor. Fiz. 67, 567 (1974) English transl.: Sov. Phys.-JETP 40, 280 (1975)
216 K.F. Alexander and E. Hantzsche: Preprint 75-4 (in German), Zentralinstitut für Elektronenphysik, Berlin, DDR (1975)
217 K.F. Alexander and W. Hintze: Preprint 76-7, Zentralinstitut für Elektronenphysik, Berlin, DDR (1976)
218 K.F. Alexander and W. Hintze: Proc. II Symp. Collective Methods Accel., Dubna, USSR, Sept. 29 - Oct. 2, 1967 (JINR, Dubna, USSR, 1977) p. 155
219 B.B. Godfrey and L.E. Thode: Annals N.Y.A.S. 251, 582 (1975)
220 B. Godfrey and L. Thode: Bull. Am. Phys. Soc. 19, 967 (1974)
221 B.B. Godfrey and L.E. Thode: IEEE Trans. Nucl. Sci. PS-3, 201 (1975)

222 A.A. Irani and N. Rostoker: Report 76-16, University of California, Irvine
223 A.A. Kolomenskii and M.A. Novitskii: Zh. Techn. Fiz. 46, 44 (1976) [English transl.: Sov. Phys.-Techn. Phys. 21, 23 (1976)]
224 A.A. Kolomenskii and M.A. Novtiskii: Kratkie Soobshcheniya po Fiz. 10, 37 (1976) [English transl.: Sov. Phys.-Lebedev Inst. Reports 10, 32 (1976)]
225 A.A. Kolomenskii and M.A. Novitskii: Proc. II Symp. Collective Methods Accel., Dubna, USSR, Sept. 29 - Oct. 2, 1976 (JINR, Dubna, USSR, 1977) p. 130
226 V.M. Bystritskii, A.N. Didenko, Yu.P. Usov, and V.N. Shustova: Proc. VII Int. Symp. Elect. Discharges and Ins. in Vac., Novosibirsk, USSR (1976) p. 362
227 B.B. Godfrey, R.J. Faehl, W.R. Shanahan, and L.E. Thode: Conf. Record-Abstracts, IEEE 1977 Int. Conf. Plasma Sci., Troy, New York, May 23-25, 1977 (IEEE Service Center, Piscataway, N.J., 1977) Abstract 5B14, p. 174
228 B.B. Godfrey: Preprint LA-UR-77-1971, Los Alamos Scientific Laboratory (1977)
229 C.L. Olson, G.W. Kuswa, D.W. Swain, and J.W. Poukey: Proc. Second Symp. Ion Sources and Formation of Ion Beams, Berkeley, California, Oct. 22-25, 1974 (Technical Information Division, Lawrence Berkeley Laboratory, Berkeley, 1974) p. III-3-1
230 A.M. Cravath and L.B. Loeb: Physics (now J. Appl. Phys.) 6, 125 (1935)
231 B.J.F. Schonland: in *Encyclopedia of Physics*, Vol. 22, ed. by S. Flugge (Springer, Berlin, 1956) p. 620
232 L.B. Loeb: Science 148, 1417 (1965)
233 D.L. Turcotte and R.S.B. Ong: J. Plasma Phys. 2, 145 (1968)
234 G.P. Mkheidze, V.I. Pulin, M.D. Raizer, and L.E. Tsopp: Zh. Eksper. I. Teor. Fiz. 63, 104 (1972) [English transl.: Sov. Phys.-JETP 36, 54 (1973)]
235 Yu.I. Abrashitov, V.S. Koidan, V.V. Konyukhov, L.M. Lagunov, V.N. Lukyanov, K.I. Mekler, and D.D. Ryutov: Zh. Eksper. I. Teor. Fiz. 66, 1324 (1974) [English transl.: Sov. Phys.-JETP 39, 647 (1974)]
236 C.N. Boyer, W.W. Destler, and H. Kim: IEEE Trans. Nucl. Sci. NS-24 #3, 1625 (1977)
237 Yu.V. Tkach, I.I. Magda, G.V. Skachek, S.S. Pushkarev, and I.P. Panchenko: Zh. Techn. Fiz. 44, 658 (1974) [English transl.: Sov. Phys.-Techn. Phys. 19, 412 (1974)]
238 J.A. Nation: Bull. Am. Phys. Soc. 19, 966 (1974)
239 J.A. Nation: J. Plasma Phys. 13, 361 (1975)
240 R. Williams, J.A. Nation, and M.E. Read: Conf. Record-Abstracts, IEEE 1977 Int. Conf. Plasma Sci., Troy, New York, May 23-25, 1977 (IEEE Service Center, Piscataway, N.J., 1977) Abstract 5B13, p. 173
241 H. Kim and H.S. Uhm: Bull. Am. Phys. Soc. 19, 966 (1974)
242 R.B. Miller, D.C. Straw, and R.F. Hoeberling: Bull. Am. Phys. Soc. 21, 1101 (1976)
243 A. Greenwald, R. Lowell, and R. Little: Bull. Am. Phys. Soc. 21, 1147 (1976)
244 A. Greenwald and R. Little: Proc. 2nd Int. Top. Conf. High Power Electron and Ion Beam Res. and Tech., Cornell University, Oct. 3-5, 1977 (LPS, Cornell University, 1978) p. 553
245 J.A. Pasour, R.K. Parker, W.O. Doggett, D. Pershing, and R.L. Gullickson: Proc. 2nd Int. Top. Conf. High Power Electron and Ion Beam Res. and Tech., Cornell University, Oct. 3-5, 1977 (LPS, Cornell University, 1978) p. 623
246 R.K. Parker, J.A. Pasour, W.O. Doggett, D.L. Pershing, and R.L. Gullickson: Bull. Am. Phys. Soc. 22, 1111 (1977)
247 C.L. Olson, J.P. VanDevender, and A. Owyoung: Abstract E10, 2nd Int. Top. Conf. High Power Electron and Ion Beam Res. and Tech., Cornell University, Oct. 3-5, 1977
248 M.J. Rhee, D.W. Hudgings, R.A. Meger, and E. Pappas: Bull. Am. Phys. Soc. 22, 1199 (1977)
249 R. Mako and A. Fisher: Bull. Am. Phys. Soc. 21, 1059 (1976)
250 R. Mako and A. Fisher: Bull. Am. Phys. Soc. 22, 1111 (1977)
251 T. Tajima, R. Mako, and A. Fisher: Bull. Am. Phys. Soc. 21, 1168 (1976)
252 T. Tajima and F. Mako: Report PPG-314, University of California, Los Angeles (1977)
253 F.C. Young and M. Friedman: Bull. Am. Phys. Soc. 19, 966 (1974)
254 F.C. Young and M. Friedman: J. Appl. Phys. 46, 2001 (1975)

255 A.A. Plyutto, P.E. Belensov, E.D. Korop, G.E. Mkheidze, V.N. Ryzhkov, K.V. Suladze, and S.M. Temchin: Zh. Eksper. I. Teor. Fiz. Pis. Red. 6, 540 (1967) [English transl.: Sov. Phys.-JETP Lett. 6, 61 (1967)]
256 A.A. Plyutto, K.V. Suladze, S.M. Temchin, and E.D. Korop: Atomnaya Energiya 27, 418 (1969) [English transl.: Sov. J. Atom. En. 27, 1197 (1969)]
257 G.P. Mkheidze, A.A. Plyutto, and E.D. Korop: Zh. Techn. Fiz. 41, 952 (1971) [English transl.: Sov. Phys.-Techn. Phys. 16, 749 (1971)]
258 V.N. Ryzhkov, A.A. Plyutto, P.E. Belensov, and A.T. Kapin: Zh. Techn. Fiz. 42, 2074 (1972) [English transl.: Sov. Phys.-Techn. Phys. 17, 1651 (1973)]
259 A.A. Plyutto, K.V. Suladze, S.M. Temchin, G.P. Mkheidze, E.D. Korop, B.A. Tskhadzy, and I.V. Golovin: Zh. Techn. Fiz. 43, 1627 (1973) [English transl.: Sov. Phys.-Techn. Phys. 18, 1028 (1973)]
260 A.A. Plyutto and A.T. Kapin: Zh. Techn. Fiz. 45, 2533 (1975) [English transl.: Sov. Phys.-Techn. Phys. 20, 1578 (1976)]
261 P.E. Belensov, A.T. Kapin, A.A. Plyutto, and V.N. Ryzhkov: Zh Techn. Fiz. 34, 2120 (1964) [English transl.: Sov. Phys.-Techn. Phys. 9, 1633 (1965)]
262 K.V. Suladze, B.A. Tskhadaya, and A.A. Plyutto: Zh. Eksper. I. Teor. Fiz. Pis. Red. 10, 282 (1969) English transl.: Sov. Phys.-JETP Lett. 10, 180 (1969)
263 G.P. Mkheidze and E.D. Korop: Zh. Techn. Fiz. 41, 873 (1971) [English transl.: Sov. Phys.-Techn. Fiz. 16, 690 (1971)]
264 K.V. Suldaze: Zh. Eksper. I. Teor. Fiz. Pis. Red. 15, 649 (1972) [English transl.: Sov. Phys.-JETP Lett. 15, 459 (1972)]
265 A.A. Plyutto and K.V. Suladze: Plazmennyye Uskoriteli (1973) p. 47
266 B.A. Tskhadaya, A.A. Plyutto, and K.V. Suladze: Zh. Techn. Fiz. 44, 1779 (1974) [English transl.: Sov. Phys.-Techn. Phys. 19, 1108 (1975)]
267 E.D. Korop and A.A. Plyutto: Zh. Techn. Fiz. 40, 2534 (1970) [English transl.: Sov. Phys.-Techn. Phys. 15, 1986 (1971)]
268 E.D. Korop and A.A. Plyutto: Zh. Techn. Fiz. 41, 1055 (1971) [English transl.: Sov. Phys.-Techn. Phys. 16, 830 (1971)]
269 A.A. Plyutto, K.V. Suladze, E.D. Korop, and V.N. Ryzhkov: Proc. V Int. Symp. Discharges and Elect. Ins. in Vacuum, Poznan, Poland (1972) p. 145
270 J.R. Kerns, C.W. Rogers, and J.G. Clark: Bull. Am. Phys. Soc. 17, 690 (1972)
271 T.E. McCann, C.W. Rogers, and D.N. Payton, III: Bull. Am. Phys. Soc. 17 690 (1972)
272 J.G. Clark, J.R. Kerns, and T.E. McCann: Bull. Am. Phys. Soc. 17, 1031 (1972)
273 J.G. Clark, J.R. Kerns, and T.E. McCann: Annals N.Y.A.S. 251, 273 (1975)
274 D.J. Johnson and J.R. Kerns: Appl. Phys. Lett. 25, 191 (1974)
275 J.R. Kerns and D.J. Johnson: J. Appl. Phys. 45, 5225 (1974)
276 J.R. Kerns and D.J. Johnson: Report AFWL-TR-75-250, Air Force Weapons Laboratory (1976)
277 R.F. Hoeberling, R.B. Miller, D.C. Straw, and D.N. Payton, III: Bull. Am. Phys. Soc. 21, 1059 (1976)
278 R.F. Hoeberling, R.B. Miller, D.C. Straw, and D.N. Payton, III: IEEE Trans. Nucl. Sci. NS-24 #3, 1662 (1977)
279 L.P. Bradley and G.W. Kuswa: Phys. Rev. Lett. 29, 1441 (1972)
280 L.P. Bradley and G.W. Kuswa: Bull. Am. Phys. Soc. 17, 980 (1972)
281 G.W. Kuswa and L.P. Bradley: Bull. Am. Phys. Soc. 17, 980 (1972)
282 G.W. Kuswa: Bull. Am. Phys. Soc. 18, 1310 (1973)
283 B. Freeman, H. Sahlin, J. Luce, O. Zucker, T. Crites, and C. Nelson: Bull. Am. Phys. Soc. 17, 1030 (1972)
284 H. Sahlin, B. Freeman, J. Luce, and O. Zucker: Bull. Am. Phys. Soc. 17, 1030 (1972)
285 B. Freeman, J. Luce, and H. Sahlin: Preprint UCRL-74158, Lawrence Livermore Laboratory (1972)
286 J.S. Luce, H.L. Sahlin, and T.R. Crites: IEEE Trans. Nucl. Sci. NS-20 #3, 336 (1973)
287 B. Freeman, R. Gullickson, O. Zucker, W. Bostick, and H. Klapper: Bull. Am. Phys. Soc. 18, 1351 (1973)
288 C.B. Mills: Report LA-5380-MS, Los Alamos Scientific Laboratory (1973)
289 B.L. Freeman, Jr.: Report UCRL-51608, Lawrence Livermore Laboratory (1974)
290 J.S. Luce: Annals N.Y.A.S. 251, 217 (1975)

291 H.L. Sahlin: Annals N.Y.A.S. 251, 238 (1975)
292 G.T. Zorn, H. Kim, and C.N. Boyer: IEEE Trans. Nucl. Sci. NS-22 #3, 1006 (1975
293 C.N. Boyer, H. Kim, and G.T. Zorn: Proc. Int. Top. Conf. E-Beam Res. and Tech. Albuquerque, New Mexico, Nov. 3-5, 1975 (National Technical Information Service, Springfield, Virginia, 1976) Vol. 2, p. 347
294 G.T. Zorn, H. Kim, and C.N. Boyer: Bull. Am. Phys. Soc. 20, 1381 (1975)
295 W.W. Destler, D.W. Hudgings, H. Kim, M. Reiser, M.J. Rhee, C.D. Strifler, and G.T. Zorn: IEEE Trans. Nucl. Sci. NS-24 #3, 1656 (1977)
296 W.W. Destler, H. Kim, and T. Dao: Bull. Am. Phys. Soc. 22, 1198 (1977)
297 R. Williams, J.A. Nation, and M.E. Read: Bull. Am. Phys. Soc. 21, 1059 (1976)
298 W.O. Doggett and W.H. Bennett: Proc. II Symp. Collective Methods Accel., Dubna USSR, Sept. 29 - Oct. 2, 1976 (JINR Dubna, USSR, 1977) p. 127
299 J.L. Adamski, P.S.P. Wei, J.R. Beymer, R.L. Guay, and R.L. Copeland: Proc. 2nd Int. Top Conf. High Power Electron and Ion Beam Res. and Tech., Cornell University, Oct. 3-5, 1977 (LPS, Cornell University, 1978) p. 497
300 P.S.P. Wei, J.J. Adamski, and J.R. Beymer: Bull. Am. Phys. Soc. 22, 1134 (1977
301 A.A. Plyutto, K.N. Kervalidze, and I.V. Kvattskhava: Atomnaya Energiya 3, 153 (1957) [English transl.: Sov. J. Atom. En. 3, 925 (1957)]
302 D.L. Morrow, J.D. Phillips, R.M. Stringfield, W.O. Doggett, and W.H. Bennett: Appl. Phys. Lett. 19, 441 (1971)
303 J.G. Linhart: Annals N.Y.A.S. 251, 234 (1975)
304 S. Peter Gary and H.W. Bloomberg: Appl. Phys. Lett. 25, 112 (1973)
305 S.L. Hsieh, H.W. Bloomberg, and S.P. Gary: J. Plasma Phys. 1, 553 (1975)
306 V.I. Kucherov: Zh.Techn. Fiz. 45, 1307 (1975) [English transl.: Sov. Phys.-Techn. Phys. 20, 317 (1976)]
307 V.I. Kucherov: Fiz. Plazmy 1, 963 (1975) [English transl.: Sov. Phys.-J. Plasm Phys. 1, 525 (1975)]
308 V.I. Kucherov: Pis. Zh. Techn. Fiz. 1 (1975) [English transl.: Sov. Phys.-Techn. Phys. Lett. 1, 224 (1975)]
309 R.F. Hoeberling: Bull. Am. Phys. Soc. 20, 1385 (1975)
310 R.J. Faehl, R.B. Miller, and B.B. Godfrey: Bull. Am. Phys. Soc. 21, 1165 (1976
311 R.B. Miller, R.J. Faehl, T.C. Genoni, and W.A. Proctor: IEEE Trans. Nucl. Sci. NS-24 #3, 1648 (1977)
312 R.B. Miller: Proc. 2nd Int. Top. Conf. High Power Electron and Ion Beam Res. and Techn., Cornell University, Oct. 3-5, 1977 (LPS, Cornell University, 1978) p. 613
313 R.B. Miller: Bull. Am. Phys. Soc. 22, 1134 (1977)
314 R.M. Johnson: Proc. Symp. ERA, Berkeley, Feb. 5-7, 1968 (National Technical Information Service, Springfield, Virginia, 1969) p. 219
315 A.A.Kolomensky and I.I. Logachev: Proc. VIII Int. Conf. High Energy Accel. (CERN, Geneva, 1971) p. 587
316 I.I. Logachev: Vestnik Akad, Nauk SSSR 3, 104 (1973) [English transl.: Vestni USSR Acad, Sci. 3, 144 (1973)]
317 A.A. Kolomenskii and I.I. Logachev: IV All-Union Conf. Charged Particle Beams, Moscow, USSR (1974)
318 A.A. Kolomenskii, I.I. Logachev, and A.A. Skryl'nik: Kratkie Soobshcheniya po Fiz. 8, 28 (1974) [English transl.: Sov. Phys.-Lebedev Inst. Reports 8, 33 (1974)]
319 A.A. Kolomenskii and I.I. Logachev: Zh. Techn. Fiz. 46, 50 (1976) [English transl.: Sov. Phys.-Techn. Phys. 21, 27 (1976)]
320 C.L. Olson: Report SLA-73-0864, Sandia Laboratories (1974)
321 A.V. Agafonov, A.A. Kolomenskii, and I.I. Logachev: Zh. Eksper. I. Teor. Fiz. Pis. Red. 22, 478 (1975) [English transl.: Sov. Phys.-JETP Lett. 22, 231 (1975)]
322 A.V. Agafonov, A.A. Kolomenskii, and I.I. Logachev: Fiz. Plaszmy 2, 953 (1976) [English transl.: Sov. J. Plasma Phys. 2, 528 (1976)]
323 A.V. Agafonov, V.G. Gapanovich, A.A. Kolomenskii, and I.I. Logachev: Proc. II Symp. Collective Methods Accel., Dubna, USSR, Sept. 29 - Oct. 2, 1976 (JINR, Dubna, USSR, 1977) p. 181
324 M. Widner, I. Alexeff, and W.D. Jones: Phys. Fluids 14, 795 (1971)
325 G.C. Goldenbaum: IEEE Trans. Nucl. Sci. NS-22 #3, 1009 (1975)

326 J.E. Crow, P.L. Auer, and J.E. Allen: J. Plasma Phys. 14, 65 (1975)
327 A.Y. Wong: Report PPG-277, University of California, Los Angeles (1976)
328 A.Y. Wong and R.L. Stenzel: Phys. Rev. Lett. 34, 727 (1975)
329 G.S. Janes, R.H. Levy, H.A. Bethe, and B.T. Feld: Phys. Fev. 145, 925 (1966)
330 J.D. Daugherty, J.E. Eninger, G.S. Janes, and R.H. Levy: Proc. First Int. Conf. Ion Sources, Saclay, June 18-20, 1969 p. 643
331 N. Rostoker: Particle Accel. 5, 93 (1973)
332 J.T. Verdeyen, W.L. Johnson, D.A. Swanson, and B.E. Cherrington: Proc. Int. Top. Conf. E-Beam Res. and Tech., Albuquerque, New Mexico, Nov. 3-5, 1975 (National Technical Information Service, Springfield, Virginia, 1976) Vol. 1, p. 657
333 W.L. Johnson, G.B. Johnson, and J.T. Verdeyen: J. Appl. Phys. 47, 4442 (1976)
334 O.A. Larent'ev: Annals N.Y.A.S. 251, 152 (1975)
335 F. Winterberg: II Nuovo Cimento 20B, 173 (1974)
336 F.E. Mills: Proc. V Int. Conf. High En. Accel., Frascati, Italy, Sept. 9-16, 1965 (Comitato Nazionale Energia Nucleare, Rome, 1966) p. 444
337 A.L. Lance: *Introduction to Microwave Theoy and Measurements* (McGraw-Hill, New York, 1964) Chap. 2
338 N.A. Krall and A.W. Trivelpiece: *Principles of Plasma Physics* (McGraw-Hill, New York, 1973) p. 140
339 R.J. Briggs: Phys. Fluids 19, 1257 (1976)
340 S.V. Yadavalli: Appl. Phys. Lett. 29, 272 (1976)
341 A.H.W. Beck: *Space Charge Waves* (Pergamon, New York, 1958) p. 103
342 W.R. Shanahan: Phys. Rev. A 16, 115 (1977)
343 R.J. Faehl and B.B. Godfrey: Preprint LA-UR-77-2362, Los Alamos Scientific Laboratory (1978)
344 V.P. Indykul, I.P. Panchenko, V.D. Shapiro, and V.I. Shevchenko: Zh. Eksper. I. Teor. Fiz. Pis. Red. 20, 153 (1974) [English transl.: Sov. Phys.-JETP Lett. 20, 65 (1974)]
345 V.P. Indykul, I.V. Panchenko, V.D. Shapiro, and V.I. Shevchenko: Fiz. Plazmy 2, 775 (1976) [English transl.: Sov. J. Plasma Phys. 2, 431 (1976)]
346 W.E. Drummond, G.I. Bourianoff, E.P. Cornet, D.E. Hasti, W.W. Rienstra, M.L. Sloan, H.V. Wong, J.R. Thompson, and J.R. Uglum: Report AFWL-TR-75-296, Parts 1 and 2, Air Force Weapons Laboratory (1976)
347 W.E. Drummond, G.I. Bourianoff, E.P. Cornet, D.E. Hasti, W.W. Rienstra, M.L. Sloan, J.R. Thompson, J.R. Uglum, and H.V. Wong: Report AFWL-TR-76-152, Air Force Weapons Laboratory (1976)
348 V.V. Velikov, A.G. Lymar, and N.A. Khizhnyak: Pis. Zh. Techn. Fiz. 1, 615 (1975) [English transl.: Sov. Phys.-Techn. Phys. Lett. 1, 276 (1975)]
349 A.N. Lebedev and K.N. Pazin: Atomnaya Energiya 41, 244 (1976) [English transl.: Sov. J. Atom. En. 41, 878 (1976)]
350 M. Friedman: Proc. 2nd Int. Top. Conf. High Power Electron and Ion Beam Res. and Techn., Cornell University, Oct. 3-5, 1977 (LPS, Cornell University, 1978) p. 533
351 M. Friedman: Bull. Am. Phys. Soc. 22, 1111 (1977)
352 A. Irani and N. Rostoker: Bull. Am. Phys. Soc. 21. 1059 (1976)
353 A. Irani and N. Rostoker: Report UCI-77-1, University of California, Irvine
354 A. Mondelli and N. Rostoker: Bull. Am. Phys. Soc. 22, 1131 (1977)
355 N. Rostoker: Proc. 2nd Int. Top. Conf. High Power Electron and Ion Beam Res. and Techn. Cornell University, Oct. 3-5, 1977 (LPS, Cornell University, 1978) p. 635
356 S. Putnam: Proc. II Symp. Collective Methods Accel., Dubna, USSR, Sept. 29 - Oct. 2, 1976 (JINR, Dubna, USSR, 1977) p. 136
357 J.W. Poukey and C.L. Olson: Bull. Am. Phys. Soc. 21, 1183 (1976)
358 C.L. Olson, J.W. Poukey, J.P. VanDevender, and J.S. Pearlman: IEEE Trans. Nucl. Sci. NS-24 #3, 1659 (1977)
359 C.L. Olson, J.P. VanDevender, J.S. Pearlman, and A. Owyoung: Bull. Am. Phys. Soc. 21, 1183 (1976)
360 M.L. Sloan: Proc. Int. Top. Conf. E-Beam Res. and Tech., Albuquerque, New Mexico, Nov. 3-5, 1975 (National Technical Information Service, Springfield, Virginia, 1976) Vol. 2, p. 334

361 E.P. Cornet: Bull. Am. Phys. Soc. 22, 1112 (1977)
362 G.I. Bourianoff: Bull. Am. Phys. Soc. 22, 1172 (1977)
363 R.B. Miller, D.C. Straw, and J.R. Kerns: Bull. Am. Phys. Soc. 19, 965 (1974)
364 R.B. Miller and D.C. Straw: J. Appl. Phys. 48, 1061 (1977)
365 R.B. Miller and D.C. Straw: Report AFWL-TR-76-156, Air Force Weapons Laboratory (1977)
366 B.B. Godfrey: IEEE Trans. Plasma Sci. PS-5 #3, 223 (1977)
367 B.B. Godfrey: Bull. Am. Phys. Soc. 21, 1183 (1976)
368 R.J. Faehl, B.B. Godfrey, B.S. Newberger, W.R. Shanahan, and L.E. Thode: IEEE Trans. Nucl. Sci. NS-24 #3, 1637 (1977)
369 B.B. Godfrey, R.J. Faehl, B.S. Newberger, W.R. Shanahan, and L.E. Thode: Proc. 2nd Int. Top. Conf. High Power Electron and Ion Beam Res. and Tech., Cornell University, Oct. 3-5, 1977 (LPS, Cornell University, 1978) p. 541
370 B.B. Godfrey, R.J. Faehl, B.S. Newberger, W.R. Shanahan, and L.E. Thode: Report LA-7148-PR, Los Alamos Scientific Laboratory (1978)
371 G.I. Bourianoff and W.W. Rienstra: Bull. Am. Phys. Soc. 21, 1101 (1976)
372 M.L. Sloan and H.V. Wong: Bull. Am. Phys. Soc. 22, 1172 (1977)
373 M. Andrews, J. Bzura, H.H. Fleischmann, and N. Rostoker: Phys. Fluids 13, 1322 (1970)
374 R.C. Davidson: *Theory of Nonneutral Plasmas* (W.A. Benjamin, Reading, Mass., 1974) p. 36
375 G. Gammel, J.A. Nation, and M.E. Read: Bull. Am. Phys. Soc. 21, 1184 (1976)
376 G. Gammel, J.A. Nation, and M.E. Read: Abstract 5B12, IEEE Int. Meeting Plasma Sci., Troy, New York (1977)
377 R. Adler, G. Gammel, J.A. Nation, M.E. Read, and R. Williams: Proc. 2nd Int. Top. Conf. High Power Electron and Ion Beam Res. and Tech., Cornell University Oct. 3-5, 1977 (LPS, Cornell University, 1978) p. 509
378 B.B. Godfrey: Preprint LA-UR-76-2646, Los Alamos Scientific Laboratory (1977)
379 B.B. Godfrey: Preprint LA-UR-77-2502, Los Alamos Scientific Laboratory (1977)
380 C.L. Olson: Report SAND-76-0292, Sandia Laboratories (1976)

Collective Ion Acceleration with Electron Rings

Uwe Schumacher

1. Introduction

There is a growing interest in the application of accelerated ions in nuclear physics, technology and medicine, which calls for the development of intense and energetic ion accelerators [1-4]. High-energy (light or heavy) ions afford the possibility of producing completely new atomic nuclei, that could not otherwise be obtained. The most fascinating goal is the search for the superheavy elements. In the wide area of technology the energetic ions are a tool for ion implantation, for radiation damage simulation in fission or fusion power reactor walls, and for the production of new materials (e.g. superconductors) and alloys. The life sciences profit from the research on high-energy ions in applying radio nuclides or accelerated ions in diagnostics or radiation therapy.

There are numerous powerful facilities specifically built for heavy-ion acceleration with some of the applications mentioned, the most advanced devices being the Super-Hilac at Berkeley (USA) and the Unilac at GSI (Darmstadt).

In all of these ordinary particle accelerators the magnitude and the structure of the electromagnetic fields are given by the conditions of vanishing current density (rot $\underline{B} = 0$) and charge density (div $\underline{E} = 0$) within the volume where the particles move. The electromagnetic fields in these "single particle" accelerators are produced by charges and currents in solid walls. The maximum electric field strength at the surface of the conducting acceleration structure is limited, however, to about 20 MV/m [5], whereas the energy gain of the particles to be accelerated is only a fraction of this value [6]. The conditions rot $\underline{B} = 0$ and div $\underline{E} = 0$ thus impose relatively strong limits on the maximum energy gain per length in ordinary accelerators. In order to reach a certain high ion energy in a linear accelerator one consequently has to build a very long and expensive device.

If the conditions rot $\underline{B} = 0$ and div $\underline{E} = 0$ are dropped, however, a wide area of new possibilities is opened for accelerators of very high energy gain per unit length or with very strong focussing.

The fields in these cases are of collective origin, being produced by a relatively dense collection of electric charges. The electric field strengths are very much higher than those produced conventionally. Two examples of these intense charge collections are strong relativistic particle beams and plasmas. The properties of the charge collections as used for particle acceleration by collective fields differ from those of plasmas in that these charge collections are non-neutral, so that there are strong electric fields, and that one particle component (e.g. the electrons) has relativistic velocity, while the other one (e.g. the ions) is at rest, the gyroradius of the relativistic component thus no longer being negligible compared with the dimension of the device.

The strong collective fields can be used to accelerate as well as to guide the charged particles in an accelerator.

The basis for collective acceleration methods was founded in 1934, when BENNETT [7,8] presented his model of the self-constricted electron beam. For a relativistic electron beam completely neutralized by ions he found the spatial distribution and the transverse Maxwellian velocity distributions as well as the well-known pinch relation for a beam in equilibrium. The basic characteristics of intense particle beams were treated by LAWSON [9-11].

The first example of a collective acceleration method was already proposed by ALFVEN and WERNHOLM [12] in 1952 for the acceleration of positive ions by the electric field near the moving focus of an intense electron beam. Owing to the very low beam intensity (and to a certain lack of focussing) the experiments did not succeed. A similar idea was put forward by HARVIE [13,14] in his statement "It is proposed that protons should be accelerated along a straight line by the electrostatic Coulomb force associated with bunches of electrons which are themselves accelerated along the same line by an externally applied electric field". In the year 1956 the concept of collective acceleration was advanced by the ideas of the three Russian physicists, BUDKER, VEKSLER and FAINBERG. BUDKER [15] applied the BENNETT model of the self-constricted relativistic beam to propose a stabilized relativistic ring that is produced by injecting the beam into a magnetic field. He suggested using this stabilized electron ring as a very strong guide field for a high-energy proton accelerator, a collective analog of a strong focussing synchrotron. The electric field within the ring rises to very high values owing to the shrinking of the ring dimensions by the (synchrotron) radiation due to the transverse electron oscillations.

FAINBERG [16] proposed plasma mechanisms for the collective acceleration of particles. The electric fields of waves excited by electron beams in plasma waveguides or those of non-linear density waves which may also travel as individual pulses (solitons), can be used to accelerate particles collectively [17].

VEKSLER [18] suggested several methods of accelerating clusters of ions and electrons by means of collective fields. The accelerating electric field is created by the interaction of a geometrically small group of particles to be accelerated with another assembly of charges (coherent "impact" acceleration) or by non-linear electromagnetic waves, or by reversing the Cerenkov and polarization losses [19,20]. VEKSLER determined the main advantages of these coherent methods of charged particle acceleration [18]: The strength of the accelerating field depends linearly on the number of particles that create the field. The synchronization of the accelerating field to the motion of the accelerated bunch is automatically fulfilled. It is possible to apply the high field strengths only at the location of the particles to be accelerated (thus avoiding spark breakdown at metallic surfaces).

It was VEKSLER together with SARANTSEV and their collaborators [21] who initiated an experimental program at Dubna with the invention of one of the collective acceleration methods, the electron ring accelerator (ERA). This work was concentrated in the Soviet Union till its announcement at the Cambridge High Energy Accelerator Conference in 1967, after which electron ring accelerator centres sprang up in many countries and new devices were developed or proposed [22].

Other mechanisms of collective ion acceleration were discovered "accidentally" when the acceleration of a relatively large number of energetic ions was observed nearly a decade ago in experiments with vacuum spark plasma sources [23-25], and in experiments with intense electron beams passing through a gas [26-28].

These phenomena of ion acceleration in linear electron beams [29-36], as well as several recent collective ion accelerator concepts that involve linear electron beams, are discussed in the accompanying article of this book by OLSON.

2. Electron Ring Acceleration

2.1 Basic Principles

In electron ring accelerators (ERA) the cluster of charges that generates the collective fields has the form of a ring of relativistic electrons. According to the proposal of VEKSLER, SARANTSEV et al. [21] the ring is formed by injecting an intense relativistic electron beam into a magnetic field. The problem of the electrostatic repulsion of the charges is solved by BUDKER's [15] concept of a partly neutralized relativistic (thus self-stabilized) ring beam. In contrast to most of the other collective acceleration schemes, the potential well is formed by a certain number of electrons, that are all contained in the ring and are not replaced by other electrons during acceleration. The acceleration of the ring can be carefully controlled. The synchronization of the accelerating mechanism, the moving potential well of the electron ring, to the ions to be accelerated is automatically fulfilled as long as the ions can stay within the ring. The rings are characterized by their "holding power", defined as the maximum electric field strength that holds the ions in the accelerated ring. If this quantity is too small and if the rings are accelerated too fast, the ions fall out of the ring and get lost. The main aim of present research on electron ring accelerators is therefore the achievement of the highest possible values of the holding power.

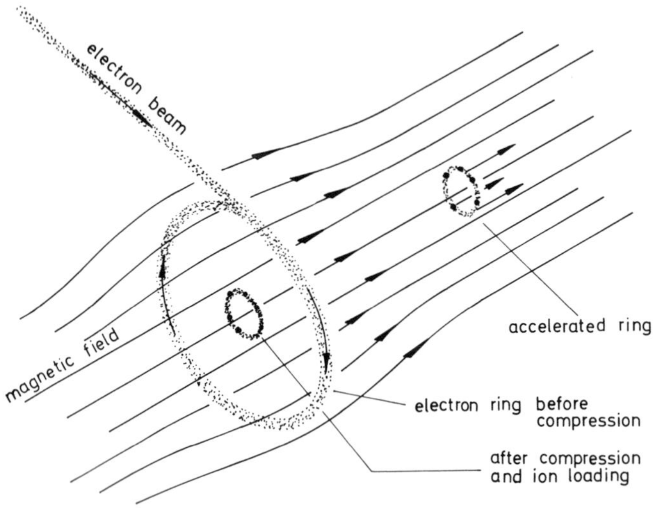

Fig.1. Schematic of an ERA using ring compression

The principle of the usual ERA concept [37-44] is illustrated in Fig.1. An unneutralized electron beam is injected into a slightly focussing magnetic field (magnetic mirror) that forms the beam into a ring and provides the axial ring focussing as long as ions are missing. In order to increase the electric field strength in the ring, it is compressed in its major and minor dimensions by increasing the magnetic field amplitude as a function of time. Since the electron energy increases during this process, the difference between the electrostatic repulsion and the magnetic attraction decreases. Moreover, to compensate this difference, the ring can now be partly neutralized by ions that are created by collisional ionization of the residua gas. The ionization is produced by the electrons just in the potential well of the electron ring. The ions have two functions: they provide the missing focussing and they are the particles that are accelerated collectively.

There are two other schemes of the electron ring accelerator, both using static magnetic fields. One method, proposed by CHRISTOPHILOS [45] and by LASLETT and SESSL [46], performs compression in the static magnetic field by a suitable choice of its shape, before loading and acceleration can start (see Fig.2).

The other method (Maryland scheme) forms the electron rings from hollow electron beams in a static magnetic field without applying compression. The electrons are injected through a cusped magnetic field structure [47-49]. It transforms the initiall longitudinally (i.e. parallel to the magnetic field axis) directed electron velocity into azimuthally oriented velocity, so that the electrons can accumulate into intens rings. After the rings are loaded with ions they can be accelerated (see Fig.3).

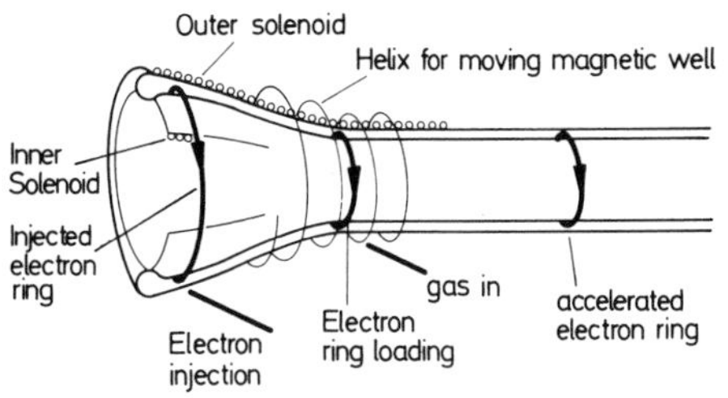

Fig.2. Schematic view of the static compressor (from [46])

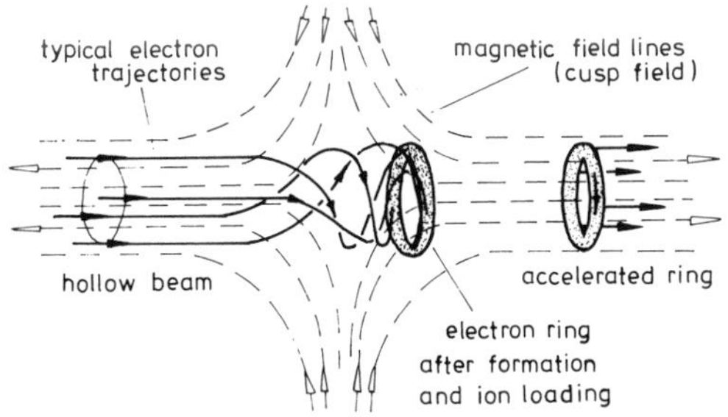

Fig.3. The Maryland scheme

2.1.1 Maximum Electric Field Strength and Holding Power

Obviously the holding power is related to and limited by the maximum electric field strength E_m in the ring prior to acceleration. E_m is determined by the electron density in the ring. For simplicity it might be approximated by the electric field strength E_{max} at the edge of a straight beam of constant electron density n_e within its circular cross-section of radius a

$$E_{max} = -\frac{en_e}{2\varepsilon_0} a \qquad \text{(MKSA units used)} \quad . \tag{1}$$

In the case of a slender ring of major radius R (a/R << 1, Fig.4) and of a total number N_e of electrons we have approximately

$$E_{max} = -\frac{e}{4\pi^2\varepsilon_0 R} \cdot \frac{N_e}{a} \quad , \tag{2}$$

which numerically reads

$$E_{max} \left[\frac{MV}{m}\right] = \frac{4.58 \cdot 10^{-12} N_e}{R[cm] \cdot a[cm]} \quad . \tag{3}$$

As expressed with the electron current

$$I_e = \frac{N_e e\beta c}{2\pi R} \quad , \tag{4}$$

where β is the electron azimuthal velocity v_φ related to the speed of light c:

$$\beta = \frac{v_\varphi}{c} \quad , \tag{5}$$

the maximum field strength is obtained from

$$E_{max} = - \frac{1}{2\pi\varepsilon_0 \beta c} \cdot \frac{I_e}{a} \quad , \tag{6}$$

which is numerically expressed as

$$E_{max}\left[\frac{MV}{m}\right] = 6 \, \frac{I_e[kA]}{\beta \cdot a[cm]} \quad , \tag{7}$$

where β is normally very near 1.

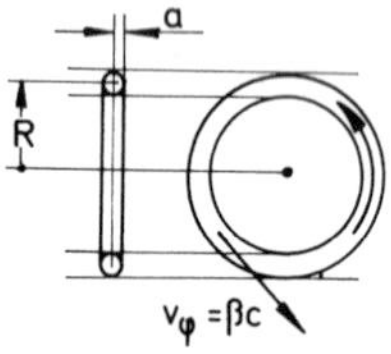

Fig.4. Electron ring geometry

For an electron ring with a major radius R = 3 cm and a minor radius a = 0.3 cm, that contains an electron number of $N_e = 10^{13}$, we estimate from (3) a maximum electric field strength of E_{max} = 50 MV/m. This value represents an increase of about an order of magnitude compared with the energy gain per unit length in conventional linear accelerators (which are limited in the accelerating field strength by sparking phenomena).

The maximum electric field strength E_{max}, however, depends on the charge distribution in the ring (so far we have assumed a uniform charge density in the minor ring cross-section). In order to compare the maximum electric field E_{max} and its axial position z_{max} for different charge distributions of elliptical beam cross-section, BOVET [50] expressed these quantities with the standard deviations σ_r and σ_z (in the radial and axial directions respectively) of these distributions as

$$\left.\begin{aligned} z_{max} &= \alpha_1 \sigma_z \quad , \\ E_{max} &= \alpha_2 \frac{eN_e}{4\pi^2 \varepsilon_0 R(\sigma_r + \sigma_z)} \quad . \end{aligned}\right\} \tag{8}$$

He finds the parameters α_1 and α_2 to be within relatively narrow limits for different electron density distribution functions (which might well describe experimentally found distributions) and for a wide range of minor axis ratio σ_r/σ_z:

$$\left.\begin{array}{l} 1.3 < \alpha_1 < 2.4 \\ 0.75 < \alpha_2 < 1.15 \end{array}\right\} \quad \text{for } 0.1 < \sigma_r/\sigma_z < 10 \quad . \tag{9}$$

The holding power E_H is related to the peak electric field E_{max} inside the ring at rest by

$$E_H = \eta E_{max} \quad . \tag{10}$$

E_H is always smaller than the peak field E_{max} owing to the fact that the electron-ion ring gets polarized by the inertial force acting on the ions when accelerating fields are applied to the ring. The factor η is calculated by MERKEL [51] and by HOFMANN [52] as dependent on the distribution of the electrons and ions in the accelerated ring or on close boundaries. η is in the range of 0.2 to 0.8 and is maximum when focussing by electric images is applied.

An example of the polarization of the ion-electron ensemble in electron rings under the action of the acceleration is given in Fig.5, where the densities of electrons and ions are numerically calculated by HOFMANN [52] and are plotted in the (r, z) cross-section as lines of constant density.

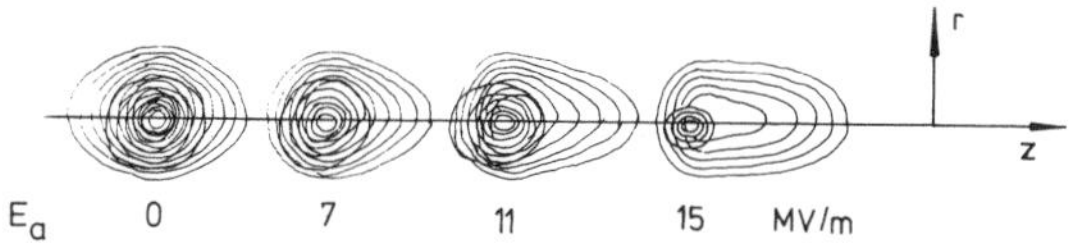

Fig.5. Density distributions for electrons and ions in electron rings for different acceleration field strengths E_a (from HOFMANN [52])

In order to increase the holding power of the electron rings the electron number N_e should be made as high as possible, while R and a should be kept as small as possible. We have to look for the fundamental limitations that are imposed on these quantities of the electron rings by the conditions of the electron beams, the ring equilibrium, its stability and its acceleration.

2.1.2 Beam Limitations

The electron rings are normally formed from unneutralized electron beams that are injected into the magnetic field. We now consider the limitations of these beams prior to and after injection into a focussing magnetic field, the space charge limit. The properties of the unneutralized beam are determined by its "perveance" $I/\tilde{U}^{3/2}$ where $\tilde{U}$ is the beam potential with respect to the cathode [53].

According to LAWSON [10,11] we introduce dimensionless parameters to describe the beam. The Budker parameter [15]

$$\nu_{Budker} = r_o\tilde{N}_e = r_o\int_0^\infty n_e(r)2\pi r dr \qquad (11)$$

is defined as the product of $\tilde{N}_e$, the number of electrons per unit length of the beam, and the classical electron radius

$$r_o = \frac{e^2}{4\pi\varepsilon_o m_e c^2} = \frac{\mu_o e^2}{4\pi m_e} \quad , \qquad (12)$$

which in numerical terms is $r_o = (2.817938 \pm 7 \cdot 10^{-6}) \cdot 10^{-13}$ cm.

For $\nu_{Budker} = 1$ and $\beta = 1$ we obtain the current

$$I = ec/r_o = 17000 \text{ A} \quad , \qquad (13)$$

which is a "natural current" since I is given by

$$I = \tilde{N}_e e\beta c \quad . \qquad (14)$$

The current $I_{AL} = \beta\gamma \cdot 17000$ A is called the ALFVEN-LAWSON current since ALFVEN [54] found this value as a critical current in cosmic ray streams.

$$\gamma = (1 - \beta^2)^{-1/2} \qquad (15)$$

is the ratio of the total energy to the rest energy of an electron that moves with a velocity βc parallel to the other electrons.

If in a current stream the longitudinal velocity should always be very much greater than the transverse velocity, so that this stream has the form of a beam, and also in order to avoid the "kink" instability, it is necessary, that $\nu_{Budker} \ll \gamma$. At the critical current I_{AL} the Larmor radius of the electrons in the magnetic field of the beam is comparable with the beam radius and, moreover the electromagnetic field energy of the beam approaches the particle kinetic energy.

The fundamental limit of electron beams is given by their space charge. LAWSON [10] calculated the spreading of a cylindrical beam of uniform cross-section due to its own space charge. He finds the radius r to be related to the axial width z by

$$r\frac{d^2r}{dz^2} = \frac{2\nu_{Budker}}{\beta^2\gamma^3} \quad . \qquad (16)$$

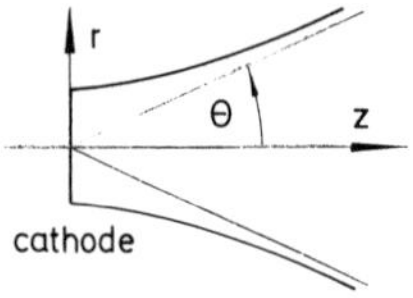

Fig.6. Beam spreading due to its own space charge

In the case of small z HARRISON [55] solved this differential equation by a hyperbola with an asymptote, the angle of which to the z-axis is given by

$$\tan \Theta = \left[\frac{2\nu_{Budker}}{\beta^2\gamma^3} \right]^{1/2} , \qquad (17)$$

as illustrated in Fig.6.

It is obvious from (17) that the beam can only exist if tan Θ is much less than unity, or

$$\nu_{Budker} << \frac{.1}{2} \beta^2\gamma^3 . \qquad (18)$$

This is the fundamental space charge limit for unneutralized beams. It is, however, not restrictive for beams that are used to form the electron rings. An example of a beam with 2 MeV electrons might demonstrate this: In this case we have $\gamma \simeq 5$ and $\beta \simeq 1$, so that we have to fulfill the condition $\nu_{Budker} << 63$. For the electron line density we thus have $\tilde{N}_e = \nu_{Budker}/r_0 << 2.23 \cdot 10^{14}\ cm^{-1}$, and for the current $I = \tilde{N}_e e\beta c << 1.07$ MA, which is far above those values used in ERA experiments.

The situation is similar for the space charge limit of a ring such as is formed from an electron beam in a focussing magnetic field (betatron field). If the betatron field of the axial component B_z is described by the magnetic field index n, defined as

$$n = - \frac{r}{B_z} \frac{\partial B_z}{\partial r} , \qquad (19)$$

and if n is equal to 0.5, then the number of electrons per unit length which can be trapped in a torus of minor and major radii a and R respectively is calculated by LAWSON [11] as

$$\nu_{Budker} = \frac{1}{4} \beta^2\gamma^3(a/R)^2 , \qquad (20)$$

where $a << R$.

The example of typical initial values for an ERA of $a/R = 0.1$ and $\gamma = 5$ (i.e. $\beta \simeq 1$) results in $\nu_{Budker} \simeq 0.3$ and $\tilde{N}_e = \nu_{Budker}/r_0 = 1.1 \cdot 10^{12}\ cm^{-1}$ or $I = \tilde{N}_e e\beta c =$ 5.3 kA. From these values it is easily concluded that this space charge limit is also far above the parameters used in ERA experiments.

2.1.3 Equilibrium Conditions

The equilibrium conditions of an electron ring are roughly approximated by those of a self-constricted stream of electrons, where the space charge is neutralized by ions, so that the repulsive force is suppressed and the attractive magnetic force holds the beam together. The conditions were first derived by BENNETT [7]. He assumes the transverse velocity distributions of the electrons and ions to be Maxwellian and thus defines the two-dimensional temperatures of electrons and ions by T_e and T_i, respectively. For the special case of vanishing ion drift velocity he finds the so-called Bennett pinch relation

$$I^2 = 8\pi/\mu_0 \cdot \tilde{N}_e k(T_i + T_e) \quad . \tag{21}$$

This is the necessary condition for the current I to deliver the beam equilibrium by its inwardly directed magnetic force.

If T_i is assumed to be zero, if the current is given by $I = \tilde{N}_e e\beta c$ and if the transverse electron temperature is expressed as $kT_e = 1/2\ \gamma m_e \cdot \langle\beta_t^2\rangle c^2$, we obtain the relation (analogous to the "perveance" of charged streams) [10,11]

$$\langle\beta_t^2\rangle/\beta^2 = \nu_{Budker}/\gamma \quad . \tag{22}$$

Since we want to define the beam by the fact that the Larmor radius of the electron in the self-field is much greater than the beam radius, we have the condition that $\beta_t \ll \beta$ or with (22)

$$\nu_{Budker}/\gamma \ll 1 \quad . \tag{23}$$

In the following we want to concentrate on systems that fulfill condition (23) since then the streaming motion predominates and we have an electron beam, although systems with $\nu_{Budker} > \gamma$ may exist as plasma streams.

In BENNETT's model [7] of the self-constricted fully neutralized electron beam an equilibrium configuration for the beam is found with a radial density distribution function proportional to $(1 + ar^2)^{-2}$. The relation of the line densities $\tilde{N}_e$ and $\tilde{N}_i$ of electrons and ions for the temperature ratio T_i/T_e is given by [51,56]

$$Z\tilde{N}_i/\tilde{N}_e = \frac{T_i/T_e + \gamma}{(1 + T_i/T_e\gamma)\gamma} \tag{24}$$

in the laboratory system, where the electrons are characterized by the relativistic factor γ, and the ions have charge Z per particle.

The ratio of the line densities of the two components of the beam is a function of the electron energy γ and the temperature ratio T_i/T_e. The limits $T_i/T_e = 0$ and $T_i/T_e = \infty$ result in the equilibrium conditions of a straight self-focussed electron beam, the Budker limits

$$1/\gamma^2 < Z\tilde{N}_i/\tilde{N}_e < 1 \quad . \tag{25}$$

Although these limits are found for a beam of special spatial distribution, they might be applied to slender electron rings (characterized by large aspect ratio R/a). In doing this, BUDKER [15] explains these limits by considering the forces in the laboratory system and in the electron system (in which the electrons are at rest). If the right inequality in (25) is fulfilled, the beam is negatively charged in the laboratory system, and there will be a potential well for the ions. In the electron system, however, where the electron line density is reduced to $\tilde{N}_e' = 1/\gamma\ \tilde{N}_e$ and the ion line density is increased by a factor of γ to $\tilde{N}_i' = \gamma\tilde{N}_i$, the beam is positively charged if the left part of (25) holds, so that here the electrons are captured in a potential well.

In the laboratory system the repulsive force of two electrons moving parallel to each other is thus greatly reduced by the magnetic attractive force, only $1/\gamma^2$ of its magnitude being uncompensated. It therefore only takes a small amount of ions, as determined by (25), to replace Coulomb repulsion by attraction. This fact should be illuminated by the forces and potential distribution of an electron beam (as an approximation to an electron ring) of constant electron density n_e within a circular cross section of radius a, of electron velocity βc in the longitudinal direction and of vanishing transverse temperature (Fig.7).

The repulsive Coulomb field of the electrons is given by

$$E_r = \begin{cases} -\dfrac{1}{2}\dfrac{en_e}{\varepsilon_0}\, r & \text{for } r \leq a \\[2ex] -\dfrac{1}{2}\dfrac{en_e}{\varepsilon_0}\dfrac{a^2}{r} & \text{for } r > a \quad , \end{cases} \tag{26}$$

with its maximum, E_{max}, at the edge of the beam (see (1))

$$E_{max} = -\frac{en_e}{2\varepsilon_0}\, a \quad . \tag{27}$$

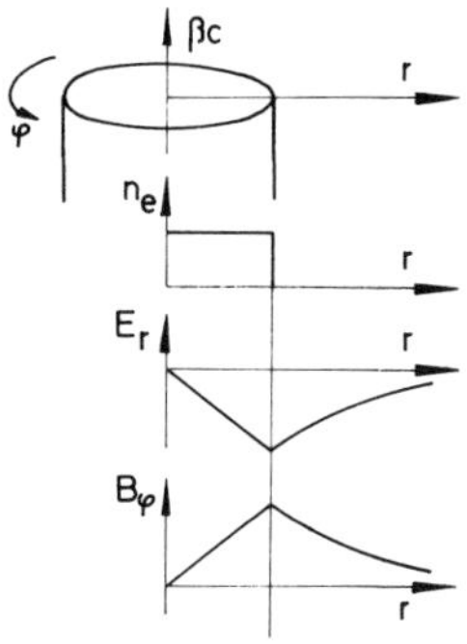

Fig.7. Radial distribution of the electric and magnetic field of an electron beam of constant density

The radial dependence of the azimuthal magnetic field component B_φ, as generated by the electron current, is determined from

$$B_\varphi = \begin{cases} \frac{1}{2}\,\beta c \mu_0 e n_e r & \text{for } r \le a \\ \frac{1}{2}\,\beta c \mu_0 e n_e \frac{a^2}{r} & \text{for } r > a \end{cases} \qquad (28)$$

so that the resulting net force F_r of the repulsive Coulomb force $-eE_r$ and the attractive Lorentz force $\beta c B_\varphi$ (with $c^2 = 1/\varepsilon_0 \mu_0$) reads

$$F_r = -\,eE_r + \beta c B_\varphi = -\,eE_r(1 - \beta^2) = -\,eE_r/\gamma^2 \quad . \qquad (29)$$

Adding a number of ions that fulfills the Budker limits (25) results in a focussing force.

2.2 Electron Ring Acceleration Mechanisms

The relativistic electron rings can be accelerated along the magnetic field lines not only with externally applied electric fields, but also by the Lorentz force if the magnetic guide field is made to expand slightly along the axis.

2.2.1 Acceleration by Electric Fields

Let us assume an external electric field of strength ε to be applied (in a direction parallel to the magnetic field) to an ion loaded electron ring of sufficiently high holding power. The rate dE_i/dz of gain of ion energy E_i per unit length in the axial direction z (i.e. the axial force on the ions) is then obtained from

$$dE_i/dz = e\varepsilon \frac{M_i}{\gamma_c m_e} \left[\frac{1 - ZN_i/N_e}{1 + N_i M_i / N_e m_e \gamma_c} \right] (1 - \alpha_c N_e) \quad , \qquad (30)$$

where M_i and m_e are the ion and electron rest mass, and the ion loading $f = ZN_i/N_e$ is the ratio of the ion to electron number times the charge state Z of the ion. γ_c is the relativistic mass factor of the electrons in the compressed state of the ring such as is adequate for acceleration. The bracket $(1-\alpha_c N_e)$ describes the energy loss of electron rings due to cavity radiation, if the electric acceleration is done by cavities. As calculated by KEIL [57], the parameter α_c is a very sensitive function of the cavity radius. For an electron number of $N_e = 3 \cdot 10^{13}$, an electric acceleration field strength of 5 MV/m and a cavity radius of 10 cm the reduction in energy gain is already 50 %. These calculations have been very well verified by measurements of the radiation loss of bunches of nearly 10^9 electrons in the SLAC (as performed by KOONTZ et al. [58]). For a wide range of electron energies the energy loss was found to be about constant, in quantitative agreement with KEIL's calculations.

It is this reduction of ion energy gain by the radiation losses that practically excludes favorable application of the method of electrically accelerating electron rings, compared to the magnetic expansion acceleration method. A large cavity bore radius, which in principle, would allow the radiation losses to be reduced, cannot be applied since it is unfavorable for low coupling impedance (as will be described later).

In the case of negligible radiation losses (for $\alpha_c N_e << 1$) and very small ion loading ($ZN_i/N_e << m_e\gamma_c/M_i$) it is obvious from (30) that the ion energy gain with this collective method of acceleration is higher by the ratio of the ion mass M_i to the relativistic electron mass $\gamma_c m_e$ than the rate of energy gain $e\varepsilon$ of the ion in the external electric field ε itself. The effective charge-to-mass ratio of the ions is increased by the factor $M_i/m_e\gamma_c$.

In order not to loose the ions from the electron ring the holding power E_H (10) of the rings should be made as high as possible. It is this value that determines the acceleration rate since the applicable external electric field ε is limited by it through

$$\varepsilon \lesssim E_H Z \frac{m_e\gamma_c}{M_i} \frac{1 + N_i M_i/N_e m_e \gamma_c}{(1 - ZN_i/N_e)(1 - \alpha_c N_e)} \quad . \tag{31}$$

2.2.2 Magnetic Expansion Acceleration

For the other acceleration method the external electric field vanishes and the axial magnetic field gradually decreases with axial distance.

The energy gain due to the magnetic expansion acceleration is similar to (30):

$$dE_i/dz = e\beta_\perp c B_r \frac{M_i}{\gamma m_e} \frac{1}{1 + N_i M_i/N_e m_e \gamma_c} \quad , \tag{32}$$

where $\beta_\perp c$ is the azimuthal electron velocity and γ the relativistic electron mass factor, both decreasing in the course of the acceleration. B_r is the radial magnetic field component responsible for the axial acceleration and limited by

$$B_r \lesssim E_H Z \frac{m_e\gamma_c}{c\beta_\perp M_i}(1 + N_i M_i/N_e m_e \gamma_c) \quad . \tag{33}$$

The magnetic expansion acceleration seems to be more effective since (as compared to (30)) the Lorentz force does not act on the ions and, moreover, the radiation losses can be made to vanish. However, since the energy given to the ions has to come from the electron kinetic energy, the electron ring expands during the acceleration, and the holding power decreases. The ring motion and the ion energy gain in this case of magnetic expansion acceleration are determined by the constancy of the energy (if radiation losses are neglected) and the canonical angular momentum of the ion-loaded electron ring.

The kinematic relations [59] between the components $\beta_\parallel$ and $\beta_\perp$ parallel and perpendicular to the magnetic field component B_z (see Fig.8) are

$$\beta^2 = \beta_{\parallel}^2 + \beta_{\perp}^2$$

or

$$1 - \frac{1}{\gamma^2} = 1 - \frac{1}{\gamma_{\parallel}^2} + \beta_{\perp}^2 \quad ,$$

which results in

$$\gamma = \frac{\gamma_{\parallel}}{(1 - \gamma_{\parallel}^2\beta_{\perp}^2)^{1/2}}$$

and

$$\gamma^2\beta_{\perp}^2 = \frac{\gamma_{\parallel}^2\beta_{\perp}^2}{1 - \gamma_{\parallel}^2\beta_{\perp}^2} \quad . \tag{34}$$

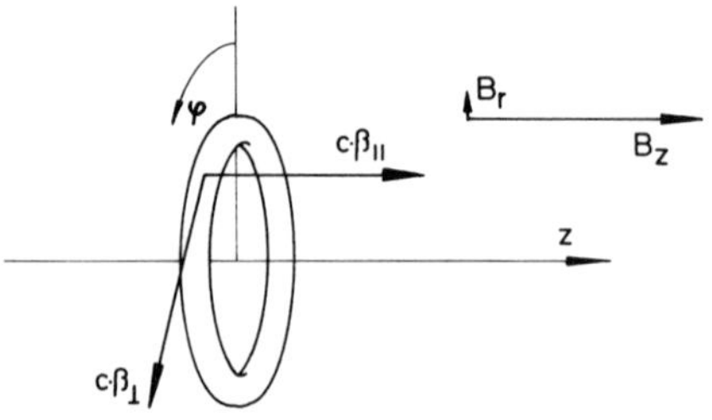

Fig.8. Electron ring with velocity and magnetic field components

The initial energy E_{in} of the ion-loaded electron ring in the compressed state (index c), as appropriate for acceleration, is

$$E_{in} = N_e m_e c^2 \gamma_c + N_i M_i c^2 \quad , \tag{35}$$

while the final energy E_f after acceleration is given by

$$E_f = N_e m_e c^2 \gamma + N_i M_i c^2 \gamma_{\parallel} \quad . \tag{36}$$

With $G_i = N_i M_i/(N_e m_e)$ and the mass ratio

$$g = \frac{G_i}{\gamma_c} = \frac{N_i M_i}{N_e m_e \gamma_c} \tag{37}$$

the energy conservation ($E_f = E_{in}$) reads

$$\gamma + \gamma_{\parallel} G_i = \gamma_c + G_i \tag{38}$$

or

$$\gamma/\gamma_c + \gamma_{\parallel} g = 1 + g \quad . \tag{39}$$

The canonical angular momentum component p_Θ is conserved in the rotational symmetric field:

$$p_\Theta = m_e \gamma \beta_\perp c R - e A_\Theta R = \text{constant}, \tag{40}$$

where R is the radius and A_Θ the azimuthal vector potential component, since for a uniform magnetic field B, which is a good approximation for our situation, the vector potential is circumferential, which gives the constancy of the flux Φ through the electron orbit

$$\Phi = \pi R^2 B = 2\pi R A_\Theta \quad , \tag{41}$$

so that from (40) with the cyclotron orbit radius R in the magnetic field B

$$R = \frac{m_e \gamma \beta_\perp c}{eB} \tag{42}$$

we obtain

$$\frac{\gamma^2 \beta_\perp^2}{\gamma_c^2 \beta_c^2} = \frac{B}{B_c} = b_1 = \frac{b_o}{\beta_c^2} \quad . \tag{43}$$

With (34) this results in

$$\gamma^2 \beta_\perp^2 = \gamma_c^2 b_o = \frac{\gamma_{\parallel}^2 \beta_\perp^2}{1 - \gamma_{\parallel}^2 \beta_\perp^2} \quad ,$$

$$\frac{1}{\beta_\perp^2} = \frac{\gamma_{\parallel}^2 (1 + \gamma_c^2 b_o)}{\gamma_c^2 b_o}$$

and

$$\gamma^2 = \gamma_{\parallel}^2 (1 + \gamma_c^2 b_o) \quad .$$

With the energy conservation equations (38) and (39) one gets for the dependence of $\gamma_{\parallel}$ on the mass ratio g and the field ratio $b_1 = B/B_c$

$$\gamma_{\parallel} = \frac{\gamma_c + G_i}{G_i + (1 + \gamma_c^2 b_o)^{1/2}} = \frac{1 + g}{(b_o + \frac{1}{\gamma_c^2})^{1/2} + g} \simeq \frac{1 + g}{b_1^{1/2} + g} \quad . \tag{44}$$

$\gamma_{\parallel}$ is proportional to the total ion energy E_t during the expansion acceleration process.

The total energy of the ions of rest mass M_i is

$$E_t = M_i \gamma_{\parallel} c^2 \quad ,$$

while their kinetic energy E_i is given by

$$E_i = (\gamma_{\parallel} - 1) M_i c^2 \simeq \frac{1 - b_1^{1/2}}{b_1^{1/2} + g} M_i c^2 \quad . \tag{45}$$

From (44) it is obvious that during the magnetic expansion acceleration - unlike during the acceleration in an electric field - the magnetic field amplitude drops and subsequently the ring radius increases.

Let us (for example in the case of vanishing mass ratio g) determine the magnetic field drop required to reach a final energy value of $\gamma_{\parallel f} = 2$. In this case the magnetic field goes down by a factor of 4, while (owing to the constancy of flux) the radius increases to twice its initial value:

$$B_f = \frac{1}{4} B_c \quad \text{and} \quad R_f = 2R_c \quad .$$

We should add the relations between the final (index f) and initial (index c; for compressed) values of γ and R, in accordance with the energy conservation

$$\gamma_{\parallel f} \gamma_{\perp f} = \gamma_{\parallel c} \gamma_{\perp c} \tag{46}$$

and the expression for the cyclotron orbit radius (42):

$$\gamma_{\parallel f} \simeq \gamma_{\parallel c} \cdot (B_c/B_f)^{1/2} \tag{47a}$$

$$\gamma_{\perp f} \simeq \gamma_{\perp c} \cdot (B_f/B_c)^{1/2} \tag{47b}$$

$$R_f \simeq R_c \cdot (B_c/B_f)^{1/2} \quad . \tag{47c}$$

Since the electron ring radius and the minor dimension increase during the process of acceleration in the expanding magnetic field, the holding power E_H gradually goes down. The magnetic field drop thus has to be designed such that the acceleration is always limited by the holding power. In seeking the limitation of the expansion rate [59] we equate the ion energy rate

$$dE_i/dz = M_i c^2 d\gamma_{\parallel}/dz \tag{48}$$

to the ion charge Ze times the holding power E_H

$$dE_i/dz = ZeE_H \quad , \tag{49}$$

where from (8) and (10)

$$E_H = \eta\alpha_2 \frac{eN_e}{4\pi^2\varepsilon_0 R(\sigma_r + \sigma_z)} \quad . \tag{50}$$

Assuming BR $(\sigma_r + \sigma_z)$ to be constant throughout the acceleration and defining

$$\lambda_1 = \frac{2\pi^2\varepsilon_0 R(\sigma_r + \sigma_z)M_i c^2 b_1}{\eta\alpha_2 e^2 N_e} \tag{51}$$

we obtain from (48)

$$(d\gamma_{\parallel}/dz)_{max} = \frac{eE_H}{M_i c^2} = \frac{b_1}{2\lambda_1} \quad . \tag{52}$$

This gives with (44) the maximum rate of the drop of B

$$(db_1/dz)_m \simeq - \frac{b_1^{3/2}}{\lambda_1} \frac{(b_1^{1/2} + g)^2}{1 + g} \tag{53}$$

with $b_1 = B_z/B_c$ and $g = N_i M_i/(N_e M_e \gamma_c)$.

For very small ion loading $(N_i M_i << N_e m_e \gamma_c)$ we have $g \simeq 0$ and

$$(db_1/dz)_m \simeq - \frac{b_1^{5/2}}{\lambda_1} \quad ,$$

which has the solution

$$B_z = B_c \left[1 + \frac{3}{2}\frac{z}{\lambda_1}\right]^{-2/3} \tag{54}$$

Typical curves of $b_1 = B_z/B_c$ and of $\gamma_{\parallel}$ (from (44)) are given in Fig.9 for three values of g.

Although the upper limit of $\gamma_{\parallel}$ seems to be relatively high, especially for low ion loading, a practical upper value of $\gamma_{\parallel}$ might be given by $\gamma_{\parallel} \simeq 1.6$.

2.3 Four Different ERA Schemes

2.3.1 ERA in Static Magnetic Fields by Hollow Beam Injection Through a Cusp

In the static ERA proposal of NELSON and KIM [60] a hollow cylindrical relativistic electron beam is created in a uniform magnetic field and injected through a cusp field configuration of rotational symmetry into an increasing solenoidal field.

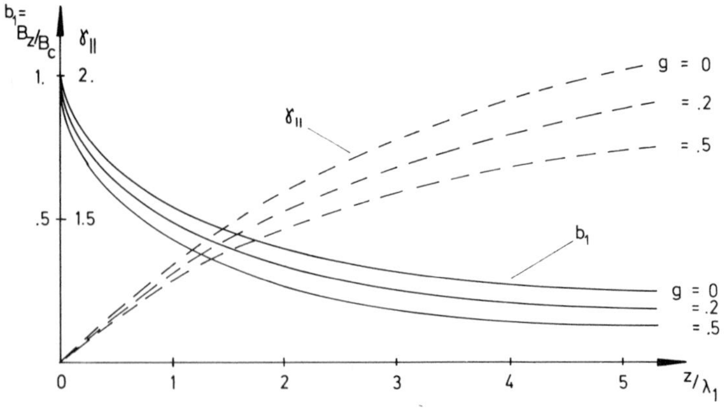

Fig.9. Magnetic field drop and energy versus axial distance

During this process the axial velocity of the electron beam is converted into rotational velocity, whereby the initially cylindrical charge distribution is converted into a ring shaped one, which is brought into equilibrium by Bennett pinching.

At rotational symmetry of the system the canonical angular momentum

$$p_\Theta = m_e \gamma R^2 \dot{\Theta} - eA_\Theta R \tag{55}$$

is conserved, where A_Θ is the azimuthal component of the vector potential and $\dot{\Theta}$ is the electron angular velocity. The relation is also known as Busch's theorem [61].

Since the magnetic field is static, the electron energy, moreover, is conserved.

For a uniform magnetic field the azimuthal vector potential component A_Θ has the simple form (see (41))

$$A_\Theta = \frac{1}{2} RB_z \quad , \tag{56}$$

and the angular velocity $\dot{\Theta}$ is related to the axial magnetic field component B_z by

$$\dot{\Theta} = \frac{eB_z}{m_e \gamma} \quad , \tag{57}$$

which is substituted in the expression for the canonical angular momentum p_Θ (55) to give

$$p_\Theta = eR^2B_z - \frac{e}{2} R^2B_z = \frac{1}{2} eR^2B_z = \frac{1}{2\pi} e\Phi \quad , \tag{58}$$

Φ being the magnetic flux through the electron orbit.

If the electrons are created in a constant magnetic field B_{zi} and if the initial angular velocity $\dot{\Theta}_i$ vanishes, the initial canonical angular momentum is related to the initial flux Φ_i by

$$p_{\Theta i} = -eR_i A_{\Theta i} = -\frac{1}{2} eR_i^2 B_{zi} = -\frac{1}{2\pi} e\Phi_i \quad .$$

As the canonical angular momentum is conserved throughout the process

$$p_\Theta = p_{\Theta i} \quad ,$$

it follows that the flux through the orbit must be reversed from the initial to the final stages:

$$\Phi = -\Phi_i \quad .$$

This principle of flux reversal by the application of a cusp magnetic field is applied in the Maryland ERA experiment [48]. Two solenoids with magnetic fields equal in magnitude but opposite in direction to each other are used. The electrons created and accelerated to high energy in one of the solenoids travel initially along the magnetic field lines (parallel to the axis z) with speed v_z near light velocity c. Passing through the cusp region the Lorentz force $v_z B_r$ turns the electrons into a circular path about the z-axis thereby slowing down the motion in the z direction (Fig.3).

In order to suppress radial oscillations generated by the $v_\Theta B_z$ force in the cusp, the width of the transition region must be very small, which in the Maryland ERA is achieved by using an iron plate between the two solenoids [49]. From (57) and $v_\Theta = R\dot{\Theta} = eB_z R/(m_e\gamma)$ the velocity v_z in the axial direction after the traversal of the cusp region is given by

$$v_z^2 = v^2 - v_\Theta^2 = v^2 - \frac{e^2 B_z^2 R^2}{(m_e\gamma)^2} \quad . \tag{59}$$

It is easily concluded that the axial velocity component v_z can be made arbitrarily small by a suitable choice of the radius R or the field magnitude B_z.

An initially cylindrical beam of (axially directed) velocity $v \simeq c$ and a pulse length of 1 ns has a length of about 30 cm. This beam is axially compressed by a factor of 30 to about 1 cm length after traversal of the cusp region, if one ends up at a final axial velocity of $v_z = 0.03$ c.

2.3.2 Ring Compression in Static Fields

The above mentioned method of electron ring acceleration in static magnetic fields calls for relatively good quality of the electron rings, which depends on the properties (current, energy and emittance) of the injected beam. In order to obtain electron rings with much higher holding powers, several methods of ring compression in static magnetic fields (without an electron energy gain) are proposed [45,46,62]. The methods are essentially the reverse of a normal magnetic expansion process (as described in Sec. 2.2.2), with the difference that the ring will not hold ions during this compression process. The ring will thus not be self-focussed, and hence the axial focussing has to be provided by external focussing (other than magnetic focussing) or by image focussing in order to achieve electron rings of desirably small minor dimensions in such a static-field compressor. CHRISTOFILOS [45] has proposed one or more alternating sections of axial deceleration and acceleration, which method calls for severe limitation of the energy spread of the injected beam. LASLETT and SESSLER [46] presented a scheme for a static-field compressor in which there is no axial ring deceleration (or acceleration) and hence no very stringent dependence on the initial energy spread. The fields of their static compressor are neither focussing nor defocussing in the axial direction. They propose the addition of a small traveling magnetic well that provides transverse focussing throughout the compression process in order to maintain the electron ring integrity. With present knowledge suitable image focussing (see Sec. 2.4.3) would probably be even better for providing the missing axial ring focussing.

Since the magnetic field in this device is static [46], the energy of an electron remains constant during the compression process, and so does its canonical angular momentum because there is no axial acceleration and the electron moves on its (circular) equilibrium orbit. Under these conditions the surface $r = r(z)$, on which the electron ring moves, is determined by

$$r(z) \cdot B_z\,[r(z),z] = \text{constant} \quad \text{and} \tag{60}$$

$$B_r\,[r(z),z] = 0 \tag{61}$$

since the force in the axial direction (z), which is proportional to B_r, must vanish.

With these conditions the magnetic field components B_z and B_r can be determined completely for points near the surface $r = r(z)$ (as an expansion in powers of the distance from the surface). Moreover, they satisfy Maxwell's equations. The focussing vanishes in the axial direction, as in the case of a ring in a uniform magnetic field, so that additional axial focussing has to be supplied. LASLETT and SESSLER [46] propose adding a moving magnetic well to the static compressor field since such a well also affords the possibility of controlling the ring position along its trajectory. The general scheme of such a static compressor is given in Fig.2. The magnetic field, as determined from the series expansion, is created by an outer and an inner solenoid, and the electron beam is injected (at the left) at large radius. Its axial motion and consequently its compression down to the loading position is controlled by the moving magnetic well, which is created by a helix with a surrounding dielectric layer and an outer conducting sheath [63]. With such a slow-wave structure the current pulse can be made slow enough.

2.3.3 Transverse ERA

The principle of transverse electron ring acceleration has been proposed and advanced theoretically and experimentally by a collaboration of groups at the Universities of Bari and Lecce [64-67]. The ion acceleration in this device is obtained by the electric field of an electron ring which drifts perpendicularly to the external magnetic field. The acceleration transverse to the magnetic field ensures axial and radial stability of the electron ring.

The axial component B_z of the magnetic field is assumed to be of the form [65]

$$B_z = B_0(1 - k_1y - s_1^2y^2) \quad , \tag{62}$$

where the three terms in the bracket represent the dipole, quadrupole and sextupole components of the magnetic field. B_z does not depend on x. An electron "ring" with its center on the x-axis (y = 0) will have an average radius of

$$r_{av} = \frac{m_e\beta\gamma c}{eB_{av}} \tag{63}$$

at an average value B_{av} of the magnetic field; but we only obtain a closed orbit in the case of vanishing k_1. For $k_1 \neq 0$, but in the approximation $k_1r_m << 1$ and $s_1^2r_m^2 << 1$, there is a drift due to the quadrupole term with a speed of

$$v_d = k_1r_mv_e/2 \quad , \tag{64}$$

where $v_e = \beta c$ is the electron velocity and

$$r_m = \frac{m_e\beta\gamma c}{eB_0} \quad .$$

The axial focussing of the electrons is provided by two separate mechanisms. A certain contribution is produced by the alternating gradient mechanism since the magnetic field alternatively increases and decreases along the orbit of the electrons owing to the linear term in the expression of the axial magnetic field component. In the case of vanishing quadratic term and on the assumption that the orbit can be approximated by a circle, the axial betatron frequency (see Sec.2.4.1), which determines the axial focussing, is found to be

$$\nu_z = 2^{-1/2}k_1r_m \quad . \tag{65}$$

Normally this quantity is small (especially during the initial phases of the acceleration) since according to (64) ν_z can be expressed by

$$\nu_z = 2^{1/2}\frac{v_d}{v_e} \quad ,$$

and the ratio of the drift velocity v_d of the electron ring to the electron speed v_e must be kept small during the phase of ion capture. Therefore the sextupole field component is intended to add a much stronger axial focussing force because the axial betatron frequency ν_z in this case can be approximated by

$$\nu_z = r_m s_1 \quad . \qquad (66)$$

The combination of the two mechanisms results in the approximate formula

$$\nu_z^2 = r_m^2(s_1^2 + k_1^2/2) \quad . \qquad (67)$$

The radial betatron frequency is always given by $\nu_r = 1$, so that the electrons are always sufficiently focussed in both transverse directions.

These focussing properties of such a drifting electron "ring" have been studied for the vacuum magnetic field configuration. The considerations do not include, however, the contributions of the self-fields of the electrons, which under certain circumstances might be advantageous for additional focussing. Although their azimuthal asymmetry in the form of klystron bunching is regarded as helpful in increasing the acceleration field strength, the collective instabilities (as described later), which might possibly set lower limits, have not yet been investigated for this device.

The principle of this transverse electron ring accelerator might be illuminated further by Fig.10, where in part a) the linear device (with the magnetic field into the drawing surface) is characterized by one electron path, indicating the drifting electron ring. The Bari-Lecce group [65,66] proposes furthermore, that this principle be used in a circular accelerator (for instance a synchrotron) with a radial dependence of the axial magnetic field component of the type given in (62). Compression of the orbits can be provided by pulsed magnetic fields.

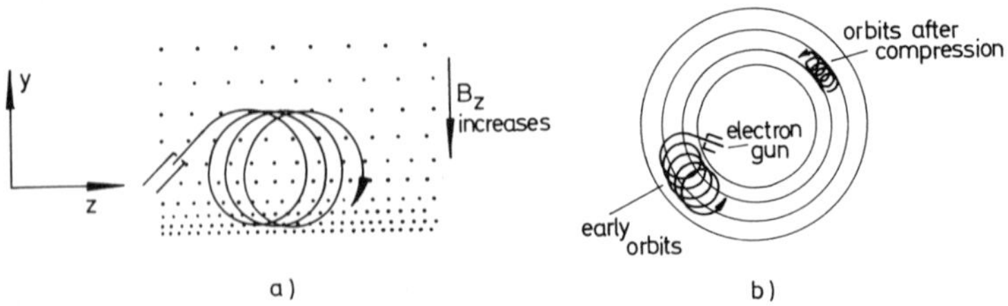

Fig.10. Principle of transverse electron ring acceleration; a) linear device, b) circular variant [65,66]

2.3.4 Ring Compression in Pulsed Magnetic Fields, Roll-out and Acceleration

In most of the experiments on electron ring acceleration (the only exceptions being the devices of Maryland and Bari-Lecce, as mentioned before) the rings are compressed just after their formation in order to increase the holding power by reducing the ring dimensions and increasing the electron energy towards its optimum value. The compression is performed by pulsing the magnetic field.

The simple formulae for the major radius R, the transverse momentum $p_\perp$, and the radial and axial betatron amplitudes a_β and b_β as functions of the magnetic field B are obtained from the conservation of the generalized azimuthal momentum (canonical angular momentum) [68] in an axially symmetric magnetic field (see (55))

$$p_\Theta = m_e \gamma R^2 \dot{\Theta} - eA_\Theta R = \text{const} \quad . \tag{68}$$

With the trajectory radius (42)

$$R = \frac{m_e \gamma \beta_\perp c}{eB} = \frac{p_\perp}{eB}$$

and the relation between the magnetic flux and the vector potential

$$\Phi = \Phi(B) = \int_S \underline{B} d\underline{f} = \oint \underline{A} d\underline{s} = 2\pi R A_\Theta$$

we obtain

$$RA_\Theta = \frac{\Phi}{2\pi} \quad \text{and} \quad Rp_\perp = m_e \gamma R^2 \dot{\Theta} = eBR^2 \quad ,$$

which is substituted in (68) to give

$$BR^2 = \frac{\Phi}{2\pi} = \frac{p_\Theta}{e} = \text{const.} \tag{69}$$

For a change of the magnetic field strength B from its initial value B_1 to its final value B_2 we obtain the ratios of the major radius R and transverse momentum $p_\perp$ as

$$\frac{R_2}{R_1} = \left(\frac{B_1}{B_2}\right)^{1/2} \left(1 + \frac{\Phi_2 - \Phi_1}{2\pi B_1 R_1^2}\right)^{1/2} \tag{70}$$

and

$$\frac{p_{\perp 2}}{p_{\perp 1}} = \left(\frac{B_2}{B_1}\right)^{1/2} \left(1 + \frac{\Phi_2 - \Phi_1}{2\pi B_1 R_1^2}\right)^{1/2} \tag{71}$$

In the special case of a homogeneous magnetic field with $\Phi = \pi R^2 B$ and $p_\Theta/e = 1/2\ BR^2 = \text{const}$ the flux through the electron orbit is conserved ($\Phi_2 = \Phi_1$), which results in the simple relations

$$R_2/R_1 = (B_1/B_2)^{1/2} \tag{72}$$

and

$$p_{\perp 2}/p_{\perp 1} = (B_2/B_1)^{1/2} \quad . \tag{73}$$

For a first orientation these relations can also be applied to situations with magnetic fields slightly deviating from homogeneity. The transformation law of the betatron amplitudes a_β and b_β, which determine the minor ring dimensions, is obtained with the conservation of the adiabatic invariant

$$\frac{p_\perp}{R} \nu a_\beta^2 = \text{constant}$$

from the transformation relations (70) and (71) to be

$$\frac{a_{\beta 2}}{a_{\beta 1}} = \left(\frac{B_1 \nu_{r1}}{B_2 \nu_{r2}}\right)^{1/2} = \left(\frac{B_1}{B_2}\right)^{1/2} \left(\frac{1 - n_1}{1 - n_2}\right)^{1/4} \tag{74}$$

and

$$\frac{b_{\beta 2}}{b_{\beta 1}} = \left(\frac{B_1 \nu_{z1}}{B_2 \nu_{z2}}\right)^{1/2} = \left(\frac{B_1}{B_2}\right)^{1/2} \left(\frac{n_1}{n_2}\right)^{1/4} , \tag{75}$$

where the betatron tunes ν (number of betatron oscillations per electron revolution (see Sec.2.4.1)) in the radial (ν_r) and axial (ν_z) directions are simplified to be determined only by the external magnetic field. With a field index n given by (19), they are obtained from

$$\nu_r^2 = 1 - n \tag{76}$$

and

$$\nu_z^2 = n \quad . \tag{77}$$

The representation of the betatron tunes by the contribution of the external magnetic field alone is normally an approximation for n-values that are not too small (say $n > 0.1$). For smaller n-values other contributions to the tunes (as described later in Sec.2.4.3) become important.

The transformation relations of the major and minor electron ring dimensions and the transverse momentum demonstrate the beneficial effect of ring compression on the maximum electric field (and the holding power) in pulsed magnetic fields. Using (8) we obtain (within the limits of the mentioned approximations)

$$\frac{E_{max2}}{E_{max1}} = \frac{B_2}{B_1}\left(1 + \frac{\Phi_2 - \Phi_1}{2\pi B_1 R_1^2}\right)^{-1/2} \cdot \frac{1 + b_{\beta 1}/a_{\beta 1}}{\left(\frac{\nu_{r1}}{\nu_{r2}}\right)^{1/2} + \frac{b_{\beta 1}}{a_{\beta 1}}\left(\frac{\nu_{z1}}{\nu_{z2}}\right)^{1/2}} , \qquad (78)$$

which indicates that the maximum electric field in the electron ring increases proportionally to the magnetic field strength for a nearly homogeneous field shape with $\Phi_2 = \Phi_1$.

The typical sequence of the different stages of an electron ring accelerator using compression in pulsed magnetic fields is the compression, the roll-out (the transition of the ring from the location of maximum compression to the acceleration section) and - via the "spill-out" - the acceleration. A schematic time behavior of the major radius R, the axial position z, the field index n, the magnetic field components B_z and B_r, the transverse electron momentum $p_\perp$ and the maximum electric field strength E_{max} is drawn for a representative case in Fig.11 (not to scale). The curves during compression are determined with (70,71) and (78) from the magnetic field time behavior. The increase of E_{max} (and $p_\perp$) is obvious.

During the roll-out process the electron ring major radius R stays about constant, and so do B_z, $p_\perp$, and E_{max}. The major change is in n and in the axial position z (which for simplicity, was assumed to be constant during compression, i.e. the compression occurred in a plane perpendicular to z). The transition of the ring during the roll-out is performed by shifting the location where the radial magnetic field component B_r vanishes since the ring position is at this location. The situation is illuminated by the axial plot of the radial magnetic field component B_r in Fig.12. At the time t_c of the maximum ring compression the ring is at the origin z = 0. The field index n is determined by the slope $\partial B_r/\partial z$ with (see (19))

$$n = -\frac{R}{B_z}\frac{\partial B_z}{\partial R} = -\frac{R}{B_z}\frac{\partial B_r}{\partial z} \quad . \qquad (79)$$

For a later time (t_2 or t_3) B_r has changed more for smaller z-values than for the larger ones, so that the position of the point with vanishing B_r (and with it the ring) wanders to higher z. Since the slope $|\partial B_r/\partial z|$ decreases, the field index n also goes down (see Fig.11). If the magnetic field pattern is made such that these curves merge into one that is always at positive B_r-values (such as are necessary for magnetic expansion acceleration), there will be a time t_{sp}, at which spill-out, the transition to the acceleration section, occurs. At this point not only B_r vanishes, but also $\partial B_r/\partial z$ (which results in n = 0). The electron ring is then free for acceleration by the monotonically increasing B_r. Since in the first section of the acceleration the slope is $\partial B_r/\partial z > 0$, we see from (79) that there is a region of negative field index (see Fig.11), which calls for additional axial focussing (see Sec.2.4.3).

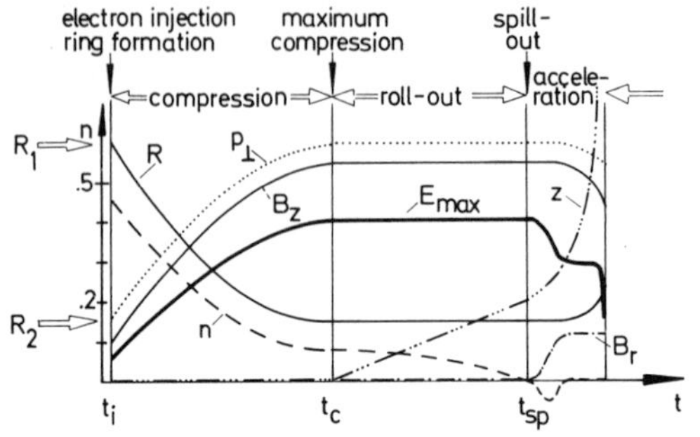

Fig.11. Representative time behavior of some electron ring quantities for a typical sequence

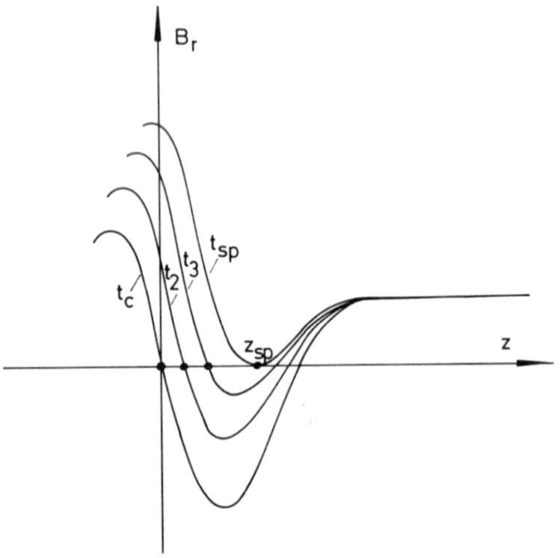

Fig.12. Axial dependence of B_r for different times during roll-out

During the acceleration (constant B_r is assumed) the maximum electric field strength E_{max} will drop to the holding power value E_H owing to the polarization of the electron ring (see (10) and Fig.11).

2.4 Single Particle Behavior

2.4.1 Betatron Tunes of Electrons in Pure Magnetic Fields

Let us consider a single electron of charge -e in a magnetic field of rotational symmetry [61]. The equations of motion in cylindrical coordinates are

$$\begin{aligned} &\frac{d}{dt}(m\dot{r}) - mr\dot{\Theta}^2 = -e(E_r + r\dot{\Theta}B_z - \dot{z}B_\Theta) \quad , \\ &\frac{1}{r}\frac{d}{dt}(mr^2\dot{\Theta}) = -e(E_\Theta + \dot{z}B_r - \dot{r}B_z) \quad , \end{aligned} \tag{80}$$

$$\frac{d}{dt}(m\dot{z}) = -e(E_z + \dot{r}B_\Theta - r\dot{\Theta}B_r) \quad , \tag{80}$$

which in the case of vanishing electric field (so that m = const) reduce to

$$\left.\begin{aligned} m\ddot{r} - mr\dot{\Theta}^2 &= -er\dot{\Theta}B_z \quad , \\ mr^2\dot{\Theta} - \frac{e}{2\pi}\Phi &= \text{const} \quad , \\ m\ddot{z} &= er\dot{\Theta}B_r \quad . \end{aligned}\right\} \tag{81}$$

The second equation describes the conservation of the canonical angular momentum ((55) with $m = m_e\gamma$). We now restrict ourselves to a magnetic field of constant gradient (field index n, (79)). In the vicinity of a radius R with $x = r - R \ll R$ the magnetic field components are approximated by

$$\left.\begin{aligned} B_r &\simeq -B_0 nz/R \text{ and } , \\ B_z &\simeq B_0(1 - nx/R) \quad , \end{aligned}\right\} \tag{82}$$

which is substituted in (81) with $r\dot{\Theta} \simeq v$ to give

$$\left.\begin{aligned} m\ddot{x} &= mv^2/2 - evB_0(1 - nx/R) \quad , \\ mrv - erB_0 x &= \text{const} \quad , \\ m\ddot{z} &= -evB_0 nz/R \quad . \end{aligned}\right\} \tag{83}$$

With the expansions

$$mv = p = p_0 + \Delta p \quad ,$$

$$1/mv = 1/p_0 \cdot (1 - \Delta p/p_0) \quad ,$$

$$1/r = 1/R \cdot (1 - x/R)$$

and the substitution

$$d\Theta = v/R \, dt$$

we obtain the linearized trajectory equations

$$\left.\begin{aligned} \frac{d^2x}{d\Theta^2} + (1 - n)x &= R\Delta p/p \quad , \\ \frac{d^2z}{d\Theta^2} + nz &= 0 \quad . \end{aligned}\right\} \tag{84}$$

Under the condition of

$$0 < n < 1 \tag{85}$$

the solutions are stable oscillations (betatron oscillations)

$$\left.\begin{aligned} x &= \Delta x + a_\beta \sin(\nu_r\Theta + \varphi_1) \\ z &= b_\beta \sin(\nu_z\Theta + \varphi_2) \end{aligned}\right\} \qquad (86)$$

with

$$\Delta x = \frac{R}{1-n}\,\frac{\Delta p}{p} \quad \text{and}$$

$$\nu_r = (1 - n)^{1/2} \qquad (87)$$

$$\nu_z = n^{1/2} \qquad \text{(see (76), (77))}. \qquad (88)$$

ν_r and ν_z are the betatron tunes, the number of betatron oscillations per tune, which are connected by the simple relation

$$\nu_r^2 + \nu_z^2 = 1 \quad . \qquad (89)$$

2.4.2 Betatron Resonances

As may be seen from Fig.11, in a representative electron ring accelerator the electron injection and the ring formation occur at a magnetic field index in the vicinit of n = 0.5. During the compression the field index drops and approaches zero at the entrance to the acceleration section. Hence several betatron resonances have to be crossed during the ring compression.

These betatron resonances can occur when the radial betatron tune ν_r (87) or the axial tune ν_z (88) is a simple fraction or when both values (ν_r and ν_z) are connecte by simple integral relations of the form [69]

$$\tilde{k}\nu_r + \tilde{\ell}\nu_z = \tilde{m} \quad , \qquad (90)$$

where $\tilde{k}$, $\tilde{\ell}$, and $\tilde{m}$ are positive or negative integers. In this equation $\tilde{m}$ indicates the harmonic order of the magnetic field's azimuthal variation that drives the resonance. If $\tilde{k}$, $\tilde{\ell}$, and $\tilde{m}$ are small, the resonance might be important. If the magnetic field has median-plane symmetry, only resonances with even values of $\tilde{\ell}$ can occur.

For a compressor geometry with median-plane symmetry the following potentially dangerous resonances have been investigated by LASLETT and PERKINS [69]:

$$\left.\begin{aligned} 2\nu_r - 2\nu_z &= 0 \quad , && \text{at } n = 0.5 \\ \nu_r + 2\nu_z &= 2 \quad , && \text{at } n = 0.36 \\ 2\nu_z &= 1 \quad , && \text{at } n = 0.25 \\ \nu_r - 2\nu_z &= 0 \quad , && \text{at } n = 0.2 \end{aligned}\right\} \qquad (91)$$

At the transition to the acceleration section a dangerous resonance of

$$\nu_r = 1 \quad , \text{at } n = 0 \tag{92}$$

can occur, which has been treated by PELLEGRINI and SESSLER [70] and others [71,72].

The traversal of a betatron resonance results in a more or less pronounced widening of the minor ring dimensions, thus lowering the holding power.

Experimental observations of single-particle resonance traversal with beam blow up have been reported from the Berkeley [73] and the Garching [74] electron ring experiments.

The traversal of the different single particle resonances might be illustrated by the "resonance diagram" in Fig.13, where the betatron tunes ν_r and ν_z are plotted as abscissae and ordinates, respectively. Due to relation (89) the trajectory of a compressor sequence is part of a circle in this diagram (heavy line), the parameter being the magnetic field index n. A typical compressor starts at about n = 0.5 and runs through the most dangerous resonances at n = 0.5, 0.36, 0.25 and 0.2 to approach finally the ν_r = 1 resonance at vanishing n.

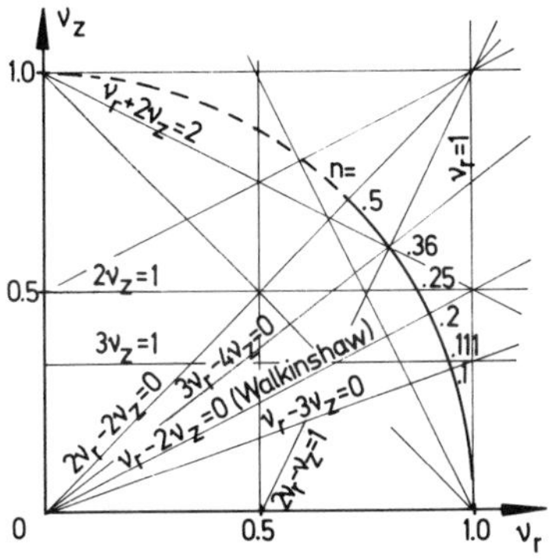

Fig.13. Resonance diagram

For the non-zero n resonances (91) LASLETT and PERKINS [69] find the following results:

α) The $2\nu_r - 2\nu_z = 0$ resonance (at n = 0.5) is driven by non-linearities, i.e. the higher radial derivatives of the magnetic field $\partial^2 B_z/\partial r^2$, $\partial^3 B_z/\partial r^3$, etc. It occurs in the absence of azimuthal bumps.

This resonance might especially be important for the electron ring research, since in most of the experiments the electron ring starts at a magnetic field index of about n = 0.5, where the initial betatron tunes in radial and axial direction are equal and the inflection of the electron beam can be performed as a three-turn inflection, as described in Sec.3.2.

Defining the radial and axial amplitudes by

$$\tilde{A}_r = \left[\left(\frac{p_r}{m_e \gamma \omega_{ce}} \right)^2 / (1 - n) + (r - R)^2 \right]^{1/2} \quad \text{and}$$

$$\tilde{A}_z = \left[\left(\frac{p_z}{m_e \gamma \omega_{ce}} \right)^2 / n + z^2 \right]^{1/2} \quad ,$$

where p_r and p_z are the momentum components and ω_{ce} is the electron gyrofrequency, it is found that the sum of the squares of the radial and axial betatron amplitudes is roughly a constant for this resonance:

$$\tilde{A}_r^2 + \tilde{A}_z^2 \simeq \text{const} . \tag{93}$$

The total growth is given by

$$G = \frac{1.6 \cdot 10^{-4} \, \tilde{A}_r^4}{R^4} \, \frac{[3 + 20b'' - 56b''^2 - 12b''']^2}{|dn/d(rev)|} \quad , \tag{94}$$

where

$$b'' = \frac{R^2}{B_z} \, \frac{\partial^2 B_z(r,t)}{\partial r^2} \Bigg/_{r = R} \tag{95}$$

and

$$b''' = \frac{R^3}{B_z} \, \frac{\partial^3 B_z(r,t)}{\partial r^3} \Bigg/_{r = R} \quad ,$$

and the growth factor equals 10^G.

$|dn/d(rev)|$ is the change of n during one electron revolution.

β) The resonance $\nu_r + 2\nu_z = 2$ (at $n = 0.36$) is driven by second harmonic variations. Just at the same field index value there is a coupling resonance $3\nu_r - 4\nu_z =$ which needs no azimuthal driving term. But because of the higher numbers $\tilde{k}(= 3)$ and $\tilde{l}(= - 4)$ this coupling resonance does not seem to be dangerous.

The growth of the $\nu_r + 2\nu_z = 2$ resonance depends on the magnitude of the Fourier coefficients that describe the second harmonic variations of the magnetic field.

γ) The $2\nu_z = 1$ resonance (at $n = 0.25$) is driven by azimuthal asymmetries (first harmonic variations) of the magnetic field. The growth rate does not depend on $\tilde{A}_r$. LASLETT and PERKINS [69] find a total growth for this resonance of

$$G = \frac{1.1 \, K^2}{|dn/d(rev)|} \quad \text{(decades)} \; , \tag{96}$$

where

$$K = \frac{1}{B_0}\left\{\left[2C_1(r,t) - r\,\frac{\partial C_1(r,t)}{\partial r}\right]^2 - \left[2S_1(r,t) - r\,\frac{\partial S_1(r,t)}{\partial r}\right]^2\right\}^{1/2}$$

with the coefficients $C_1(r,t)$ and $S_1(r,t)$ obtained from the representation of the midplane magnetic field as

$$B_z = B_0(r) + \sum_{\tilde{m}} [C_{\tilde{m}}(r,t)\cos\tilde{m}\Theta + S_{\tilde{m}}(r,t)\sin\tilde{m}\Theta] \quad .$$

δ) The $\nu_r - 2\nu_z = 0$ resonance (at n = 0.2) is often referred to as the Walkinshaw resonance [75]. It is regarded as potentially dangerous in electron ring accelerators: It may lead to axial amplitudes which are twice the original radial amplitude before traversal of this resonance, because the radial and axial amplitudes for this resonance are related by [76]

$$\tilde{A}_z^2 + 4\tilde{A}_r^2 \simeq \text{const} \tag{97}$$

for small and moderate amplitudes.

The total growth of this resonance is found to be [69,77,78]

$$G = \frac{0.96}{|dn/d(rev)|}\left(\frac{\tilde{A}_r b''}{R}\right)^2 \quad \text{(decades)} \ , \tag{98}$$

where b" is the (relative) second radial derivative of the axial magnetic field component on the equilibrium radius (95).

The growth can be kept small as long as b" is very small during traversal through this resonance. Since

$$b'' = \frac{R^2}{B_z}\left.\frac{\partial^2 B_z(r,t)}{\partial r^2}\right/_{r=R} = n(1+n) - r\,\frac{\partial n}{\partial r} \simeq 0.24 - R\left.\frac{\partial n}{\partial r}\right/_{r=R} \ , \tag{99}$$

the derivative of n should be made such that

$$\left.\frac{\partial n}{\partial r}\right/_{r=R} \simeq \frac{0.24}{R} \quad .$$

Normally this approach is difficult to obtain (for technical reasons). One therefore tries to increase simultaneously the field index change per revolution dn/d(rev) in order to reduce the growth. But even for a relatively high value of dn/d(rev), of about 3×10^{-4} (as was obtained with the first Garching compressor) the Walkinshaw resonance occurs. This is obvious from the numerical calculations of the axial amplitude growth for different initial radial betatron amplitudes $\tilde{A}_r$ in Fig.14.

Since all "total growth" formulae of the mentioned resonances are inversely proportional to dn/d(rev), the growths are reduced appreciably if the resonances are crossed rapidly enough. If the growth is very large, however, as in the demonstrated case of the n = 0.2 resonance, so that it is limited only by the initial radial betatron amplitude, the final axial amplitudes might not be reduced by faster resonance crossing.

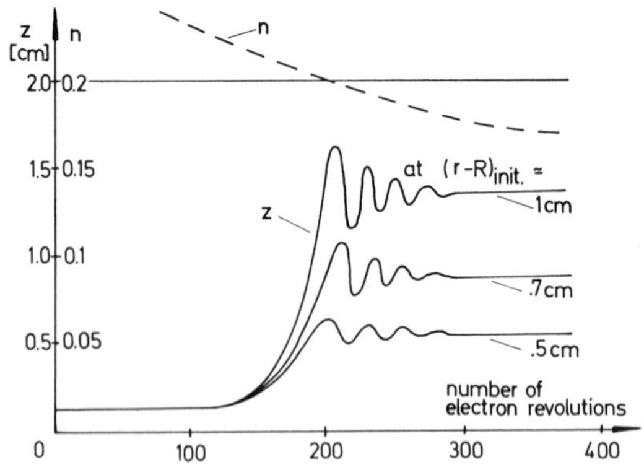

Fig.14. Axial growth at n = .2 resonance traversal for the first Garching compressor (calculations kindly performed by W.A. Perkins, unpublished)

As an example of the relative importance of the non-zero n single-particle resonances the growth rates of the axial betatron amplitude (in decades per electron revolution) from the calculations of LASLETT and PERKINS [69,78] are plotted in Fig.15 as a function of n for electrons injected into a magnetic field that is constant in time.

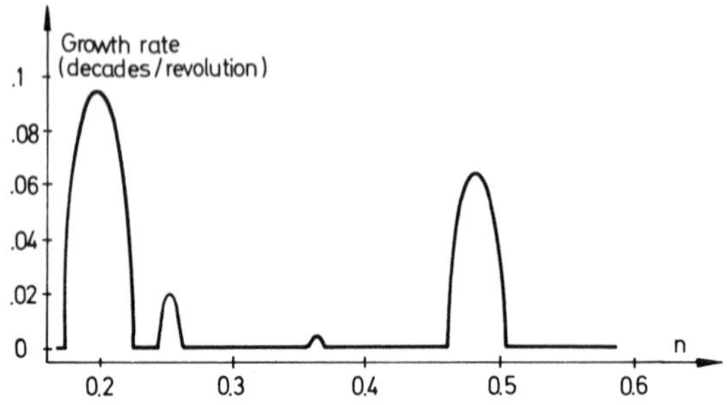

Fig.15. Growth rate of axial betatron amplitude for different field index values n (from ref.[69])

The numerical results of Fig.15 demonstrate directly that there are regions of growth near n = 0.20, 0.25, 0.36, and 0.50. Since the axial motion of the electron is modified by its radial motion of appreciable amplitude, the peaks obtained are not exactly centered about the n values stated. The peaks around n = 0.25 and 0.36 do not appear for an azimuthally symmetric field.

So far we have assumed median-plane symmetry of the magnetic field and hence have had to deal only with resonances of even $\tilde{\ell}$-values in (90). If we have no median-plane symmetry, odd $\tilde{\ell}$-value resonances such as

$$\left.\begin{aligned} \nu_r - 3\nu_z &= 0 \ , \quad \text{at } n = 0.1 \\ 3\nu_z &= 1 \ , \quad \text{at } n = 0.11 \end{aligned}\right\} \qquad (100)$$

might occur, the growth rate of which has not yet been calculated. Preliminary numerical calculations of LASLETT [79] indicated the occurrence of resonant effects at $\nu_z = 1/3$ when an azimuthal magnetic field B_φ is present. But further computational study of the potentially strong single-particle resonances (100) seems to be desirable [79].

For very small field index n, which is obtained at the transition of the electron ring from the roll-out region into the acceleration section and also during the acceleration the $\nu_r = 1$ integer resonance can be excited, since then the betatron oscillation frequencies of the individual electrons are incoherently crossing this resonance due to the focussing action of the ions being created in the electron ring. In contrast to the incoherent oscillation frequency the coherent radial oscillation frequency of the ring as a whole remains always below unity.

The crossing of the integer resonance $\nu_r = 1$ can cause a large increase of the minor ring dimensions, resulting in a decrease of the electric field of the ring. The integral resonance $\nu_r = 1$ is driven by a first harmonic disturbance of the B_z-field with an amplitude of $(\Delta B_z)_1$. If there is a second harmonic disturbance of the magnetic field index of amplitude $(\Delta n)_2$ the half-integral resonance

$$2\nu_r = 2$$

may occur, which increases the amplitude of certain electrons parametrically in the vicinity of $\nu_r = 1$.

The Hamiltonian [80]

$$H = 1/2x'^2 + 1/2\nu_r^2 x^2 - (\Delta B_z)_1 \frac{R}{B_z} x \cos\Theta + \frac{1}{2}(\Delta n)_2\, x^2 \cos 2\Theta \ , \qquad (101)$$

where $x' = dx/d\Theta$ is the momentum conjugate to x, yields the radial equation of motion for an electron (see (84) with $\Delta p = 0$ and $\nu_r^2 = 1 - n$)

$$d^2x/d\Theta^2 + \nu_r^2 x = -(\Delta B_z)_1 R/B_z \cos\Theta + (\Delta n)_2\, x \cos 2\Theta \ . \qquad (102)$$

The first term on the right produces a coherent and stationary displacement of the entire electron ring. In the single-particle approximation, where self-fields (which up to now we have not taken into account) are neglected, the ring performs a coherent $\tilde{m} = 1$ oscillation with the frequency [81,82]

$$S_k = (1 - \nu_r)\omega_{ce} \tag{103}$$

and the (stationary) amplitude

$$\tilde{A} = \frac{R}{B_z} \frac{(\Delta B_z)_1}{2(1 - \nu_r)} \quad . \tag{104}$$

If ν_r approaches 1, the frequency S_k gets very small, while the amplitude goes up. There is, however, the possibility to reduce ν_r^2 by collective fields (see next section) such that $1 - \nu_r$ can be increased in order to keep $\tilde{A}$ small.

For the slow crossing of the incoherent integral resonance at $\nu_r = 1$ PELLEGRINI and SESSLER [70] find the resonant incoherent increase Δa of the small radial electron ring dimension if there is a spread in the radial betatron frequency that leads to Landau damping

$$\Delta a/a \simeq R/a \cdot \omega_{ce}^2 \frac{(\Delta B_z)_1}{B_z} \cdot \left[\frac{\pi}{\omega_{ce} u}\right]^{1/2} \Delta^2/\Lambda^2 \quad . \tag{105}$$

Here the resonance is crossed at a relative speed of $u = \omega_{ce}^2 \; d(<\nu_r^2>)/dt$ with the average frequency $<\nu_r>$ being at $<\nu_r> = 1$.

The spread in the square of the frequency in the electron ring is given by

$$\Delta^2 = N_e^{-1} \sum (\nu_r^2 - <\nu_r^2>)$$

and the shift in the square of the frequency, as induced by the ions is given by

$$\Lambda^2 = <\nu_r^2> - (1 - S_k/\omega_{ce})^2 \quad .$$

Here the ratio is typically of the order of $\Delta^2/\Lambda^2 \simeq 10^{-2}$ (see also [82]).

S_k also includes the contributions of ions and images, as discussed in the next section.

The $\nu_r = 1$ resonance can be avoided for small values of the ion loading.

The half-integer resonance $2\nu_r = 2$ is driven by the second term on the right side of (102) within a stop band of width $\Delta\nu_r \simeq (\Delta n)_2/2$. The radial amplitude change Δa due to passing through this resonance at a speed of u normalized to a is calculated by SYMON [80] to be

$$\Delta a/a \simeq \omega_{ce}^2 \frac{(\Delta n)_2}{4} \left[\frac{2\pi}{\omega_{ce} u}\right]^{1/2} \quad . \tag{106}$$

In order to avoid a pronounced amplitude increase, the field gradient bump $(\Delta n)_2$ at a given speed u has to be made very small, and since $(\Delta n)_2$ is caused by a non-circularity of the field coils or because these are not exactly centered on the axis the coil tolerances have to be small, this being in the range of 10^{-3} to 10^{-4} for typical ERA designs [80,82].

Since these tolerances are not easy to obtain, the crossing of this half-integer resonance should be avoided by applying additional contributions to the betatron tunes. In recent detailed calculations HOFMANN [149] found that the minor radius growth depends on the second harmonic of the field gradient bump $(\Delta n)_2$ as well as on the field bump $(\Delta B_z)_2/B_z$.

2.4.3 Ion and Image Focussing

When calculating the tunes in Sec.2.4.1, we only took the forces of the outer magnetic field into account, which certainly dominate if the magnetic field index is in the range of 0.05 to 0.95. For smaller field index values the forces of the electrons, the ions and the image charges and currents become more and more important. The betatron tunes, i.e. the betatron frequencies of the electrons normalized to the electron gyrofrequency, are a direct expression of the focussing forces. Their squares have been calculated by LASLETT [83-85] and by IVANOV et al. [86] (see also [87-89]):

$$\nu_r^2 = (1 - n) + \mu \left\{ \frac{2R^2}{\sigma_a(\sigma_a + \sigma_b)} \left(f - \frac{1}{\gamma^2}\right) + (1 - f)\frac{P}{2} - 4\left[\frac{(1 - f)\varepsilon_{1,E}}{(S_E - 1)^2} - \frac{\varepsilon_{1,M}}{(S_M - 1)^2}\right] - n\left[\left(1 - \frac{f}{2}\right)P + (1 - f)K - \bar{L}\right]\right\} \tag{107}$$

and

$$\nu_z^2 = n + \mu \left\{ \frac{2R^2}{\sigma_a(\sigma_a + \sigma_b)} \left(f - \frac{1}{\gamma^2}\right) - (1 - f)\frac{P}{2} + 4\left[\frac{(1 - f)\varepsilon_{1,E}}{(S_E - 1)^2} - \frac{\varepsilon_{1,M}}{(S_M - 1)^2}\right] + n\left[\left(1 - \frac{f}{2}\right)P + (1 - f)K - \bar{L}\right]\right\} , \tag{108}$$

with σ_a and σ_b being the minor electron ring radial and axial dimensions (rms), which replace $a/\sqrt{2}$ and $b/\sqrt{2}$, respectively, in the cited references [90,91], and

$$\mu = \frac{N_e r_o}{\gamma 2\pi R} = \frac{\nu_{Budker}}{\gamma}$$

$$f = Z_i N_i / N_e \qquad \text{(ion loading fraction)}$$

$$P = 2 \ln \left[\frac{8\sqrt{2}R}{\sigma_a + \sigma_b}\right]$$

$$K \simeq 1/(S_E - 1)$$

$$\bar{L} \simeq 1/(S_M - 1) \qquad \text{and}$$

S_E = radius of the cylinder for electric images divided by R
S_M = radius of the cylinder for magnetic images divided by R
$\varepsilon_{1,E}$ and $\varepsilon_{1,M}$ are the electrostatic and magnetostatic image field coefficients respectively [89].

The first term arises from the external field (see (87,88)). The first term in the curly brackets, which is proportional to f, describes the ion focussing, while the electron focussing is expressed in the sum term proportional to $1/\gamma^2$. The term $(1 - f)P/2$ originates from the toroidicity of the electron ring, leading to focussing in the radial direction but defocussing in the axial direction. The first squar bracket is the image field term, which can give a positive contribution to focussing in the axial direction, but a negative one in the radial direction.

In order to obtain the necessary positive contribution to axial focussing from the images, the image currents have to be suppressed since they have an axially defocussing effect in contrast to the image charges.

The last term represents the (normally negligible) applied-field correction due to the self-fields.

The sum of the squares of the tunes is given by the relation

$$\nu_r^2 + \nu_z^2 = 1 + \frac{4\mu R^2}{\sigma_a(\sigma_a + \sigma_b)} \left(f - \frac{1}{\gamma^2}\right) , \qquad (109)$$

which differs from unity (see (89)) only in the effect of the "direct" self-fields acting in the region where div $\underline{E}$ and rot $\underline{B}$ do not vanish [84].

Equations (107) and (108) show that for a normal electron ring experimental sequence with n finally approaching zero there is always focussing in the radial direc tion, while in the axial direction the electron and toroidal defocussing can only be overcome by ion or image focussing. For the ion focussing we have to have at least $f > 1/\gamma^2$, which calls for an ion loading fraction of $f > 1.1 \cdot 10^{-3}$ at $\gamma = 30$. The image focussing, however, turns out to be more effective if - as has been discussed the image currents can be avoided ($\varepsilon_{1,M} = 0$). Such a structure is the image cylinder called "squirrel cage" [92,93], a device consisting of several conducting strips extending inside or outside the ring (but relatively near to it) all along the compression and acceleration sections (Fig.16). Since the azimuthal image currents are suppressed by the gaps between the strips, the defocussing effect is excluded.

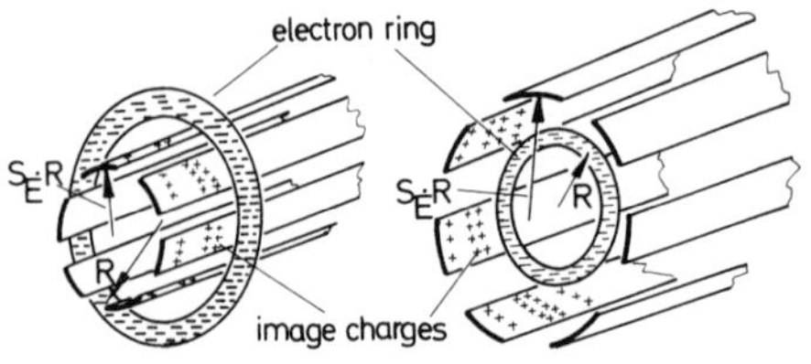

Fig.16. Inner (left) and outer (right) squirrel cage

Calculations of the pertinent image coefficients $\varepsilon_{1,E}$ and $\varepsilon_{1,M}$ have been performed by HOFMANN [89], who, moreover, was able to demonstrate the positive effect of the squirrel cage focussing on the holding power (see Sec.2.1.1), being a factor of 2 to 3 more than that of pure ion focussing.

2.4.4 Ring Integrity Conditions at Spill-out and Subsequent Acceleration

In order to maintain the equilibrium of the electron ring (loaded with ions) during compression and acceleration, it is necessary to keep the betatron tunes ν_r and ν_z always real and positive. During spill-out and the subsequent acceleration the ring, moreover, has to have adequate self-focussing quality to avoid its destruction by shearing forces that are produced by magnetic field changes over the ring cross-section.

Although these ring integrity conditions were calculated by PELLEGRINI and SESSLER [94] with mainly ion focussing in mind, we shall follow them here, modifying their results slightly in also including the image focussing terms. We also replace the ring minor dimensions a and b by $\sqrt{2}\,\sigma_a$ and $\sqrt{2}\,\sigma_b$, respectively. Examples are given for the Garching experiment [95-97], which is built to fulfill these conditions carefully.

In Sec.2.2.2 (33) we saw that there is a limit for the radial magnetic field component B_r in a magnetic expansion acceleration structure above which the ions cannot keep up with the electrons. If the ions are only supplying the ring self-focussing, the ring will immediately lose its integrity. In the case of image focussing, however, the integrity can be maintained even if the ions are lost [98]. But ion acceleration then certainly cannot be accomplished.

There are more restrictions on B_r and its derivatives in the vicinity of the spill-out region. The radial and axial variations of B_r in this region exert non-elastic forces on the electrons. These forces tend to pull the ring apart in the axial direction unless they are counteracted by sufficiently large self-focussing forces. Hence there are upper limits on $\partial^2 B_r/\partial z^2$ and on $\partial B_r/\partial r$ for electron rings of given self-focussing quality in order to maintain its integrity up to, at and behind spill.

We consider first the case of vanishing energy spread of the electrons ($\Delta E/E = 0$).

Let us describe the axial dependence of B_r prior to spill-out by

$$B_r(z) = \frac{\partial B_r}{\partial z}(z_e)\,(z - z_e) + \frac{1}{2}\,\frac{\partial^2 B_r}{\partial z^2}(z_e)\,(z - z_e)^2 \quad , \tag{110}$$

where we introduce $\xi = z - z_e$ as the deviation of the particle from the equilibrium axial position z_e, where $B_r = 0$ (see Fig.12).

From the equation of motion in the axial direction one finds the potential

$$\tilde{V} = \frac{1}{2} Q_s^2 \xi^2 - \frac{eR^2}{m_e \gamma c} \left[\frac{\partial B_r}{\partial z} (z_e) \frac{\xi^2}{2} - \frac{\partial^2 B_r}{\partial z^2} (z_e) \frac{\xi^3}{6} \right] , \tag{111}$$

where Q_s^2 is proportional to the ring self-focussing gradient.

If we include the external magnetic focussing by $\nu_z^2 = Q_s^2 + n$, using the magnetic field index

$$n = - \frac{R}{B_z(z_e)} \frac{\partial B_r}{\partial z} (z_e) ,$$

we obtain

$$\tilde{V} = \frac{1}{2} \nu_z^2 \xi^2 - \frac{R}{6} \left[\frac{\frac{\partial^2 B_r}{\partial z^2} (z_e)}{B_z (z_e)} \right] \xi^3 , \tag{112}$$

which has its maximum (as plotted in Fig.17) at the axial position

$$\xi_{max} = \frac{2\nu_z^2}{R} \frac{B_z(z_e)}{\partial^2 B_r/\partial z^2 (z_e)} .$$

The axial ring dimensions have to be smaller than ξ_{max},

$$\sqrt{2}\, \sigma_a < \xi_{max} ,$$

or the ring self-focussing has to be high enough, so that the ring is not pulled apart.

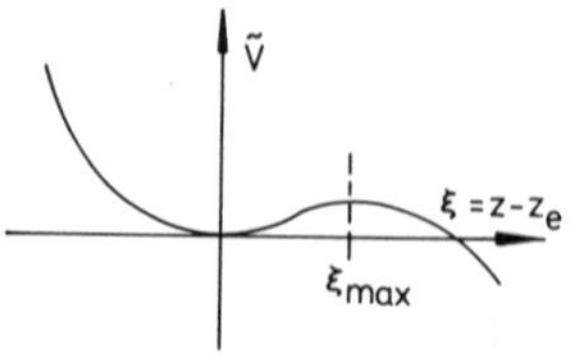

Fig.17. The axial dependence of the potential $\tilde{V}$

Since at spill $\partial B_r/\partial_z = 0$ (see Sec. 2.3.4) and $z_e = z_{sp}$ we have $\nu_z^2 = Q_s^2$ and

$$Q_s^2 > \frac{R\sigma_a}{\sqrt{2}} \frac{\frac{\partial^2 B_r}{\partial z^2} (z_{sp})}{B_z (z_{sp})} . \tag{113}$$

In the cited example [95-97] with the parameters $R = 2.5$ cm, $\sigma_a = \sigma_b = 0.5$ cm, $\partial^2 B_r/\partial z^2 = 1$ G/cm^2, $\gamma = 30$ and $B_z = 2 \cdot 10^4$ G we have to ask for a self-focussing of

$$Q_s^2 > 5 \cdot 10^{-5} \quad , \tag{114}$$

which can easily be fulfilled.

If there is no image focussing, we get from

$$Q_s^2 \simeq \frac{N_e r_e}{\gamma 2\pi R} \left[\frac{R^2}{\sigma_a^2} \left(f - \frac{1}{\gamma^2}\right) - (1 - f) \ln \frac{4\sqrt{2}R}{\sigma_a} \right]$$

that with $N_e = 5 \cdot 10^{12}$ we must have

$$f = Z\, N_i/N_e > 0.14 \quad ,$$

this high value of ion loading being necessary mainly to compensate for the toroidal defocussing.

Using image focussing with a squirrel cage [89] at a distance of 0.5 cm from the ring, the condition (114) can be fulfilled very well.

For nonvanishing energy spread ($\Delta E/E \neq 0$) of the electrons there is a spread in their equilibrium radii. Since B_r varies radially, there is again a force tending to tear the ring apart.

In this case we have correspondingly

$$\tilde{V} = \frac{1}{2} \nu_z^2 \xi^2 - \frac{R}{6} \frac{\frac{\partial^2 B_r}{\partial z^2}(z_{sp})}{B_z(z_{sp})} \xi^3 - \frac{R^2}{B_z(z_{sp})} \frac{\partial B_r}{\partial r}(z_{sp}) \frac{\Delta E}{E} \xi$$

and

$$\xi_{max} > \sqrt{2}\, \sigma_a + \sqrt{2} \frac{R}{\sigma_a} \frac{\Delta E}{E} \frac{\partial B_r/\partial r(z_{sp})}{\partial^2 B_r/\partial z^2(z_{sp})} \quad ,$$

and we obtain the condition

$$Q_s^2 > \frac{R\sigma_a}{\sqrt{2}} \frac{\partial^2 B_r/\partial z^2(z_{sp})}{B_z(z_{sp})} + \frac{R^2}{\sqrt{2}\sigma_a} \frac{\Delta E}{E} \frac{\partial B_r/\partial r(z_{sp})}{B_z(z_{sp})} \quad . \tag{115}$$

Even for the moderate ring quality, as given by the quantities σ_a, Q_s^2 and $\Delta E/E$, the ring integrity can be maintained by designing the magnetic field at spill-out to be very smooth (small values of $\partial^2 B_r/\partial z^2\, (z_{sp})$ and $\partial B_r/\partial r\, (z_{sp})$).

With the same data as above and with $\partial B_r/\partial r = 0.5$ G/cm and $\Delta E/E = 0.05$ we have to ask that

$$Q_s^2 > 6 \cdot 10^{-5} \quad ,$$

which can be achieved by the same squirrel cage focussing device.

At post-spill there is a region of defocussing from the outer magnetic field (see Sec. 2.3.4 and Fig. 11), so that the corresponding conditions demand a small value of $\partial B_r/\partial z$, and we must fulfill the two conditions [94]

$$\left.\begin{aligned} Q_s^2 &> \frac{R}{B_z}\frac{\partial B_r}{\partial z} \quad \text{and} \\ Q_s^2 &> \frac{R}{\sqrt{2}\sigma_a}\frac{\Delta E}{E}\left[\frac{2B_r}{B_z} + \frac{R}{B_z}\frac{\partial B_r}{\partial r}\right] + \frac{R}{B_z}\frac{\partial B_r}{\partial z} \quad . \end{aligned}\right\} \qquad (116)$$

With the same data as before, adding only $B_r = 5$ G and $\partial B_r/\partial z \lesssim 1$ G/cm, we have to require that

$$Q_s^2 > 3 \cdot 10^{-4} \quad ,$$

which calls for at least moderate electron ring quality, as can be achieved with the electron ring data mentioned and a squirrel cage structure 0.5 cm from the ring.

2.5 Collective Instabilities

In numerous devices with circular particle beams (synchrotrons, storage rings) containing a high number of particles, several collective instabilities have been observed. The collective fields produced by a large number of particles and the surrounding boundaries normally drive the particle motion so that the field amplitudes are enhanced and the instability grows up to a saturation value determined by nonlinearities or up to beam destruction.

In intense electron rings with their high number of particles and strong collective fields the investigation and suppression of collective instabilities is one of the major objectives. The instabilities may occur in the longitudinal direction by electromagnetic bunching forces (negative mass instability) or in the transverse direction by the forces of the eddy currents in resistive boundaries (resistive wall instability) or by the attractive forces of ions (electron ion instability). We restrict ourselves to these three types of collective instabilities and neglect the coupling of longitudinal and transverse motion for simplicity.

2.5.1 Negative Mass Instability

The longitudinal collective instability (or negative mass instability) has its origin in the space charge forces acting upon the azimuthal motion of circulating electron beams. Before its experimental observation it had been discovered by NIELSEN, SESSLER and SYMON [99] and by KOLOMENSKI and LEBEDEV [100]. Qualitatively, the mechanism of this instability can be understood from Fig.18. We assume a uniform ring of circulating monoenergetic electrons with a small density perturbation (representing an excess charge, top in Fig.18). An electron in front of the extra charge (particle 1) will gain energy from the repulsive force acting on it. Since the electron speed is near c, so that for an electron ring (with $0 \leq n \leq 1$) the angular frequency ω decreases with its momentum p ($d\omega/dp < 0$)(we are always above the transition energy), the particle goes to a greater orbit radius, thereby decreasing its revolution frequency and thus approaching the excess charge.

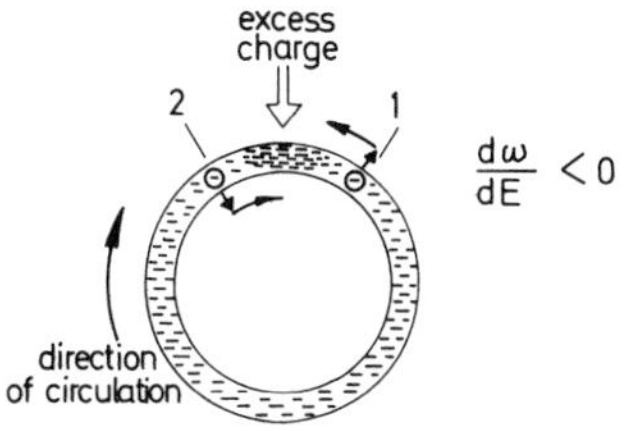

Fig.18. Mechanism of negative mass instability

For a particle behind the density perturbation (particle 2) we have a similar situation: This particle loses energy on repulsion from the excess charge and drops to a smaller orbit radius, thereby increasing its revolution frequency, so that this electron, too, approaches the excess charge. An initial fluctuation in the azimuthal density distribution thus leads to bunching of the electrons on the circumference. This instability is called the negative mass instability since this mechanism leads to a negative inertia term in the wave equation describing the motion of the perturbation [99].

The situation is very similar to the interacting particles in the circulating rings of planet Saturn, the stability properties of which were already analyzed in a famous essay by James Clerk MAXWELL [101]. As in the electron ring the particles in Saturn's rings have a negative angular inertia ($d\omega/dE < 0$). However, the forces between individual particles are attractive (in contrast to those in electron and other charged particle beams), so that a single ring is stable, and instabilities may arise only from the interaction of adjacent rings.

The theory of the negative mass instability has been described in several articles [102-110], an excellent survey on this subject having been given by SESSLER [111] in his Erice lectures.

So far we have assumed a monoenergetic particle distribution for the electron ring. If there is energy spread of the electrons in the ring, however, the associated spread of the revolution frequency tends to smear out the excess charges and to smooth the density distribution. We consequently have a threshold for the onset of growth of the negative mass instability that depends on the electron energy spread and the ring surroundings.

The threshold value is obtained from the linear theory [99,102] in treating the particle motion by means of the one-dimensional Vlasov equation in cylindrical coordinates for the canonical variables $\Theta = 2\pi(p_\Theta - p_0)$, where p_Θ is the canonical angular momentum with the mean value p_0:

$$\frac{\partial\tilde{\psi}}{\partial t} + \dot{\Theta}\,\frac{\partial\tilde{\psi}}{\partial\Theta} + 2\pi e\langle RE_\Theta\rangle\,\frac{\partial\tilde{\psi}}{\partial W} = 0 \quad . \tag{117}$$

$\langle RE_\Theta\rangle$ is the average value of the product of the radius and the azimuthal electric field that acts on the particles. $\tilde{\psi}$ is thought of as being composed of a constant (in azimuth and time) part $\tilde{\psi}_0(W)$ and infinitesimal perturbation $\tilde{\psi}_1(W)e^{i(\tilde{m}\Theta - \omega t)}$:

$$\tilde{\psi}(W,\Theta,t) = \tilde{\psi}_0(W) + \tilde{\psi}_1(W)e^{i(\tilde{m}\Theta - \omega t)} \quad . \tag{118}$$

Combining (117) and (118) and linearizing yields

$$\tilde{\psi}_1(W)e^{i(\tilde{m}\Theta - \omega t)} = -\frac{2\pi i e\langle RE_\Theta\rangle}{(\omega - \tilde{m}\dot{\Theta})}\,\frac{d\tilde{\psi}_0}{dW} \quad . \tag{119}$$

With the total electron number in the ring

$$N_e = 2\pi R\int\tilde{\psi}_0(W)dW \quad ,$$

the perturbed charge density per unit length λ_2

$$\lambda_2 = e\int\tilde{\psi}_1(W)dW$$

and the particle distribution function

$$f_0(W) = 2\pi R\tilde{\psi}_0(W)/N_e$$

we obtain

$$\lambda_2 e^{i(\tilde{m}\Theta - \omega t)} = -2\pi i e^2\langle RE_\Theta\rangle\int\frac{d\tilde{\psi}_0}{dW}\,\frac{dW}{\omega - \tilde{m}\dot{\Theta}} \quad .$$

By introducing the quantities U and V (which are normally used throughout all theoretical papers on the negative mass instability) NEIL and SESSLER [102] write the dispersion relation in the form

$$- 1 = (U - iV)\tilde{I} \quad \text{with} \qquad (120)$$

$$\tilde{I} = \int \frac{df_0}{dW} \frac{dW}{\omega - \tilde{m}\dot{\Theta}} \quad .$$

The quantities U and V are connected to the electromagnetic fields that are induced by the electron ring perturbation in the surrounding medium. These fields act on the electron ring itself. U and V depend on the geometry and the electromagnetic properties of the surrounding material (vacuum walls, shielding materials, diagnostics, image cylinders etc.).

RUGGIERO and VACCARO [103] introduce new dimensionless quantities U' and V', which only depend on the shape of the distribution function in momentum spread $\Delta p/p$ (see also [114]). They normalize the dispersion relation in writing (in linear approximation) [102]

$$\dot{\Theta} = 2\pi f_0 + k_0 W \quad ,$$

where f_0 is the revolution frequency of a central electron,

$$k_0 = 2\pi f_0 \, df_0/dE, \; \delta = \Delta E/2f_0 \quad ,$$

and ΔE is the full energy spread at the half-height of the distribution function; and they obtain

$$U' - iV' = \frac{U - iV}{\tilde{m}|k_0|\delta^2} \quad . \qquad (121)$$

For beams with reasonable distribution functions (no large off-centre tails, no discontinuous derivatives and without a central plateau) RUGGIERO and VACCARO's [103] results are summarized [108] by the stability requirement

$$|U' - iV'| \lesssim 0.7 \quad . \qquad (122)$$

From this relation KEIL and SCHNELL [108] obtain a convenient form of the threshold condition for the negative mass instability. If $\langle E_{\Theta,\tilde{m}}\rangle$ is the $\tilde{m}$-th harmonic average retarding electric field strength at the ring, generated by the amplitude $I_{\tilde{m}}$ of the $\tilde{m}$-th harmonic of the ring current, $2\pi R e \langle E_{\Theta,\tilde{m}}\rangle$ is the average energy loss per revolution. And $\langle E_{\Theta,\tilde{m}}\rangle$ is related to U and V by [102]

$$\langle E_{\Theta,\tilde{m}}\rangle = - \frac{i}{2\pi Re} (U - iV) \quad .$$

On the other hand this field strength is related to the perturbed current $I_{\tilde{m}}$ by

$$\langle E_{\Theta,\tilde{m}}\rangle = \frac{\oint E_{\Theta,\tilde{m}} ds}{2\pi R} = - Z_{\tilde{m}} I_{\tilde{m}} \quad , \qquad (123)$$

where the longitudinal coupling impedance $Z_{\tilde{m}}$ was introduced by SESSLER and VACCARO [112]. From the combination of all these relations we obtain the KEIL-SCHNELL criterion [108] as the maximum number of electrons N_e up to which the effect of the electromagnetic bunching forces is compensated by the dispersing action (Landau damping [113], produced by an energy spread $\Delta E/E$, defined as FWHM)

$$N_e \leq |\tilde{\eta}| \frac{\gamma R Z_0}{2\beta^3 r_0 \left|\frac{Z_{\tilde{m}}}{\tilde{m}}\right|} \left(\frac{\Delta E}{E}\right)^2 . \tag{124}$$

$\tilde{\eta}$ is the chromaticity (relative change in revolution frequency per unit momentum change), which for the electron ring is

$$\tilde{\eta} = 1/\nu_r^2 - 1/\gamma^2 .$$

Z_0 is the impedance of free space,

$$Z_0 = \sqrt{\mu_0/\varepsilon_0} \simeq 120\ \pi \text{ ohms} \simeq 377 \text{ ohms}.$$

The longitudinal coupling impedance $Z_{\tilde{m}}$ turns out to be one of the most important quantities for electron ring accelerators since its value determines directly the maximum attainable electron number and consequently the maximum holding power in the electron rings. We see this in combining (124) with (2) and (10) and eliminating a by using $a/R \simeq 0.7\ \Delta E/E$ (Gaussian energy distribution assumed) and R by (42):

$$E_H \leq \eta|\tilde{\eta}| \frac{c}{1.4\pi\beta^4} \cdot B \cdot \Delta E/E \cdot \frac{1}{|Z_{\tilde{m}}/\tilde{m}Z_0|} \tag{125}$$

or numerically (with $\beta \simeq 1$)

$$E_H[\text{MV/m}] \leq \eta|\tilde{\eta}| \cdot 6.81 \cdot B[\text{kG}] \cdot \Delta E/E \cdot \frac{1}{|Z_{\tilde{m}}/\tilde{m}Z_0|} . \tag{126}$$

The holding power thus turns out to be inversely proportional to the coupling impedance $|Z_{\tilde{m}}/\tilde{m}|$.

If we are above the threshold for negative mass instability as given by (124), the growth rate of the instability for impedances similar to that of a ring in free space is given by the linear theory [102,115]

$$1/\tau = \omega_{ce} \text{Im} \left[\frac{i\tilde{m} I_e Z_{\tilde{m}}}{2\pi\beta^2\gamma U_0} \right]^{1/2} , \tag{127}$$

where $I_e = N_e e\omega_{ce}/2\pi$ is the circulating current and $U_0 = mc^2/e$ (0.511 MV for electrons). The formula is derived on the assumption that $1/\tau << \tilde{m}\omega_{ce}$, and that the coupling impedance $Z_{\tilde{m}}$ is far above the threshold impedance $Z_{\tilde{m}thr}$ as obtained from (124). Since one has $1/\tau \to 0$ at the threshold, MÖHL [115] proposes to correct (127) by a factor $(1 - (Z_{\tilde{m}thr}/Z_{\tilde{m}})^2)^{1/2}$. The growth rate $1/\tau$ increases with the mode number $\tilde{m}$. For very large mode numbers $\tilde{m}$, however, the threshold condition becomes incorrect since then the electric field varies too rapidly over the minor beam cross-section. MÖHL [115] introduces a cut-off at the mode number

$$\tilde{m}_c = \pi/2 \cdot R/a \tag{128}$$

such that the wavelength of the highest unstable mode is equal to twice the minor beam diameter. He finds (for typical electron ring parameters) extremely short growth times which already decrease to the electron revolution time for mode numbers $\tilde{m}$ between 10 and 20.

These results suggest the need to work with electron densities which are below the threshold, which value can be increased by reducing the coupling impedance $Z_{\tilde{m}}$.

There are numerous calculations [116-119] and also measurements [120,121] of the negative mass coupling impedance for different models and geometries. We want to mention only the results of two typical geometries that use highly conductive walls near the ring to suppress the azimuthal electric fields at the location of the electron ring. For infinitely large walls at a distance h from the ring orthogonal to the ring axis BOVET and PELLEGRINI [68] (see also [122]) find the following expres-sion for the coupling impedance:

$$|Z_{\tilde{m}}/\tilde{m}| = \begin{cases} Z_0/2\ [1/\gamma^2(1 + 2\ \ln(2h/\pi a)) + (h/\pi R)^2] & \text{for } h \ll R \\ Z_0/2\gamma^2(1 + 2\ \ln(8R/a)) & \text{for } h \gg R\ , \end{cases} \tag{129}$$

if the effect of coherent radiation on the negative mass instability is neglected. $|Z_{\tilde{m}}/\tilde{m}|$ can be kept small as long as h/R is small, i.e. the walls should be brought as near to the ring as possible [109,110], as has been verified in the experiments [107].

A very useful calculation of the longitudinal coupling impedance $|Z_{\tilde{m}}/\tilde{m}|$ has been performed by FALTENS and LASLETT [119] for the electron ring in cylindrical geometry. For a stationary electron ring circulating near an inner conducting tube they found a strongly reduced maximum value of $|Z_{\tilde{m}}/\tilde{m}|$ that might be very advantageous for compressed electron rings. If the electron ring radius is R and the inner tube radius R_{IN}, the maximum coupling impedance is approximately given by [119]

$$(|Z_{\tilde{m}}|/\tilde{m})_{max} \simeq 300\ \frac{R - R_{IN}}{R_{IN}}\ \text{ohm} \tag{130}$$

in the range of $R_{IN}/30 \leq R - R_{IN} \leq R_{IN}/3$.

Its maximum value could thus be restricted to 60 ohms, for instance, if the ring is brought close to the tube, so that $(R - R_{IN})/R_{IN} = 1/5$.

The result is similar for an electron ring near an outer conducting tube. In this case, however, the quality factor of this device for higher-order resonant modes has to be made small to avoid resonant peaks of the coupling impedance.

LASLETT [79] uses this advantageous value of the coupling impedance to estimate the maximum attainable holding power in an electron ring accelerator. He assumes an electron ring to be situated at a distance $h = R - R_{IN} = 4\sigma_a$ from an inner conducting tube, where σ_a is the standard deviation of the radial minor ring dimension, and gets

$$|Z_{\tilde{m}}|/\tilde{m} = 300\ h/(R - h) \simeq 1200\ \sigma_a/R \text{ ohm} \quad .$$

We arrive immediately at LASLETT's result if we insert this value of the longitudinal coupling impedance in (126) and use $\Delta E/E \leq 2.36\ \sigma_a/R$ (assuming negligible betatron oscillation amplitude), $Z_0 = 377$ ohm, $|\tilde{\eta}| = 1$ and $\eta \simeq 0.8$:

$$E_H[\text{MV/m}] \leq 4.04\ B[\text{kG}] \quad . \tag{131}$$

The numerical factor is slightly different from LASLETT's result [79] because of different choice of safety factors and relations between the ring aspect ratio and the relative energy spread.

Equation (131) indicates, however, that for a typical magnetic field strength of B = 20 kG an effective acceleration field strength (holding power) of about 80 MV/m could be obtained. This holding power can be increased linearly with the applied magnetic field strength B.

So far we have only treated the linear negative mass theory and neglected radiation effects. The non-linear negative mass instability theory was developed by PELLEGRINI and SESSLER [123,124] and a computational study was made by SHCHINOV et al. [125], who found that the instability saturates at a relatively low level of oscillation energy accompanied by an increase of the beam energy spread up to the threshold value. The linear theory of the radiative instability was treated by BONCH-OSMOLOVSKI and PERELSTEIN [126], and the non-linear theory of this radiation instability by BONCH-OSMOLOVSKI et al. [127].

2.5.2 Resistive Wall Instability

For the suppression of the negative mass instability during the electron ring compression it is advantageous to have highly-conducting side plates nearby [107,109, 118]. These walls, however, cannot be applied since, owing to their high electrical conductivity, they prevent penetration of the pulsed magnetic fields necessary for electron ring compression. One therefore normally uses poorly-conducting side plates in order to combine the property of the pulsed field penetration with that of the negative mass coupling impedance reduction. In this case, however, there is the possibility of driving a transverse coherent motion of the ring as a whole. The coherent transverse instability occurs at a frequency of $(M - \nu)\omega_{ce}$, where M is a positive integer, and ν is the radial (or axial) betatron tune. It is driven by the forces of the currents induced in the imperfectly-conducting layers by the transverse electron ring displacement.

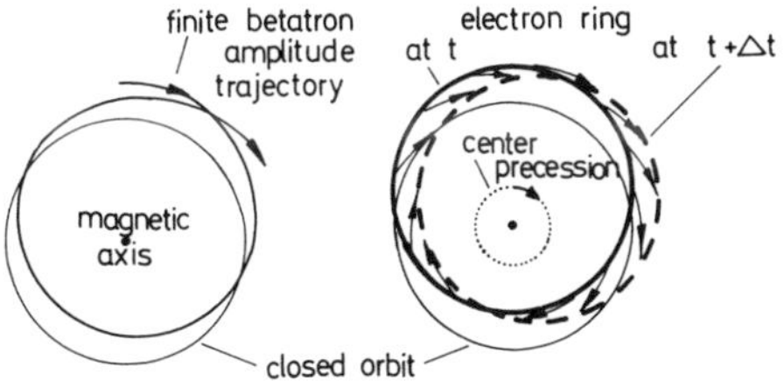

Fig.19. Particle trajectories at coherent radial motion

In the following we concentrate on the radial coherent motion of the most important mode M = 1. The coherent motion of the electron ring is elucidated in Fig.19. The trajectory of one particle performing radial betatron oscillations around its closed orbit is drawn on the left. On the right we assume that we have a displacement of the electron ring as a whole at the time t. All individual electrons in the ring perform betatron oscillations around the closed orbit as in the left part of the figure, their trajectories being indicated by arrows. After a short time Δt we see that the electrons again form a ring of the same major radius (dotted line) but with a displaced center (see arrow). The ring as a whole thus performs a precessional motion around the magnetic field axis. The ring center moves on a circle.

LAMBERTSON and LASLETT [128] calculated that radial oscillations of the beam current could induce wall currents (in imperfectly conducting side plates) which produce axially directed magnetic fields that lead to radially directed Lorentz forces in quadrature with the radial displacement.

The result of their calculation is demonstrated by Fig.20, where the transversely oscillating ring beam (of radius R) is considered in a straightened-out geometry with coordinates x, y, z. The position of the beam current I is described by

$$x_b = \tilde{A}\cos(y/R - S_k t) \quad ,$$

where

$$S_k = (1 - \nu_r)\beta c/R = (1 - \nu_r)\omega_{ce}$$

is the collective frequency of the first radial mode.

They find the induced current I_y per unit width in each of the thin side plates with a surface resistance of R_s (ohms per square) to be

$$I_y = -\frac{\mu_o S_k I}{2\pi R_s}\tilde{A}\,\frac{x - x_b}{(x - x_b)^2 + h^2}\sin(y/R - S_k t) \quad .$$

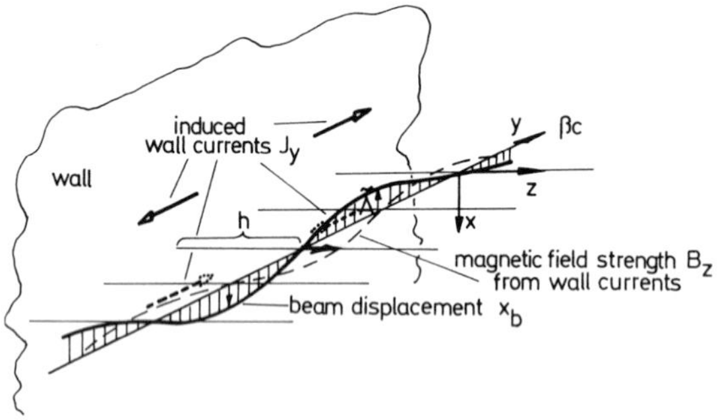

Fig.20. Beam displacement and induced magnetic field strength of the resistive wall instability

These current distributions in both side plates lead to a magnetic field component at the beam (z = 0) of

$$B_z(z = 0) = -\frac{\mu_o^2 I}{4\pi h R_s}\dot{x}_b = \frac{\mu_o^2 S_k I}{4\pi h R_s}\, ix \quad ,$$

if a complex notation ($e^{-iS_k t}$) is used.

The circular character of the orbit is taken into account by the inclusion of a dimensionless correction factor f(h/R) which is between 0.5 and 1.

Corresponding results are obtained if the ring beam is situated between coaxial circular tubes. LAMBERTSON and LASLETT [128], moreover, show that the effect of wall currents arising from the motion of electrostatically induced charges may be neglected compared with the magnetic forces.

From the magnetic field alone they obtain the e-folding growth rate of a collective radial oscillation (in the absence of Landau damping suppression) [128,129]:

$$1/\tau_G = V = \frac{N_e r_o \beta c \pi}{\nu_r \gamma h^2 24} f(h/R)\, \frac{D_1}{1 + D_1^2} \tag{132}$$

with

$$D_1 = \frac{Z_o h S_k}{2R_s R\omega_{ce}} \quad .$$

For typical electron ring parameters and boundaries this would lead to relatively high growth rates, unless there is Landau damping as characterized by a spread ΔS_k in the collective frequency S_k. This frequency spread ΔS_k is the full width at half the maximum value of the distribution of S_k. It is calculated from

$$\Delta S_k = |E \cdot \partial S_k/\partial E \cdot \Delta E/E + \partial S_k/\partial a^2 \cdot \Delta a^2|$$

with the Landau damping coefficient [79,130] (for the first radial collective mode M = 1),

$$E\partial S_k/\partial E = - 1/\beta^2 \, [S_k(1/\nu_r^2 - 1/\gamma^2) + \frac{R\omega_{ce}\partial\nu_r/\partial r}{\nu_r^2}] \tag{133}$$

and

$$\partial\nu_r/\partial r = - \partial n/\partial r \cdot 1/2\nu_r \quad .$$

The first term in the expression for ΔS_k arises from the energy spread $\Delta E/E$ of the electrons affecting the collective frequency spread. This is normally dominant. The second term describing the phase mixing due to the radial betatron amplitude spread Δa^2 is normally very small and should be neglected here.

From the theory of transverse collective instabilities [131,132] it follows that a coasting beam is stable with respect to self-amplification of a small amplitude transverse oscillation as long as

$$\Delta S_k \geq 2|U + V + iV| \quad , \tag{134}$$

where traditionally the magnitude of the self-fields created by a transverse beam oscillation is denoted by the coefficient U + V + iV. If U is assumed to be small, we should require at least that

$$\Delta S_k \geq 2V = 2/\tau_G \quad . \tag{135}$$

Following SCHNELL and ZOTTER [133], we give a simplified criterion for transverse stability of an electron ring. From the original theory [131] the quantity U + V + iV, expressed in MKS units, is given by

$$U + (1 + i)V = \frac{N_e r_o}{\mu_o \nu_r \gamma} \, \frac{\langle E + vxB \rangle}{I\Delta} \quad ,$$

where <E + vxB> is the average deflecting field around the circumference of the ring, and $I\Delta$ is the beam's perturbed dipole moment, which induces the field.

Now a transverse coupling impedance [133,135] is defined as

$$Z_T = i \, \frac{\int_0^{2\pi R} [E + vxB]ds}{\beta I\Delta} \, [\text{ohm/m}] \tag{136}$$

and dimensionless quantities U' and V' are introduced [134] by

$$U' + iV' = 2\,\frac{U + V + iV}{\Delta S_k} \quad .$$

From the work of HÜBNER, RUGGIERO and VACCARO [132,134] one obtains the stability criterion

$$|U' + iV'| \le 2F/\pi$$

with the form factor F being in the vicinity of unity, so that we find [133]

$$N_e \le F\,\frac{2R\mu_o\nu_r\gamma}{\beta r_o|Z_T|}\,\Delta S_k \quad . \tag{137}$$

For a given spread in the collective frequency and a given transverse coupling impedance one thus gets an upper limit for the electron number in the ring.

Since there is not very much information about the transverse coupling impedance, SCHNELL and ZOTTER [133] replace it by its relation to the longitudinal coupling impedance Z_L, about which more is known ((123), $Z_L = Z_{\tilde{m}}$).

In the special case of a pipe (with surface impedance) surrounding the beam at a radius b_p the relation is [135]

$$Z_T = \frac{Z_L}{|1 - \nu_r|}\,2R/b_p^2 = \frac{Z_L 2c}{S_k b_p^2}$$

(for different geometries one has to add a form factor depending on the cross-sectional geometry). The final stability criterion thus reads (with F = 1)

$$N_e \le \frac{\mu_o\nu_r(1 - \nu_r)\gamma b_p^2}{\beta r_o|Z_L|}\,\Delta S_k \quad . \tag{138}$$

The general impression concerning the situation of electron rings as regards the resistive wall instability is that with a suitable choice of the frequency spread ΔS_k it might always be possible to stay below the threshold for the electron number in contrast to the situation for the negative mass instability.

2.5.3 Ion-Electron Transverse Instabilities

The forces driving the resistive wall instability are produced by the magnetic field of currents which are induced in the imperfectly-conducting layers by the transverse electron ring displacement. These Lorentz forces are in quadrature with the radial (or axial) displacement.

If, however, we have no resistive walls nearby, exactly the same electron ring behavior may occur [136] when replacing the Lorentz forces by electrostatic forces from (a small number of) ions in the electron ring. This just occurs - if we again restrict ourselves to the first radial mode - at the resonance of the coherent transverse electron motion (near the frequency $S_k = (1 - \nu_r)\omega_{ce}$) with the ion motion at the ion bouncing frequency $\omega_B = Q_i\omega_{ce}$ with

$$Q_i^2 = \frac{r_o m_e N_e ZR}{M_i \sigma_a(\sigma_a + \sigma_b)\pi} \quad . \tag{139}$$

At this resonance the electrostatic forces are in quadrature with the radial electron ring displacement (as in Fig.20). There is, however, in a certain region in the vicinity of this resonance the possibility of unstable transverse collective oscillations of the electron ring against the ions. These oscillations were investigated for linear electron-ion beams [15,137] and considered with respect to the electron ring accelerator already at the time of the Berkeley ERA Symposium [138-140]. Extensive studies for the dipole and the quadrupole modes were done by KOSHKAREV and ZENKEVICH [141] and by LASLETT [143] and others [144,145].

The dipole mode appears to have been observed in the Bevatron [142]. After the observation of strong collective oscillations with a frequency near $S_k = (1 - \nu_r)\omega_{ce}$ in the Garching ERA experiment [96,98] DOMMASCHK [146] found out the dangerously unstable situation of the ion-electron rings near the end of electron ring compression (when the above-mentioned resonance occurs) and performed the relevant calculations, including the beneficial effect of an azimuthal magnetic field.

The stability behavior of the dipole resonance is obtained by solving the averaged, coupled equations of the displacements of the electron (x_e) and ion (x_i) beam from the equilibrium orbit, assuming linearity of the mutual forces and negligible momentum spread [141,146]

$$\left.\begin{aligned} &(\partial/\partial t + \omega_{ce}\partial/\partial\varphi)^2 x_e + \omega_{ce}^2 Q_e^2 x_e = \omega_{ce}^2 Q_1^2 x_i \quad , \\ &\partial^2/\partial t^2 x_i + \omega_{ce}^2 Q_i^2 x_i = \omega_{ce}^2 Q_i^2 x_e \quad , \end{aligned}\right\} \tag{140}$$

with

$$Q_e^2 = \nu^2 + Q_1^2 \quad . \tag{141}$$

The coupling coefficients Q_1 and Q_i between the electrons and the ions have simple physical meanings: Q_i is the ion bouncing frequency (139) and Q_1 correspondingly the electron bouncing frequency in the potential well formed by the ion ring:

$$Q_1^2 = \frac{r_o R N_i Z}{\gamma\pi\sigma_a(\sigma_a + \sigma_b)} = \frac{N_i M_i}{N_e m_e \gamma} Q_i^2 = g Q_i^2 \quad . \tag{142}$$

ν is equal to the radial betatron tune ν_r and the axial betatron tune ν_z (107,108), respectively, if we treat the dipole resonance in the radial and axial directions.

Expecting a solution of the differential equations (140) of the form

$$x_e = x_{eo}\exp[i(\tilde{m}\Theta - \omega t)] \quad ,$$

$$x_i = x_{io}\exp[i(\tilde{m}\Theta - \omega t)] \quad ,$$

we obtain the following dispersion equation

$$[Q_e^2 - (\tilde{m} - \omega/\omega_{ce})^2][Q_i^2 - (\omega/\omega_{ce})^2] = Q_i^2 Q_1^2 \quad , \tag{143}$$

which gives stable solutions only for $\tilde{m} = 0$. For $\tilde{m} \geq 1$ there are unstable areas in the Q_i-Q_1 diagram, as calculated by KOSHKAREV and ZENKEVICH [141] as well as by LASLETT [143] and DOMMASCHK [146].

In the case of the first radial mode ($\tilde{m} = 1$) the frequency Re(ω), normalized to $S_k = \omega_{ce}(1 - \nu_r)$, and the relative growth rate Im(ω)/Re(ω) of the ion-electron oscillations are plotted in a normalized diagram in Fig.21, with Q_i over $1 - \nu_r$ as abscissa and $Q_1^2/\nu_r Q_i$ as ordinate, which is valid for ν_r near to 1, as calculated by DOMMASCHK [146,82].

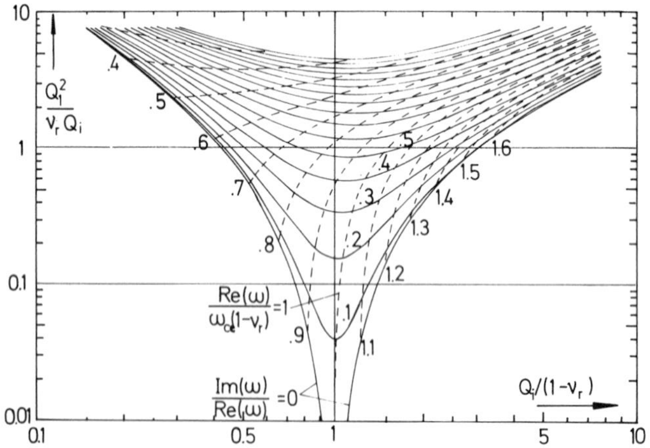

Fig.21. The instability diagram for the first radial ion-electron mode (from [146])

To get a direct impression of the instability regions in a diagram with linear (and not normalized) scales of Q_i and Q_1 for the ring during its compression (where the magnetic field index n and the tunes ν change) we have plotted four typical instability diagrams in Fig.22 [44]. They include the first two dipole modes $\tilde{m} = 1$ and $\tilde{m} = 2$ as well as the quadrupole mode (Q) [141,143,146]. The parameter ν (the betatron tune) describes the focussing. It is set equal to ν_r and ν_z for radial and axial ion-electron oscillations, respectively. At the beginning of the electron ring compression we have both tunes about equal ($\nu_r \simeq \nu_z \simeq 0.7$), so that we have a situation illuminated by the cases b) or c). Since at this time $N_i \simeq 0$ we have $Q_1 \simeq 0$ and since normally $Q_i < 0.1$, we are in a relatively safe situation. If the field index n now decreases, the instability diagram a) is typical of the axial mode, while case d) represents the situation for the radial mode. For the axial ion-electron oscillations there does not seem to be a dangerous instability area nearby (except for the quadrupole mode). From case d), however, it is obvious that for the radial mode we have to run through the $\tilde{m} = 1$ instability area, as DOMMASCHK [146] was first to point out. There are some experiments [96,98] that seem to support this theory.

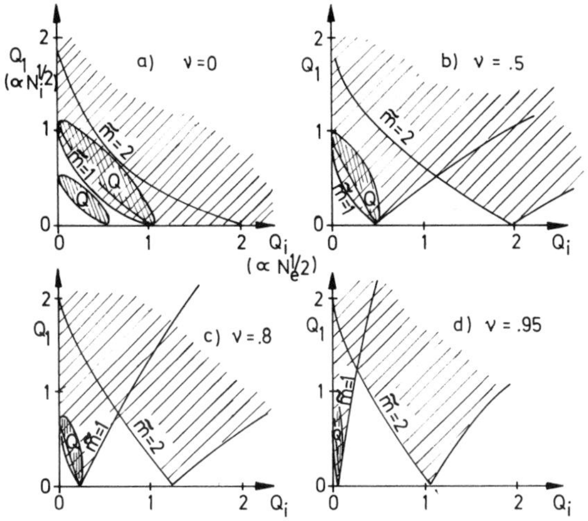

Fig.22. Instability regions in the Q_i, Q_1 plane (from [141,146])

Since the stable area in these diagrams is limited, the ion-electron instability can obviously impose severe constraints on an effective electron ring acceleration rate. There are, however, several possibilities of avoiding this ion-electron coherent resonant motion. One way to get round the instability is to work with optimum electron rings (having high Q_i) to perform very fast electron ring compression at low gas pressure (such that Q_1 stays about zero) down to a very low magnetic field index, such that $(1 - \nu_r) \ll Q_i$, and then to perform the ion loading.

Another powerful method is to use Landau damping, which has been neglected in the considerations above. Landau damping, however, can only suppress the ion-electron instability if there is frequency spread both in the electrons and in the ions [141]. If the full width at half maximum in the frequency of the collective electron and ion motion is denoted by Δ_e and Δ_i, respectively, LASLETT, SESSLER and MÖHL [145] find as a sufficient condition for stabilization of the ion-electron dipole mode by Landau damping, as applied for the first radial mode [82,147]

$$\left[\Delta_e^2 - \left(\frac{2U}{\nu_r}\right)^2\right]^{1/2} \cdot \Delta_i \geq \frac{Q_1^2 \cdot Q_i}{\nu_r} \tag{144}$$

(in the vicinity of the resonance).

The frequency spread Δ_e of the electrons must thus be larger than the coherent frequency shift $2U/\nu_r$, as induced by the collective fields:

$$\Delta_e > 2U/\nu_r \ .$$

Simultaneously the corresponding frequency spread of the ions, Δ_i, must be non-zero to ensure stability.

Δ_e can be estimated from the Landau damping coefficient (133) $E/S_k \cdot \partial S_k/\partial E$, from Q_i and the energy spread $\Delta E/E$ to be

$$\Delta_e \simeq E/S_k \cdot \partial S_k/\partial E \cdot \Delta E/E \cdot Q_i \ .$$

Since $E/S_k \cdot \partial S_k/\partial E$ can be made about equal to 2 [82,98] and assuming $\Delta E/E \simeq 0.05$, we arrive at $\Delta_e \simeq 0.1\ Q_i$, which might be sufficient to fulfill the above-mentioned condition. For smaller values of the Landau damping coefficient or the energy spread, however, the frequency spread Δ_e may be smaller than the minimum spread that is necessary to fulfill (144), so that no damping of the ion-electron instability is possible.

In a detailed calculation of the nonlinearity of the ion oscillations in the electron ring potential DOMMASCHK [146] estimated the ion spread Δ_i for a typical ERA [82] to be $\Delta_i \simeq 0.2\ Q_i$, which might be sufficient to fulfill the condition (144).

In an experimental investigation [98] we varied the Landau damping coefficient, assuming the ion spread Δ_i to be constant, and found - although not all experimental results could be explained - better stabilization of the ion-electron dipole oscillation for higher Landau damping coefficients.

For the transition through the instability region of the first radial ion-electron mode in the very slow process of ionization of heavy nuclei prior to the acceleration, however, HOFMANN [147] found that the Landau damping might not be sufficient to suppress the ion-electron instabilities. The inclusion of nonlinearities in the transverse motion of the ions, however, might be helpful [147].

In their calculations on the transverse two-stream instability LASLETT et al. [145] find the stability area in the Q_i, Q_1 diagram for ion-electron oscillations not significantly improved by the inclusion of images.

In the case of an electron ring loaded with two species of ions OTT et al.[148] find no pronounced displacement of the instability region in the Q_i, Q_1 diagram. For small ion loading fractions the instability region splits into two regions corresponding to the separate interactions of each ion species with the electrons.

2.5.4 Azimuthal Field Application

BONCH-OSMOLOVSKI et al. [92] proposed the application of an additional azimuthal magnetic field B_φ as a method of focussing an ion loaded electron ring in a linear collective accelerator. In several papers the benefical effect of such an ancillary field on the betatron tunes (including the possibility of avoiding the $\nu_r = 1$ resonance) [136,150-153] and the enhancement of Landau damping coefficients for collective transverse instabilities [154-156] has been pointed out. For the case of vanishing magnetic field index n the possibility of stabilizing the dipole mode [157,158] and the quadrupole mode [159] was treated theoretically. A review of the application of an azimuthal magnetic field component in collective ion accelerators is given by BONCH-OSMOLOVSKI et al. [160].

DOMMASCHK [146] treated the important case of non-zero field index, where a resonance between the precessional mode of the ring and the ion bouncing frequency leads to strong collective transverse ion-electron oscillations (see Sec.2.5.3).

The trajectory of an electron is determined by the focussing of an external magnetic field B_z (with a field index n), on which an azimuthal magnetic field B_φ and the repulsion of the other electrons in the ring are superposed. We assume the absence of ions and nearby walls, which would contribute electric and magnetic forces. If we define

$$Q^2 = \frac{r_o N_e}{2\pi\gamma^3} (R/\sigma_a)^2$$

(from (107) and (108) with $\sigma_a = \sigma_b$) and

$$\alpha = B_\varphi / B_z \quad ,$$

the expansion of the equations of motion about the equilibrium orbit results in [152]

$$\left.\begin{aligned} \partial^2 x/\partial\varphi^2 + (1 - n - Q^2)x &= \alpha\partial z/\partial\varphi \quad , \\ \partial^2 z/\partial\varphi^2 + (n - Q^2)z &= -\alpha\partial x/\partial\varphi \quad , \end{aligned}\right\} \qquad (145)$$

which leads with $x = \tilde{A}e^{i\nu\varphi}$, $z = \tilde{B}e^{i\nu\varphi}$ to the algebraic equation

$$\nu^4 - (1 - 2Q^2 + \alpha^2)\nu^2 + (1 - n - Q^2)(n - Q^2) = 0 \quad .$$

The solutions ν_A^2 and ν_B^2 are connected by the relationships [152]

$$\left.\begin{aligned} \nu_A^2 + \nu_B^2 &= (1 + \alpha^2 - 2Q^2) \quad , \\ \nu_A^2 \cdot \nu_B^2 &= (Q^2 - 1/2)^2 - (n - 1/2)^2 \quad , \end{aligned}\right\} \qquad (146)$$

which in the case of vanishing space charge $Q^2 = 0$ reduce to

$$\nu_A^2 + \nu_B^2 = 1 + \alpha^2 \quad ,$$
$$\nu_A^2 \nu_B^2 = n(1 - n) \quad , \tag{147}$$

and hence, since $0 < n < 1$,

$$\nu_{A,B}^2 = \frac{1 + \alpha^2}{2} \pm \left[\left(\frac{1 + \alpha^2}{2} \right)^2 - n(1 - n) \right]^{1/2} = 1/2 \left\{ 1 + \alpha^2 \pm \left[(1 - 2n)^2 + 2\alpha^2 + \alpha^4 \right]^{1/2} \right\} . \tag{148}$$

From these last equations it is obvious that the addition of an azimuthal magneti field component allows the frequencies ν_A and ν_B of the stable betatron oscillation to be controlled. Specifically, it is possible to avoid the crossing of the dangerous $\nu = 1$ resonance since $\nu_A^2 > 1$ and $\nu_B^2 > 0$ can easily be achieved by having $\alpha^2 > n(1 - n)$. Moreover, a B_φ field gives the possibility of adjusting (and even appreciable increasing) the Landau damping coefficients for transverse collective instabilities [154,155] since these coefficients in the case of $\alpha \neq 0$ are obtained from (133) by inserting for the radial derivatives of the two eigenfrequencies ν_A and ν_B in question

$$\partial\nu/\partial_r = \frac{1}{2\nu} \cdot \frac{(1 - 2n)\partial n/\partial r + 2(1 - n)\nu^2\alpha^2/r}{1 - 2\nu^2 + \alpha^2} \quad . \tag{149}$$

This potentially attractive method of applying an azimuthal magnetic field component, however, is only restricted to the first part of the electron ring acceleration, where the axial velocity is small. At higher axial speed the ion-electron ring gets polarized in the radial direction owing to the Lorentz force $e\beta_z cB_\varphi$, which has at least to be limited by the holding power. HOFMANN [161] showed that under present-day experimental conditions the attainable energy would be limited to about 1 MeV per nucleon.

This severe limitation could eventually be removed by applying a multipole magnetic field with azimuthally varying B_r and B_φ field components, as proposed by BONCH-OSMOLOVSKI and DAVILOV [162]. Such a multipole magnetic field configuration is superimposed on the axial magnetic field and is produced by a system of axially directed conductors with opposite currents on a cylinder inside or outside of the electron ring. This method affords the possibility of stabilizing the transverse and even the negative mass instabilities by increasing the appropriate Landau damping coefficients. The feasibility of this new proposal, however, is not yet certain.

2.6 Optimum Electron Rings

Since the range of application of the principle of collective ion acceleration with electron rings is relatively wide (high-energy accelerators for light nuclei or low-energy heavy-ion accelerators with electric or with magnetic expansion acceleration) there cannot be a simple, single performance criterion.

The development in understanding the ERA from theory and experiment has been accompanied by the search for its performance characteristics [68,79,90,91,163]. They may be briefly summarized.

For an electron ring in the uniform external magnetic field at the end of its compression there are several constraints:

1) Axial focussing must exist, $\nu_z^2 > 0$ (108).

2) The threshold for the occurrence of the negative mass instability must be sufficiently high (124).

3) There should be enough Landau damping to stay below the limit given by the resistive wall instability (138).

4) One has to work within the stable areas of the electron-ion collective instability.

If at first we assume negligible betatron oscillation amplitudes (compared with the spread in closed orbit radii resulting from the electron energy spread), we obtain a holding power E_H, as given by (126), in fulfilling the condition imposed by the negative mass instability. For a magnetic field strength of B = 20 kG, $\tilde{\eta} = 1$, and an energy spread of as large as $\Delta E/E = 0.1$ we arrive at

$$E_H[\mathrm{MV/m}] \leq \eta \cdot \frac{13.62}{|Z_{\tilde{m}}/\tilde{m}Z_0|} \quad . \tag{150}$$

Since for an electric acceleration structure nearby walls have to be avoided (in order to obtain low values of the coupling impedance $|Z_{\tilde{m}}/\tilde{m}Z_0|$) and even effective image focussing might be impossible, we have about $|Z_{\tilde{m}}/\tilde{m}Z_0| \simeq 0.5$ and $\eta \approx 0.2$ [52] and arrive at a holding power value as small as

$$E_H \lesssim 5.4 \text{ MV/m} \quad .$$

In contrast to this small value for the holding power in electric field acceleration structures, the acceleration in a magnetic expansion structure looks quite promising as a collective ion acceleration method since here we can use nearby conductors to arrive at low values of the coupling impedance. Using the maximum coupling impedance as obtained by FALTENS and LASLETT [119], we obtain with (131)

$$E_H \lesssim 80 \text{ MV/m}$$

under the same conditions since for this structure, where strong image focussing can be applied, we have $\eta \simeq 0.8$ [52].

If magnetic field strength values B higher than 20 kG can be applied, the holding power value can even be increased (linearly with B).

It is still an open question, if the maximum values of the holding power, as just mentioned, may be increased even more by applying the multipole azimuthal magnetic field component [162], as described in the preceding section.

3. Experiments and Results

3.1 Electron Beam Generators

Two types of electron beam generators are used for present-day electron ring experiments, namely electrostatic and inductive generators. Both use a high voltage pulse applied across a diode such that electrons are emitted by field emission from the cathode and accelerated to the anode. In the electrostatic device the full voltage is applied to the anode-cathode gap from a Blumlein or transmission line, while in the inductive generator the electrons are further accelerated by induced fields after their passage through the anode region. A schematic of the electrostatic generator using a Marx generator to charge a coaxial Blumlein line [164,165] is given in Fig.23.

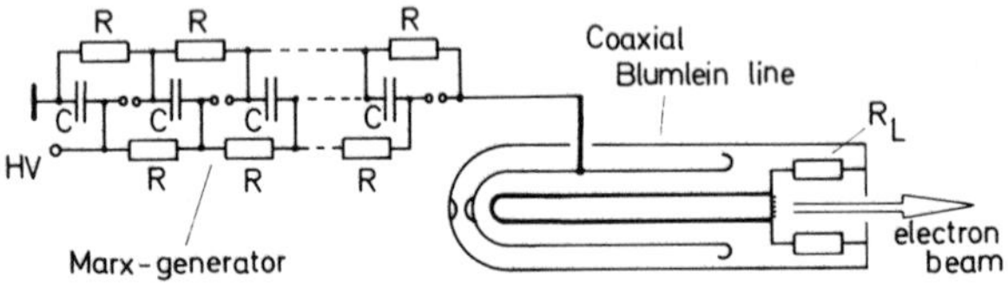

Fig.23. Electron beam generator using Marx generator, Blumlein line and field emission diode

These devices are applied for the production of intense relativistic electron beams [166-169,48]; in some special cases the Blumlein circuit is omitted and only a Marx generator with a large number of stages is used [170,96]. The condition of very small emittance of the electron beam, as is necessary for high quality electron rings, is matched by a special design of the cathode anode region, taking the space charge of the beam for electrostatic imaging into account [171-173].

Inductive generators differ from the device just mentioned in that the electrons are accelerated to higher energies by induced electric fields in a linear accelerating structure just when they have left the anode region. The induction accelerator principle was introduced by CHRISTOFILOS et al. [174], as demonstrated in Fig.24. The electron beam path is surrounded by a ring of ferromagnetic material (laminated iron core or ferrite). If a current pulse from any source (a cable is drawn) leads to a flux change $d\Phi/dt$ through the cross-section area of the ferrite, an axial electric field is induced along the axis of the ferrite ring of the integral value

$$\oint \underline{E}\, d\underline{\ell} = - d\Phi/dt \ ,$$

which corresponds to the energy gain of the electron on passing through the ring. The induction accelerator principle is not only used for the Astron Linear Accelerator [175], but also for the electron injectors of the Berkeley [165,176] and Dubna [177] electron ring experiments.

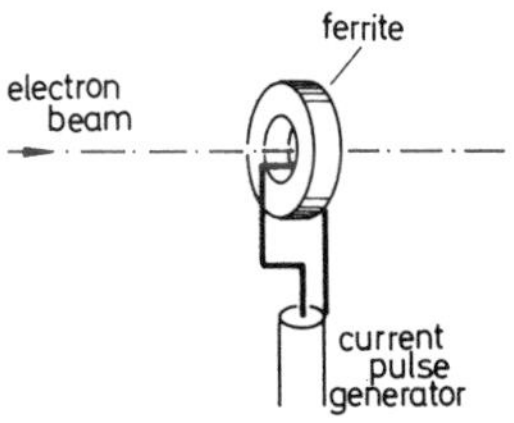

Fig.24. Induction accelerator principle [174,165]

3.2 Beam Injection and Inflection

The beam injection is different for the static field experiment with its cusp section [48] and the pulsed field compressor experiments. The formation of the electron ring in the static field accelerator is done by the injection of a hollow electron beam through a cusped magnetic field. The hollow electron beam is formed by the field emission of electrons from a ring shaped cathode. The particle trajectories in simplified cusp geometries were computed and analyzed [47,60,178] and found to be of use for the production of intense rotating electron rings for collective ion acceleration [48,179]. The first experimental work on the injection of intense relativistic electron beams through a magnetic cusp [180,181] demonstrated the feasibility of this concept. In recent experiments and calculations [182,183] a threshold energy of the electrons was found for their transmission through the cusp, as was expected. The finite width of the cusp transition leads to an electron ring radius oscillation downstream from the cusp transition, which results in a widening of the minor ring dimensions.

In the pulsed field compressor experiments the electron beam is injected in the plane of the electron ring tangentially to it. The device is such that the electrons do not feel the compressor field until they leave the injection snout. One has to take care that they do not hit the snout during the subsequent revolutions, this being achieved by setting the closed orbit radius smaller than the injection radius and reducing the coherent radial betatron oscillations of the electrons by a fast changing magnetic field. This inflection field is opposite to the main compressor field. It can be applied independently of the azimuth as in one Dubna version [184]. LAMBERTSON [37] proposed to install the inflection field only on the azimuthal section opposite the injection snout, so that the closed orbit is shifted away from the snout till the betatron oscillations are reduced. He drew attention to the fact that if the injection occurs at a field index around n = .5 (with $\nu_r \simeq 2/3$) beam stacking may be very effective. Analytical and numerical calculations [185,122,186] showed that the optimum location of the inflector is not at 180° from the snout but around 135°, where the electron trajectory crosses the closed orbit (see Fig.19). Inflection calculations in the presence of an azimuthal magnetic field [186] and in the case of axial inflection [187] were performed.

BOVET [188] studied the spiral injection into an electron ring compressor in the form of multiturn injection in the longitudinal phase plane. This method calls for a very steep magnetic field rise since the incoming electrons with their energy rising with time should all be injected on their own closed orbit at a fixed injection radius. Owing to the large field increase during one electron revolution the spiral step can be made larger than (or equal to) the beam size plus septum width, so that the electrons do not hit the snout. The method has the advantage that beam stacking with large momentum spread $\Delta p/p$ to prevent the negative mass instability is possible. Probably because of the technical constraints it has not been tried out yet.

An electrostatic type inflector system has been studied at Nagoya University [189].

3.3 Ion Loading and Ionization

The ion loading in nearly all devices under consideration occurs through impact ionization by the electrons (and thus within their potential well) of atoms, molecules or clusters [190,191]. In early calculations of the ionizing probabilities by successive electron impact and the stripping time for heavy elements LEVY [192] and JANES [193] use the following expression for the ionization cross-section [194]:

$$\sigma = \frac{\pi e^4}{4E} \sum_{j=1}^{j=\tilde{N}} 1/I_j \, \ell n(E/I_j) \quad ,$$

where E is the energy of the incident electron and the sum is taken over the $\tilde{N}$ bound electrons each with binding energy I_j. They plotted the dependence of the concentrations of the different ionization levels as a function of time for different elements, and they found that their results depend linearly on the electron density, but that they are only very weakly influenced by the electron energy.

SALOP [195] found out that Auger processes have a significant influence in speeding up the ionization rates in the ionization of noble gases by a relativistic electron ring. He took both direct continuum ionization and inner shell excitation followed by Auger ionizing transitions into account. Recently SIEBERT et al. [197] even found better agreement with the relevant experiments [198]. The calculations (using the rate equations) show that, if one starts from a certain number of neutral and avoids further penetration of neutrals into the electron ring, there are always a few neighbouring ionization states present. Owing to the decreasing ionization cross-sections in subsequent steps it takes a long time to obtain a certain high ionization state: For typical electron ring parameters it might take about 50 msec to fully ionize argon and nearly 1 sec for xenon [195,196].

The conditions of avoiding further penetration of neutrals into the electron ring after the formation of the first ion loading is difficult to fulfill experimentally if the ion loading is done simply by filling neutral gas into the vacuum vessel (as done so far). Therefore the method of cluster injection [190] is proposed for defined ion loading with a well collimated cluster beam. It turns out that the necessary density can easily be achieved [199,82].

3.4 Diagnostics

The diagnostics of relativistic electron rings uses the specific properties of these rings: the high electron energy for X-ray production when targets are hit, the high electron current resulting in strong electric and magnetic self-fields and the radiation emitted from the circulating electrons or the excitation of ions.

The electron beam prior to its injection into the compressor is diagnosed by current collectors (e.g. Faraday cups), by magnetic probes, and Rogowski belts for the current measurement from the surrounding magnetic field, or by other non-intercepting methods such as the low-inductance and low-resistance current monitor measuring the return current in the beam tube, developed by AVERY et al. [200]. All these different diagnostic methods are in very good agreement as long as the electrons are prevented from hitting the probes [201].

Magnetic pick-up loops can also be applied very well to determine the electron number, the ring radius and its compression velocity during the electron ring compression [202-204]. Normally, the self-field of the electron ring is a small fraction of the compressor magnetic field at a distance from the ring where its disturbing influence on the ring behavior can be assumed to be negligible. The voltage induced by the compressor field is therefore cancelled by that of an additional probe, located at a distance from the electron ring where only a comparable amount from the compressor field is induced. This configuration allows the electron ring current and the total electron number to be determined in a relatively narrow region of radial and axial ring position. It can hardly be applied for the large axial displacement during the roll-out of the electron ring. A long extended probe on the compressor axis, however, that integrates the axial magnetic field component $B_z(z)$ of the electron ring almost all along the axis is independent of the axial ring position and therefore suitable for the roll-out phase [205]. Since the integral

$$\int_{-\infty}^{+\infty} B_z(z)dz = \mu_0 I$$

depends linearly on the electron ring current I, the electron number N_e is found to depend only linearly on the electron radius (in contrast to a single loop on axis) and is independent of the axial ring position:

$$N_e = \frac{2\pi R}{\mu_0 ec} \int_{-\infty}^{+\infty} B_z(z)dz \quad . \tag{151}$$

Such a device has been applied successfully in the Garching ERA experiment [205, 97]. It is a magnetic field probe with a helical winding on the compressor axis extending beyond the roll-out region. The compression field is cancelled with the same probe by a part with the opposite orientation of the helical winding. Because of the weak dependence on the radius R (in (151)) the electron number can be determined with this probe nearly all through the compression and roll-out of the ring [97] and even in the very first part of the acceleration. It cannot be applied, however, for the higher axial electron velocities in the acceleration section since then the time resolution of the helical probe with its relatively high inductance is not sufficient.

In the acceleration section small, low-inductance single-turn loops measuring the axial magnetic field component are used [205]. If such a probe is located at an axial position z_0, and if an electron ring of major radius R and negligible minor dimensions passes by with an axial speed of dz/dt, the probe measures a voltage proportional to the magnetic field derivative of [205]

$$dB_z(z)/dt = -3 \frac{\mu_0 ecN_e}{4\pi R^2} \cdot \frac{(z - z_0)/R^2}{[1 + (z - z_0)^2/R^2]^{5/2}} \cdot dz/dt \quad . \tag{152}$$

The induced voltage of the probe affords the possibility of determining some characteristic properties of the accelerated rings. The electron number is found from the maximum of $B_z(z)$, which is at $z = z_0$. The axial ring speed $v_z = dz/dt$ is obtained from the extremum values of dB_z/dt, which in the case of negligible d^2z/dt^2 are located at

$$z_m = z_0 \pm R/2 \quad .$$

Another value is obtained from the time of flight of the ring between the two extrema, which are separated just by the amount R of the electron ring radius. A further possibility of determining dz/dt is afforded by the slope of dB_z/dt at $z = z_0$ (152). Certainly the application of two or more probes at different axial positions allows independent axial speed measurements by the time of flight between different probes since the zero crossing of the dB_z/dt signal accurately gives the time when the ring is at the probe location z_0.

Two or more probes, moreover, allow evaluation of the acceleration d^2z/dt^2 (and with it the holding power) of the electron ring. The holding power is obtained from the acceleration measurement at a certain location in the ERA section for loaded as well as unloaded rings. If the acceleration increases along the acceleration section, the holding power of the electron rings is determined from the maximum acceleration up to which the difference in the ring inertia persists. This method has already been applied in the Garching ERA experiments [206,207].

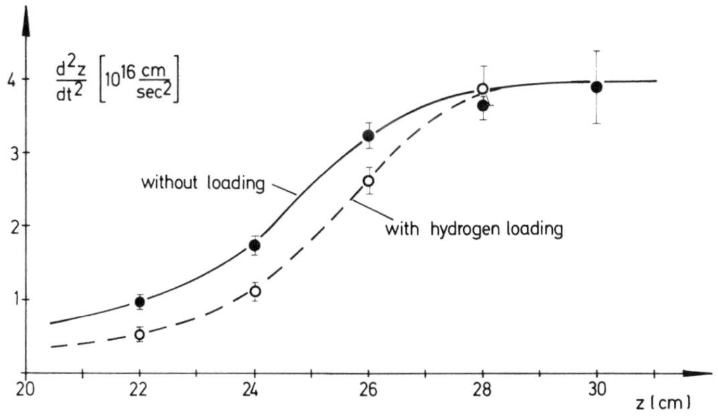

Fig.25. The electron ring acceleration versus z with and without hydrogen loading [206,207]

Figure 25 gives an example of the measurements of electron ring acceleration versus the axial distance z from the compression plane for ion loaded and unloaded rings. The values of d^2z/dt^2 are quite different for all $z < 26$ cm, corresponding to an acceleration of $d^2z/dt^2 < 3 \cdot 10^{16}$ cm sec^{-2} of the unloaded ring. Since the acceleration d^2z/dt^2 of the pure electron ring is related to the applied radial magnetic field component B_r by

$$\left.\begin{array}{l} B_r[G] = 5.12141 \cdot 10^{-17}\, d^2z/dt^2 \ [\mathrm{cm\ sec^{-2}}] \\ (\text{for } \beta_\perp = v_\varphi/c \simeq 1) \quad , \end{array}\right\} \tag{153}$$

this results in a maximum applicable B_r value of $B_r = 1.5$ G and a holding power of about $E_H = 3$ MV/m. The method, however, is limited to relatively low axial ring velocities (of about $v_z = 1.5 \cdot 10^9$ cm/sec) since later on the signal quality is impaired by the finite time resolution of the probes, cables and oscilloscopes used.

Although the magnetic probes are small loops on the compressor axis, it is very important to take care not to disturb the magnetic field configuration especially in the acceleration section. It was not until a special construction of these probes [205] was applied that the ring was not affected by their presence.

Optical diagnostic methods such as the observation of synchrotron radiation or of scattered laser light normally have the smallest interference with the electron ring.

Since the electrons move on circular orbits in a magnetic field, they emit synchrotron radiation which can be observed in the visible and near infrared spectral region near the end of the ring compression. The radiation occurs tangentially to the circular electron orbit into a relatively narrow cone. SCHWINGER [208] calculated the synchrotron radiation based upon classical electrodynamics, and several papers have treated experiments with this type of radiation and described its properties [209-213].

The instantaneous power $P_\lambda(\lambda,\psi)$ radiated per unit wavelength and radian by an electron in a circular orbit of radius R is [211]

$$P_\lambda(\lambda,\psi) = 27/(32\pi^3) \cdot e^2c/R^3 \cdot (\lambda_c/\lambda)^4\gamma^8[1 + (\gamma\psi)^2]^2 \cdot \left\{ K^2_{2/3}(\tilde{\xi}) + \frac{(\gamma\psi)^2}{1 + (\gamma\psi)^2} K^2_{1/3}(\tilde{\xi}) \right\} , \quad (154)$$

where ψ is the angle between the direction of the emitted light and the plane of the electron orbit (elevation angle).

The "critical wavelength" λ_c is given by

$$\lambda_c = 4\pi R/(3\gamma^3)$$

or in convenient units for electron rings

$$\lambda_c[\mu m] = 5590 \cdot R[cm]/(E[MeV])^3 \quad .$$

The argument $\tilde{\xi}$ of the modified Bessel functions is defined by

$$\tilde{\xi} \equiv \lambda_c/(2\lambda) \cdot [1 + (\gamma\psi)^2]^{3/2} \quad .$$

One concludes that the angular width of the radiated power is in the vicinity of γ^{-1}, so that the radiation of an electron ring occurs into a disc of angular width γ^{-1} around the electron ring plane. Under typical electron ring conditions one can assume uncorrelated emission in the visible and near infrared spectral regions.

The small emission angle of the synchrotron radiation allows the cross-section of the electron ring to be observed, so that no Abel inversion is necessary. The minor ring dimensions and density distributions can thus easily be evaluated. An example of the axial (relative) synchrotron light intensity distribution of the first Garching electron ring experiment [214] is given in Fig.26 for two different wavelength regions.

The electron ring cross-section was even imaged photographically in some of the experiments at Berkeley [73,215] and Dubna [216]. The integration of $P_\lambda(\lambda,\psi)$ over the elevation angle ψ gives the spectral distribution $P_\lambda(\lambda)$ of the synchrotron radiation [208-211]

$$P_\lambda(\lambda) = 3^{5/2}/16\pi^2 \cdot e^2c/R^3 \cdot \gamma^7 \cdot (\lambda_c/\lambda)^3 \int_{\lambda_c/\lambda}^{\infty} K_{5/3}(\zeta)d\zeta \quad . \tag{155}$$

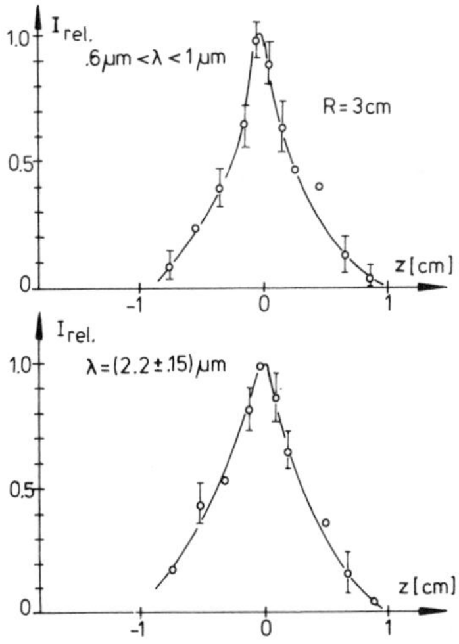

Fig.26. Axial profile of the compressed electron ring [214]

The spectrum has a maximum at 0.42 λ_c and a full width at half-maximum of 0.84 λ_c and drops exponentially for $\lambda \ll \lambda_c$ and quite slowly (proportionally to about $\lambda^{-7/3}$) for $\lambda \gg \lambda_c$ [210].

An expression of the asymptotic behavior for $\lambda \gg \lambda_c$ (corresponding to [210]) in convenient units for electron ring accelerators is

$$P_\lambda(\lambda)\ [W/(\mu m \cdot electron)] \simeq 9 \cdot 10^{-10}\ (R[cm])^{-2/3} \cdot (\lambda[\mu m])^{-7/3} \quad , \tag{156}$$

which is independent of the electron energy, while for the spectral distribution $P_\lambda(\lambda)$ near its peak at $\lambda \simeq \lambda_c/2$ the radiated power increases with the seventh power of the electron energy:

$$P_\lambda[W/(\mu m \cdot electron)] \simeq 9 \cdot 10^{-21}\ \gamma^7\ (R[cm])^{-3} \quad . \tag{157}$$

For typical electron ring parameters the spectral distribution $P_\lambda(\lambda)$ is plotted in Fig.27a and 27b for different electron energies E and different ring radii R. One easily concludes from these plots that the radiated power depends very sensitively on the electron energy in the visible spectral region. At energies of 15 MeV and more and radii around 2.5 cm, however, one is already very near the maximum of the spectral emission in the near infrared region. The determination of the (absolute) spectral power distribution gives a very good value of the electron energy. An example of a measured spectrum of the synchrotron radiation, as obtained from our first Garching electron ring experiment [214] is given in Fig.28. The absolute values are found from a calibration with a tungsten strip lamp. The measured points are very well matched by the calculated curve belonging to an energy of 12.6 MeV, while curves with only about half an MeV difference from this value can clearly be excluded.

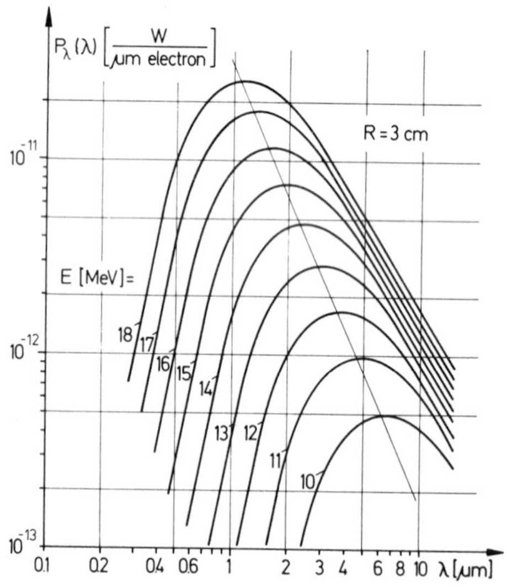

Fig.27a. Synchrotron radiation spectrum for different electron energies

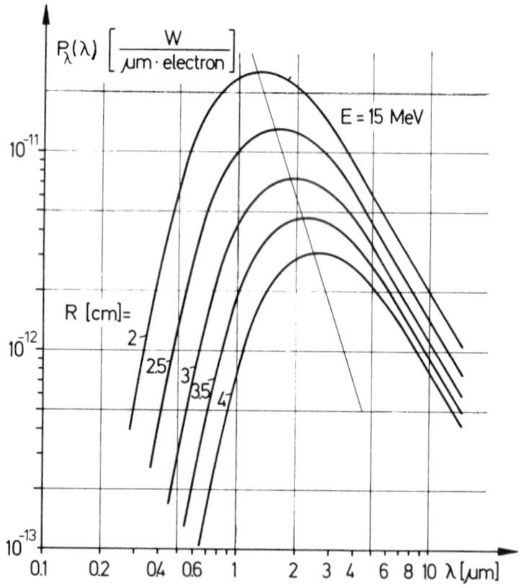

Fig.27b. Synchrotron radiation spectrum for different electron ring radii

The diagnostic possibilities offered by the angular intensity distribution (154) [213] or the polarization of the radiation [208,209] have not yet been explored in electron ring experimental devices.

The radial intensity distribution of the synchrotron radiation from the electron ring gives information about its minor radial extent, modified by the electron energy distribution. It is very difficult to distinguish between the contribution of the betatron oscillations or the closed orbit spread (as caused by the electron energy distribution) to the minor radial ring width. It seems to be very important to know the amounts of these two contributions to the radial minor dimension in order to find the mechanism (betatron resonances or collective longitudinal instability) that leads to an increase in the minor radial electron ring dimension and a subsequent drop of the holding power.

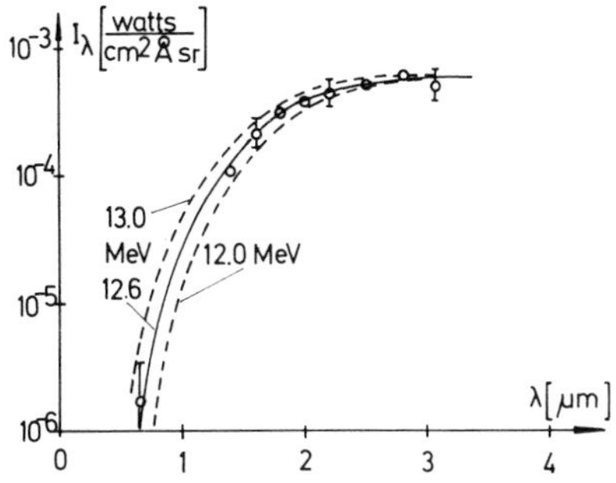

Fig.28. Calculated and measured spectra of the synchrotron radiation [214]

PETERSON and RECHEN [217] proposed to analyze the momentum and betatron amplitude distributions in a circulating beam by a two-probe method where one probe clips the beam at the inner radius and the other at the outer radius. The distributions in particle momentum and in amplitude of radial betatron oscillations are obtained from the gradient function of the X-ray signals, as demonstrated in some of the Berkeley experiments [218]. The clipping of the beam at a probe, however, might interfere with the ring, possibly causing changes of the quantities to be measured. Therefore another method is proposed that takes advantage of the energy sensitivity and the narrow angular distribution of the light scattered from relativistic electrons [219]. The interference of this method with the electron ring is almost negligible. The distributions in particle momentum and in radial betatron oscillation amplitude can be measured separately, the betatron oscillation distribution from the angular radiation distribution of the scattered light and the particle momentum spread from the scattering spectrum.

We assume the case of the energy of the light being very small compared with the electron energy. A photon with frequency ω_0 and wave vector $\underline{k}_0$ hits an electron of momentum vector $\underline{p}$ at an angle Θ, so that the polarization vector is perpendicular to $\underline{p}$ (see Fig.29). The frequency ω_1 of the scattered radiation is then given in the laboratory system by

$$\omega_1 = \omega_0 \frac{1 + \beta \cos \Theta}{1 - \beta \cos \chi} \quad . \tag{158}$$

Since β, the electron velocity relative to the speed c of light, is very close to 1 for relativistic electrons, the maximum energy $\hbar\omega_1$ of the scattered photons can be very much higher than the energy of the incident light. For backward scattering from a head-on collision of the photon and the electron ($\Theta = 0$, $\chi = 0$), for instance, the factor is

$$\omega_1/\omega_0 = 1 + \beta/(1 - \beta) = \gamma^2(1 + \beta)^2 \simeq 4\gamma^2 \quad (\text{for } \gamma >> 1) \quad .$$

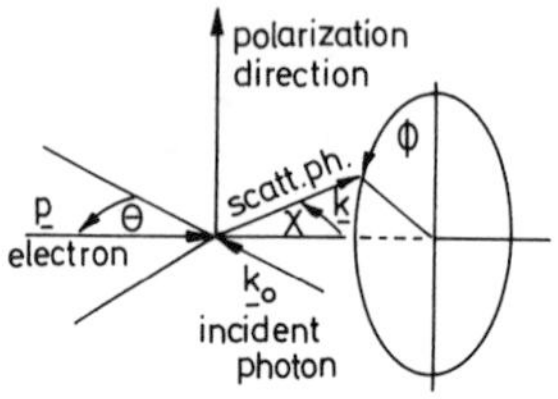

Fig.29. Scattering geometry

The scattered photons from ruby laser light (λ_0 = 6943 Å) on γ = 30 electrons (as in a typical ERA) may thus be found in the 2 Å wavelength region! The scattered light wavelength depends very sensitively on the electron energy, so that the electron energy can be deduced from the scattering spectrum. Moreover, for relativistic electrons (with $\gamma > 1$) the scattering occurs into a relatively narrow cone around the direction of the electron momentum p. The differential photon scattering cross-section [219-221], taken at $\Theta = \pi/2$ (normal incidence), is

$$d\sigma/d\Omega(\chi,\phi) = r_0^2 \left[\frac{1}{\gamma^2(1 - \beta\cos\chi)^2} - \frac{\sin^2\chi\cos^2\phi}{\gamma^4(1 - \beta\cos\chi)^4} \right] \tag{159}$$

with $d\Omega = \sin\chi\, d\chi\, d\phi$, the integration of which over ϕ and χ results in the Thomson formula

$$\sigma = 8/3\ \pi r_0^2 = \sigma_{Thomson} \quad .$$

The cross-section (for $\phi = 0$) becomes zero at an angle of

$$\chi_0 = \text{arc}\cos\beta = \text{arc}\sin 1/\gamma \simeq 1/\gamma \quad (\text{for } \gamma \gg 1) \quad ,$$

which is as small as $\chi_0 = 1.91^0$ for $\gamma = 30$. This narrow angular distribution can be used to determine the radial angular betatron oscillation distribution and hence the radial betatron amplitudes. An example of the radial angular distribution of the scattered light for parabolic radial betatron distributions of different maximum angles ψ_{max} is given in Fig.30, where ψ is the elevation angle against the tangent. It turns out that as long as ψ_{max} is larger than $\chi_0 \simeq 1/\gamma$ the measured angular spread $\Delta\psi$ (FWHM) is a very unique (linear) function of the angular standard deviation, that is nearly independent of the assumed angular distribution function [219].

A similar behavior is found for the spectral distribution depending on the electron energy distribution. An example of the power scattered into a large solid angle for parabolic distributions of different standard deviation widths σ_γ/γ_0 is given in Fig.31, which demonstrates the sensitive dependence of the spectrum on the electron energy distribution. Here, too, we find a nearly linear dependence of the difference of the characteristic maximum occurring frequency ω_{max} and the frequency where the maximum scattered power is found on the standard deviation widths σ_γ/γ_0 that is nearly independent of the assumed electron energy distribution function [219].

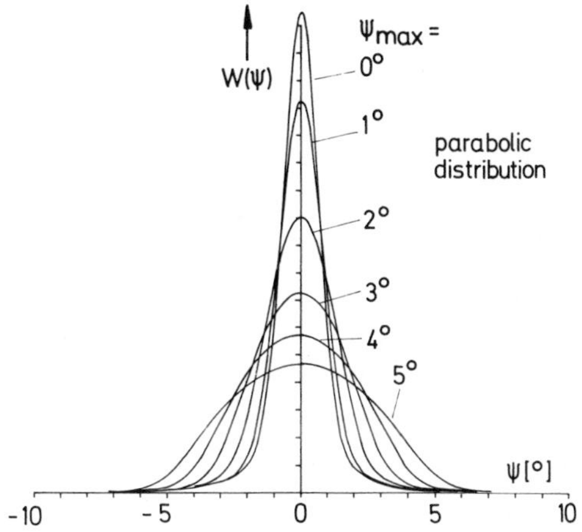

Fig.30. Angular scattering distributions [219]

Although the total number of scattered photons for typical electron rings should be at least in the range of 10^5 to 10^6, they have not been found in the recent Garching experiments [222].

There are several other optical diagnostic methods that make use of the radiation emitted by the ions in the electron ring [223,224]. These methods are very useful, in principle, since they give information not only about the ring characteristics but also about the ion loading and the degree of their ionization. It seems, however, as if the loading fraction and the atomic mass of the ions have to be very high [223]. In an experiment at the Berkeley device SCHMIEDER [225] was able to prove the emission of K and L-X-rays from xenon.

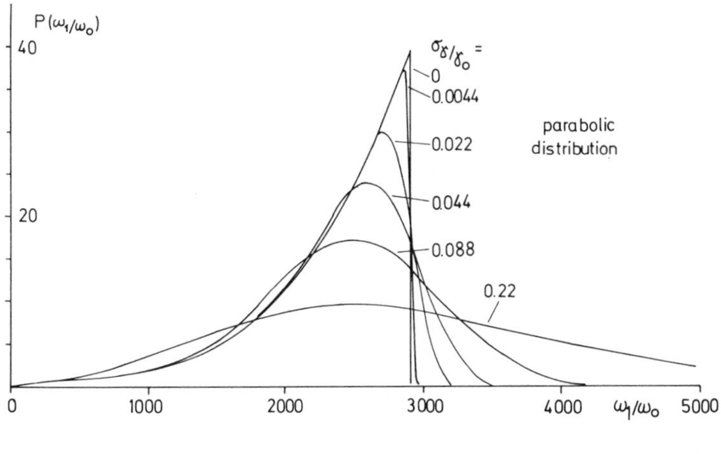

Fig.31. Scattering spectra for different energy spreads [219]

Finally, we should mention the diagnostics for the collectively accelerated ions. These ions can be detected by nuclear tracks in material such as plastics [226,227] or by nuclear reactions [228]. The nuclear track registration is very well investigated and described by PARETZKE [226]. This method is based on the latent tracks produced by the deceleration of ions in dielectric materials and widened (and thus made visible) by an etching process. The nuclear track method has been used to detect the collectively accelerated He ions in the Garching electron ring accelerator experiment [206,207].

The Dubna ERA group was the first to apply nuclear reactions for the detection of the energy and particle number of collectively accelerated ions [229]. The method was to detect α particles with the nuclei of ^{63}Cu by measuring the induced activity on a copper target being hit by the beam of α particles. The nuclear reaction method of recording the collectively accelerated ions calls for ion energies that are normally in the range of at least a few MeV/nucleon. An excellent review of the nuclear reaction measurements of ions in the range of interest for electron ring accelerators is given in [228].

3.5 Static Field Experiments and Electron Ring Stopping

The principle of the acceleration of electron rings in static magnetic fields underlying the Maryland ERA concept is given in Fig.3 and in Sec.2.3.1 [43,47-49,60,62, 178,179]. In the University of Maryland experiment the electrons are emitted from a hollow cathode parallel to the magnetic field lines and successfully passed through the cusp region, where most of the longitudinal velocity is transferred into azimuthal velocity. Recent observations, however, indicate strong radial beam blow-up downstream from the cusp region. This was successfully suppressed by positioning an energy spreading foil behind the cusp region in order to avoid the negative mass instability [230,231]. This type of collective instability, the theory of which has recently been extended to a cylindrical electron sheath by UHM [232], is regarded as a possible candidate for the observed beam blow-up. Before ion loading can occur [233], however, the electron layer must be decelerated down to relatively low axial velocities since at present the minimum axial speeds are still in the range of β_z = 0.2 to 0.3 [231]. The electron ring stopping can be accomplished by several different mechanisms, by interaction with resistive material in the vicinity of the ring [234-237], by the passage through a gas cloud [238] or by a fast-pulse coil trapping system [231]. KALNINS, KIM and LINHART [234] treated the problem of the interaction of an electron ring with a resistive wire loop. The electron ring induces image currents in these loops which owing to the finite resistivity of the loops and subsequently the phase shift, leads to a decelerating axial force on the ring. MERKEL [235] and HERRMANN [236] determined the decelerating force on the motion of electron rings close to conducting cylinders, which is due to the ohmic losses in the walls. The conditions can be specified to match the range of requirements of the Maryland experiment [237]. In an acceleration structure, however, an undesirable run-away situation might occur [236] since the decelerating force has a maximum at a certain electron ring axial speed which depends on the resistivity and the geometry of the cylinder. The effect of the deceleration of the electron rings by ions [238] is based on the creation of ions in the potential well of the electron ring which predominantly derive their potential energy from the ring as long as they cannot be captured in the ring, thus decelerating the ring.

The experimental verification of the electron ring stopping by one of these methods or a combination of them seems to be a vital problem of this ERA device. The experiments at Maryland are accompanied by intense efforts at theoretical description of the electron rings [239-243].

The transverse ERA concept [64-67] (see Sec. 2.3.3 and Fig.10) is being studied experimentally and theoretically at the Universities of Bari and Lecce. The experiment ANEL 1 is a static magnetic field ERA with a length of 50 cm and a major radius of about 5 cm. The electron energy is 100 keV for the time being and will be increased up to 2 MeV later on. The device ANEL 2 is a collective synchrotron in which the electron ring will move along a circular path.

3.6 Pulsed Field Compressor Experiments

There are numerous experimental activities in the area of pulsed field ERAs in the Soviet Union (at the Joint Institute of Nuclear Research (JINR) at Dubna, at the Institute for Theoretical and Experimental Physics (ITEP) at Moscow, and at the University of Tomsk), in the USA (work being done at Berkeley), in Japan (at the Institute of Plasma Physics at Nagoya University) and in Europe (at Saclay, at Kernforschungszentrum Karlsruhe and at Max-Planck-Institut für Plasmaphysik at Garching).

The experiments on collective ion acceleration at JINR at Dubna already started in the late fifties under the leadership of the late VEKSLER [18,20]. The first publication was of a talk presented at the Cambridge Accelerator Conference [21] in 1967.

The initial experimental device [244-247] consisted of an induction linac [177] that injects an electron current of about 100 A at a kinetic energy of 1.5 MeV into a pulsed magnetic field at an injection radius of 40 cm, the pulsed field being generated by three nested pairs of coils. This device yielded successful α particle acceleration [229] which, although not repeated, encouraged further intensive experimental activity in Dubna. The new experimental device consists of a new induction accelerator (SILUND) and a compressor made of four nested pairs of coils with a 1 m long solenoid for heavy-ion acceleration in an expanding magnetic field [246-248,184,216]. The vacuum vessel is a tapered stainless steel chamber placed near the electron ring in order to reduce the coupling impedance to about $|Z_{\tilde{m}}/\tilde{m}| \lesssim 6\ \Omega$. The first results obtained with this device are very promising [216]: The electron number captured in a ring was more than 10^{13}, and the minor ring dimensions at the end of the compression (R = 4 cm) were found to be as small as 2 to 3 mm. These rings could even be brought into the expansion acceleration section. However, ion acceleration has not yet been observed since the ion loading due to high background gas pressure was too high and, moreover, the radial magnetic field component B_r chosen was obviously too large. The next experimental findings from this collective heavy-ion accelerator prototype will certainly be interesting.

Parallel to the expansion acceleration structure, there is the development of a superconducting RF acceleration section for an electron ring accelerator at Dubna [249].

The experiments on electron rings at the Institute of Theoretical and Experimental Physics (ITEP) at Moscow are mainly concentrated on the investigation of collective effects and instabilities (such as the ion-electron instability) during the compression of the electron ring [250-252]. The compressor consists of six nested pairs of coils and has the character of a model experiment [252]. In the future a static magnetic field will also be used [250]. The experimental investigations at ITEP are accompanied by extensive theoretical work relating to the ERA.

The experiments at the Institute for Nuclear Physics at Tomsk (USSR) are performed with a device (TREK) that is built for obtaining intense relativistic electron rings in a magnetic mirror field [253]. The electrons are injected at an energy of 2 MeV and a current of 50 kA. The inflection is provided by the image fields of the intense beam induced in the cylindrical metal surface that surrounds the ring beam. The ring is compressed by a sequence of 4 nested pairs of coils and strong ion focussing is used.

The device is mainly used for analyzing the most dangerous instabilities to obtain very intense electron beam rings, which are thought to solve the problem of compact ion accelerators (ERA) as well as that of controlled nuclear fusion.

The electron ring research conducted at the Institute of Plasma Physics of Nagoya University (Japan) is concentrated on the investigation of electron ring formation and compression and especially beam-wall interaction [254,255]. The chamber, which allows a very good vacuum to be obtained is made of stainless steel and is of such a simple form that it can be described by the pill-box model. A 2 MeV, 1 kA electron beam is injected, the inflection is performed using an electric inflector [189], but only two coils are prepared to compress the ring radially by a factor of about 2 [255] since the later stage of compression seems to present little difficulty.

The extensive theoretical and experimental work at the Lawrence Berkeley Laboratory, which started just after the Cambridge Accelerator Conference [21], has very much influenced the research on ERA. Many of the Berkeley results have already been incorporated in this report, so that only a brief description will follow. It is to be regretted that the experiments were terminated about two years ago.

The compressor devices used in their experiments consisted of about 2 to 4 pairs of coils consecutively pulsed with a compression time of the order of several hundred μsec [256,257]. Most of the experiments concentrated on the initial phases of electron ring compression because of indications that an early longitudinal instability might be responsible for the production of RF radiation [104,107], for the nonlinearity of the trapped electron current (about a factor of 2 reduction at full current injected) [218], and for the increased momentum spread with injected current [217,218]. A principal objective of all the experiments therefore was the reduction of the longitudinal coupling impedance of the device in order to trap higher electron numbers without occurrence of the negative mass instability. As a result of the effort electron rings with electric fields close to the theoretical expectation were created [258,259] in a slightly focussing magnetic field (unfortunately without the possibility of magnetic expansion acceleration).

The main goal of the electron ring experiments at Saclay using an electron beam with a current of about 1 kA at an energy of 2.5 MeV [260,261] is to use the electron ring for producing a very short electron pulse [262] for physical-chemistry studies.

The Karlsruhe electron-ring accelerator experiments (unfortunately terminated a few years ago) were performed with a pulsed magnetic field compressor [263-267] relatively similar to the Berkeley device. They concentrated on investigations of the single-particle resonances and of the longitudinal collective instability of the electron ring in a vacuum chamber with conducting side plates as well as on the development of electron beam diodes that deliver high currents of electrons with an instantaneous energy spread in order to suppress the collective instabilities [268]. The developments were extended to new ERA compressors [269] and cluster injection devices for controlled ion loading of the electron rings [122,190].

The Garching electron-ring experimental devices differ from all others (that also use pulsed fields) in their extremely short compression time (about two orders of magnitude less than used elsewhere), which was chosen in order to cross resonances quickly and to relax vacuum requirements. The first Garching device [270-272] was a fast three-stage compression arrangement in which electron rings with 5×10^{12} electrons were compressed to minor dimensions smaller than 0.5 cm [96,173]. To investigate still faster compression (in order to avoid the observed single-particle resonances [74]) as well as the problems of roll-out and expansion acceleration, a single-stage, fast-compressor experiment was made up of a single-turn coil that featured very fast compression [95,96]. In this device it was shown that only if there is adequate image focussing (from a "squirrel cage" structure [89]) electron rings can be accelerated as integral entities [98,273]. At the end of 1974 the collective acceleration of light elements such as hydrogen and helium to a relatively low energy (~ 200 keV/nucleon) was demonstrated [206,207]. The holding power obtained was in the range of 3 to 4 MV/m and could only be slightly increased to about 5 MV/m in the following [97,274]. The main aim of the experiments was the study of the electron ring properties as influenced by collective instabilities such as the ion-electron transverse instability being studied in this device [98,273,274] and the negative-mass instability being investigated in an accompanying experiment using a stationary magnetic field [275].

Since the compression time of the single-stage compressor experiment was of the order of 10 μs only, but very much longer times are needed to load the electron ring with heavy ions of sufficiently high ionization, the new experiment ("Pustarex") [82,276,277] will combine fast pulsed compression with a long ionization interval by using static magnetic fields that hold the ring in a "waiting-room" for ion loading and thereafter allow axial magnetic expansion acceleration.

4. Conclusions

The experiments and calculations related to intense relativistic electron rings - as described in the preceding sections - have demonstrated the advantage and potential of the electron ring accelerator for collectively accelerating ions. The feasibility of this concept was shown in the Dubna [229] and Garching [206,207] electron ring experiments with collective acceleration of light nuclei, and progress has recently been made in the Maryland ERA experiments [279].

Although the holding power in most of these experiments was more than one order of magnitude below the values that should be possible with such devices, the acceleration rates already exceeded those of conventional linear ion accelerators. The quality of the electron rings in past experiments has mainly suffered from poor properties of the injected electron beams and still imperfect surroundings (relatively high coupling impedance) of the electron ring. Being optimized in many respects, the new devices of JINR at Dubna [216,246-248], MPI für Plasmaphysik at Garching [82,277] and the University of Maryland ERA [179,279] will certainly allow substantial progress towards higher values of the holding power of the electron rings and hence higher acceleration rates. It is, however, still an open question whether the maximum value of 80 MV/m, as should be possible for optimized electron rings (with respect to collective instabilities, see Sec.2.5) in a magnetic field of 20 kG, is attainable. Half this value (or even less) would already be very attractive for a collective ion accelerator. Moreover, the magnitude of the magnetic field could be increased by a factor of three to about B = 60 kG, thus trebling the holding power ((126) and (131)) for a very powerful electron ring accelerator.

The intense relativistic electron ring with its high intrinsic electric field which exceeds all values which can be obtained using metallic conductors is not only a strong and cheap vehicle for accelerating the ions; the electron ring itself has extraordinary properties which might lead to applications that have not yet been exploited [136]:

The electron ring collects a very high charge in a relatively small volume. Typical electron rings [97,258] with a total electron number of $N_e = 5 \cdot 10^{12}$ at a major radius of 2.5 cm and minor half-axis of 0.5 cm carry a charge of $Q \approx 8 \cdot 10^{-6}$ Cb. It would be necessary to apply a voltage of as much as about 3 MV to charge a conductive torus of the same dimensions.

The non-accelerated ring can have a very long life-time, determined by the synchrotron radiation time constant. According to BUDKER's original idea [15], the electron ring field strength even increases under certain conditions when the electrons lose energy by synchrotron radiation. Since the spectrum of the synchrotron radiation is continuous and exactly calculable, as long as the radiation is incoherent, the electron rings might serve as intense radiation standards (as the electron synchrotron in the X-ray region) in the visible and the neighbouring spectral regions. An electron ring in a magnetic field of 60 kG would have a major radius of about 1.7 cm, and the electrons would have an energy of about 26 MeV, such that the critical wavelength λ_c (see Sec. 3.4) is in the visible region at about 0.55 μm = 5500 Å. The radiated power P_λ with its maximum in the ultraviolet spectral region at about 2700 Å would be as much as $P_\lambda \approx 2.2$ nW/μm for each electron (157) or $P_\lambda \approx$ 11 kW/μm for $N_e = 5 \cdot 10^{12}$ electrons in the ring, which is a very powerful source.

The compressed electron ring can, moreover, be used for research on the ionization processes and on the spectroscopy of highly-stripped ions [215,225]. The advantages of the electron ring in this respect result from the strong electric field of the ring that holds the ions together in a relatively small volume where they nearly continuously interact with the electrons. The electric field, furthermore, ejects the low-energy electrons (being produced by the ionization process) from the ring so that there is practically no recombination and high ionization states can be reached. First measurements in an electron ring as a spectroscopic source for ion study were reported by SCHMIEDER [225]. The ionization of heavy atoms by electron impact takes, however, a very long time since the ionization cross-sections at MeV energies of the electrons are very small [195,197]. The possibility of injecting electron beams of lower energy parallel to the magnetic field lines through the electron ring, for which the ionization cross-section is more than two orders of magnitude higher, has not yet been exploited. A conceptual compressor design study for intense electron rings specifically devoted to the spectroscopy of highly stripped ions is the subject of a paper by the Berkeley ERA group [280].

The unloaded electron ring, if accelerated to high axial speed, might serve to study short-pulse chemistry [262]. At a speed of $\beta_z \approx 0.2$ an electron ring with an axial extent of 2b = 1 cm will deposit an electron number of about $N_e = 5 \cdot 10^{12}$ (i.e. a charge of about $8 \cdot 10^{-6}$ Cb) in about $1.6 \cdot 10^{-10}$ s.

The main field of application of the electron ring accelerator will certainly be defined by the applications of the collectively accelerated ions. The applications of accelerated ions in nuclear physics, technology and medicine are so numerous that they cannot be presented in detail here. Their use in atomic physics (inner shell excitation by fast ions, production of highly stripped ions and X-ray measurements) and in nuclear physics (nuclear excitation, nuclear reaction mechanisms, production of nuclides and new elements) have to consider the characteristics of the electron ring accelerator in view of the very low duty cycle and the relatively large energy spread of the ions. While the large energy spread imposes restrictions on the choice of the nuclear reactions to be investigated, the low duty cycle offers the possibility of applying special diagnostics, such as the time-of-flight method, because of the short-pulse high-flux ion pulses from the ERA.

Besides the potential applicability of electron ring accelerators, especially at ion energies of the order of a few times ten MeV (in principle, all types of ions can be accelerated collectively), other important uses of collectively accelerated ions (of even lower energies) are in the field of solid-state and environmental physics [3,4], mainly for investigating chemical and structural changes in materials. These include methods of analyzing the material by ions backscattered from the surface and providing information on the elemental composition of the surface, ion-excited X-ray analysis of environmental samples, charged particle activation analysis and material analysis by means of nuclear reactions.

Finally, relativistic electron rings and beams have a strong bearing on the field of fusion research. Some devices such as electron or proton layers, which are investigated as one approach to a fusion reactor, are very similar to relativistic electron rings such as used for collective ion acceleration. For fusion devices, however, very high currents of tens of amperes of proton or deuteron beams with very high power are injected into magnetic mirror systems to heat plasmas and produce fusion reactions. Recently, another approach, known as "inertial confinement fusion", requires extremely high-power electron or ion beams in order to heat and compress pellets containing deuterium and tritium and thus to produce a net power yield from fusion reactions. Although it is obvious that electron ring accelerators will not be capable of producing the desired high ion beam power for this application, several questions such as beam transport, beam acceleration and collective instabilities might be answered with the support of electron ring research.

It is hoped, however, that the main potential of electron rings to serve as powerful vehicles for collective ion acceleration will soon be demonstrated by the new electron ring experiments to allow development of a compact and powerful collective ion accelerator using electron rings.

Acknowledgements

The author wishes to express his gratitude to Professor A. Schlüter, who stimulated him to write this survey. He should like to thank him and the members of the Garching ERA group, especially Drs. C. Andelfinger, W. Dommaschk, W. Herrmann, I. Hofmann, P. Merkel and M. Ulrich for numerous discussions and excellent collaboration in the course of the experiments. He is indebted to K. Klug for carefully typing the manuscript.

List of Important Symbols

$\underline{A}$, A	Vector potential
A_Θ	Azimuthal vector potential component
$\tilde{A}_r$, $\tilde{A}_z$	Radial and axial amplitude
a	Small radial ring half axis
a_β	Radial betatron amplitude
$\underline{B}$, B	Magnetic field strength
B_c	Magnetic field strength at compressed ring
B_r, B_Θ, B_z	Magnetic field strength components in cylindrical coordinates
b	Small axial ring half axis
b_o, b_1	Magnetic field ratio abbreviations
b_β	Axial betatron amplitude
c	Speed of light, $c = (2.99792458 \pm 1.2 \cdot 10^{-8}) \cdot 10^{10}$ cm sec^{-1}
$\underline{E}$	Electric field strength
E_f	Final energy
E_H	Holding power
E_i	Ion energy
E_{in}	Initial energy
E_m, E_{max}	Maximum electric field strength
E_r, E_Θ, E_z	Electric field strength components in cylindrical coordinates
E_t	Total ion energy, $E_t = E_i + M_i c^2$
$\Delta E/E$	Relative electron energy spread
e	Elementary charge, $e = (1.602189 \pm 5 \cdot 10^{-6}) \cdot 10^{-19}$ A sec
F_r	Radial force component
f	Ion loading fraction, $f = ZN_i/N_e$
f_o	Electron revolution frequency
G	Growth
G_i, g	Abbreviations
h	Distance of the beam from a wall
I	Current
I_e	Electron current
k	Boltzmann constant, $k = (1.38066 \pm 5 \cdot 10^{-5}) \cdot 10^{-3}$ JK^{-1}
$\tilde{k}$, $\tilde{\ell}$, $\tilde{m}$	Integers
M	Mode number
M_i	Ion mass
m	Electron mass, $m = \gamma m_e$
$\tilde{m}$	Mode number
m_e	Electron rest mass, $m_e = (9.10995 \pm 5 \cdot 10^{-5}) \cdot 10^{-28}$ g
N_e	Total electron number
$\tilde{N}_e$	Electron line density

N_i	Total ion number
$\tilde{N}_i$	Ion line density
n	Magnetic field index
n_b	Beam density
n_e	Electron density
P	Radiation power
p	Momentum
$p_\perp$	Transverse electron momentum
p_Θ	Canonical angular momentum
$p_{\Theta i}$	Initial value of the canonical angular momentum
Q_1, Q_i	Coupling coefficients between electrons and ions
Q_s^2	Non-external axial focussing, $Q_s^2 = \nu_z^2 - n$
R	Major ring radius
R_s	Surface resistance
r	Radius
r_0	Classical electron radius, $r_0 = \mu_0 e^2/4\pi m_e = (2.817938 \pm 7\cdot10^{-6})\cdot10^{-13}$cm
S_E, S_M	Radius of electric or magnetic images divided by R, respectively
S_k	Coherent oscillation frequency, $S_k = (1 - \nu_r)\omega_{ce}$
T_e	Electron temperature
T_i	Ion temperature
t	Time
$\tilde{U}$	Electron beam potential with respect to the cathode
U, V	Parameters describing the electromagnetic fields related to the collective instabilities
$\tilde{V}$	Potential
v	Particle velocity
v_r, v_ϕ, v_z	Velocity components in cylindrical coordinates
x	Radial deviation from major ring radius, $x = r - R$
Z	Ion charge
Z_0	Impedance of free space, $Z_0 = (\mu_0/\varepsilon_0)^{1/2} = (\varepsilon_0 c)^{-1} = \mu_0 c = 376.732\ \Omega$
$Z_{\tilde{m}}$	Coupling impedance ($\tilde{m}$-th mode)
z	Axial position
α	Ratio of azimuthal to axial magnetic field component
α_c	Radiation loss parameter
α_1, α_2	Parameters describing the holding power function
$\underline{\beta}$, β, β_e	Electron speed relative to the speed of light c
β_t, $\beta_\perp$	Electron speed transverse to B divided by c
$\beta_\parallel$	Electron speed parallel to B divided by c
γ	Relativistic factor
γ_c	Relativistic factor of the electrons in the compressed state
$\gamma_\parallel$, $\gamma_\perp$	Relativistic factor with respect to the magnetic field direction

ε	External electric field strength
ε_0	Dielectric constant (permittivity), $\varepsilon_0 = (\mu_0 c^2)^{-1} = (8.85418782 \pm 7 \cdot 10^{-8}) \cdot 10^{-14}$ Asec $\cdot$ (Vcm)$^{-1}$
$\varepsilon_{1,E}$; $\varepsilon_{1,M}$	Image field coefficients, electrostatic and magnetostatic, respectively
η	Ratio of the holding power to the maximum electric field strength in the electron ring
$\tilde{\eta}$	Chromaticity
Θ	Azimuth
$\dot{\Theta}$	Angular velocity
λ	Wave length
λ_c	Critical wave length
λ_1	Abbreviation
μ	Abbreviation, $\mu = \nu_{Budker}/\gamma$
μ_0	Permeability, $\mu_0 = 4\pi \cdot 10^{-9}$ H/cm = 1.25663706144 $\cdot 10^{-8}$ Vsec $\cdot$ (Acm)$^{-1}$
ν	Betatron tune
ν_{Budker}	Budker parameter
ν_r, ν_z	Radial and axial betatron tune
ξ	Deviation from equilibrium axial position
σ_r, σ_z	Radial and axial small ring dimensions, standard deviations
Φ	Magnetic flux
ω	Angular frequency
ω_{ce}	Electron gyrofrequency

The numerical values of the physical constants are from [278].

References

1 G.N. Flerov, V.S. Barasenkov: Wissenschaft u. Fortschritt 25, 10 (1975)
K. Treml: Phys.uns.Zeit 6, 169 (1976)

2 J.A. Martin: *Cyclotrons and Their Applications*, Birkhäuser Verlag, Basel, (1975) p.574

3 J.F. Ziegler (ed.): *New Uses of Ion Accelerators*, Plenum Press, New York, London (1975)

4 GSI-Arbeitstagung, Gesellschaft für Schwerionenforschung, Report GSI-P2, Darmstadt, Nov.1975

5 A. Schempp, P. Feigl. H. Klein: GSI-Report PB-4-75, Darmstadt (1975)

6 B. Piosczyk, G. Hochschild, J.E. Vetter, E. Jaeschke, R. Repnow, Th. Walcher: IEEE Trans.Nucl.Sci. NS-22, 1172 (1975)

7 W.H. Bennett: Phys.Rev.45, 890 (1934)

8 W.H. Bennett: Phys.Rev.98, 1584 (1955)

9 J.D. Lawson: J.Electron.Control 3, 587 (1957)

10 J.D. Lawson: J.Electron.Control 5, 146 (1958)

11 J.D. Lawson: J.Nucl.Energy, Part C: Plasma Phys.1, 31 (1959)

12 H. Alfvén, O. Wernholm: Arkiv Fysik 5, 175 (1952)

13 R.B.R. Harvie: United Kingdom Atomic Energy Research Establishment Internal Memorandum No.AERE GM/87, March 1951 (unpublished)

14 J.D. Lawson: Part.Accelerators 1, 41 (1970)

15 G.I. Budker: Proc. CERN Symposium on High-Energy Accelerators and Pion Physics, Geneva, Vol.1, (1956) p.68; Atomnaya Energiya 1, 9 (1956), English translation: Sov.Atomic Energy 1, 673 (1956)

16 Ya.B. Fainberg: Proc.CERN Symposium on High-Energy Accelerators and Pion Physics, Geneva, Vol.1, (1956) p.84; Atomnaya Energiya 6, 431 (1959)
17 Ya.B. Fainberg: Part.Accelerators 6, 95 (1975)
18 V.I. Veksler: Proc.CERN Symposium on High-Energy Accelerators and Pion Physics, Geneva, Vol.1, (1956) p.80
19 I.E. Tamm: J.Phys.1, 439 (1939)
20 V.I. Veksler: Atomnaya Energiya 2, 427 (1957), Engl.transl.: Sov.Atomic Energy 2, 525 (1957)
21 V.I. Veksler, V.P. Sarantsev, A.G. Bonch-Osmolovskii, G.I. Dolbilov, G.A. Ivanov, I.N. Ivanov, M.L. Iovnovich, I.V. Koshunov, A.B. Kuznetsov, V.G. Mahankov, E.A. Perelshtein, V.P. Rashevskii, K.A. Reshetnikova, N.B. Rubin, S.B. Rubin, P.I. Ryltsev, O.I. Yarkovoi: Atomnaya Energiya 24, 317 (1968); Proc.6th Int. Conf.on High-Energy Accelerators, Cambridge, Mass., USA, (1967) (Cambridge Electron Accelerator Report No.CEAL-2000, 1967), p.289; Report JINR-P-3440-2, Dubna (USSR), 1967
22 A.A. Kolomenskii: Part.Accelerators 5, 73 (1973)
23 A.A. Plyutto: Sov.Phys.JETP 12, 1106 (1961)
24 A.A. Plyutto, P.E. Belensov, E.D. Korop, G.P. Mcheidze, V.N. Ruchkov, K.V. Suladze, S.M. Temchin: JETP Lett.6, 61 (1967)
25 A.A. Plyutto, K.V. Suladze, S.M. Temchin, E.D. Korop: Atomnaya Energiya 27, 418 (1969)
26 S.E. Graybill, J.R. Uglum: J.Appl.Phys.41, 236 (1970)
27 S.E. Graybill, W.H. McNeill, J.R. Uglum: Rec.11th Symp.on Electron, Ion and Laser Beam Technology (San Francisco Press, Calif., 1971) p.577
28 J. Rander, B. Ecker, G. Yonas, D. Drickey: Phys.Rev.Lett.24, 283 (1970)
29 C.L. Olson: Part.Accelerators 6, 107 (1975)
30 G. Yonas: Part.Accelerators 5, 81 (1973)
31 N. Rostoker: Proc.7th Int.Conf.on High-Energy Accelerators, Yerevan (USSR), (1969) p.509
32 J.R. Uglum, S.E. Graybill, W.H. McNeill: Bull.Am.Phys.Soc.14, 1047 (1969)
33 S.D. Putnam: Phys.Rev.Lett.25, 1129 (1970)
34 S.D. Putnam: IEEE Trans.Nucl.Sci.NS-18, 496 (1971)
35 J. Wachtel, D. Eastlund: Bull.Amer.Phys.Soc.14, 1047 (1969)
36 C.L. Olson: Proc.9th Int.Conf.on High-Energy Accelerators, Stanford (USA), (1974) p.272
37 Proc.Symp.on Electron Ring Accelerators, Lawrence Radiation Laboratory, Berkeley Cal., USA, Report UCRL-18103 (1968)
38 A.M. Sessler: Report UCRL-19242, Lawrence Radiation Laboratory, Berkeley, Cal., USA; Proc.7th Int.Conf.on High-Energy Accelerators, Yerevan (USSR), (1969) p.431
39 H. Schopper: Phys.Bl.24, 201 and 255 (1968)
40 D. Keefe: IEEE Trans.Nucl.Sci.NS-16, 25 (1969); Part.Accelerators 1, 1 (1970); Sci.Am.226, 22 (1972)
41 L.J. Laslett, A.M. Sessler: Comments Nuclear Particle Phys.3, 93 (1969); 4, 211 (1970)
42 J.M. Peterson: Report LBL-373 Preprint (1971); LBL-704 Preprint, Lawrence Berkel Laboratory, Berkeley, Cal., USA, (1972)
43 J.D. Lawson: Part.Accelerators 3, 21 (1972)
M.P. Reiser: NTvNat.38, 219 (1972)
44 U. Schumacher: Report IPP 0/20, MPI für Plasmaphysik, Garching, (1973)
45 N.C. Christofilos: Technical Note UCID-15378, Lawrence Radiation Laboratory, Livermore, Cal., USA, (1968); Phys.Rev.Lett.22, 830 (1969); IEEE Trans.Nucl.Sci. NS-16, 1039 (1969)
46 L.J. Laslett, A.M. Sessler: IEEE Trans.Nucl.Sci.NS-16, 1034 (1969)
47 J.G. Kalnins, H. Kim, D.L. Nelson: IEEE Trans.Nucl.Sci.NS-18, 473 (1971)
48 M.P. Reiser: IEEE Trans.Nucl.Sci.NS-18, 460 (1971)
49 M.P. Reiser: IEEE Trans.Nucl.Sci.NS-19, 280 (1972)
50 C. Bovet: Report ERAN-88, Lawrence Berkeley Laboratory, Berkeley, Cal., USA, (1970)
51 P. Merkel: Report IPP 0/18, MPI für Plasmaphysik, Garching, (1973)
52 I. Hofmann: Report IPP 0/21, MPI für Plasmaphysik, Garching, (1974); Proc.IXth Int.Conf.on High-Energy Accelerators, Stanford, USA, 1974, p.245

53 J.R. Pierce: *The Theory and Design of Electron Beams* (Van Nostrand, New York, 1949)
54 H. Alfvén: Phys.Rev.55, 425 (1939)
55 E.R. Harrison: J.Electron.Control 4, 193 (1958)
56 W. Kegel, P. Merkel: unpublished report, MPI für Plasmaphysik, Garching (1968)
W. Kegel: Plasma Phys.12, 105 (1970)
57 E. Keil: Report CERN-ISR-TH/68-49, CERN, Geneva, (1968)
58 R.F. Koontz, G.A. Loew, R.H. Miller: Proc.8th Int.Conf.on High-Energy Accelerators, CERN, Geneva, 1971, p.491
59 W.B. Lewis: Ref.[37] 195 (1968)
60 D.L. Nelson, H. Kim: Proc.VIIth Int.Conf.on High-Energy Accelerators, Yerevan, USSR, Vol.II, (1969), p.540
61 H. Bruck: *Accélérateurs Circulaires de Particules* (Bibliothèque des Sciences et Techniques Nucléaires, Paris 1966)
62 R.E. Berg, H. Kim, M.P. Reiser, G.T. Zorn: Phys.Rev.Lett.22, 419 (1969)
63 G.E. Dombrowski: Report ERAN-24, Lawrence Radiation Laboratory, Berkeley, Cal., USA, (1968)
64 G. Brautti, T. Clauser, M. Leo: Proc.4th Work Meeting on Electron Ring Accelerators, Report IPP 0/3, 26, MPI für Plasmaphysik, Garching (1971)
65 I. Boscolo, R. Coisson, M. Leo, A. Luches, S. Mongelli, G. Brautti, T. Clauser, A. Rainò: IEEE Trans.Nucl.Sci.NS-19, Number 2, 287 (1972)
66 I. Boscolo, R. Coisson, M. Leo, A. Luches, G. Brautti, A. Rainò: Proc.2nd Int. Conf.on Ion Sources, Vienna, (1972) p.654
67 G. Brautti, I. Boscolo, R. Coisson, M. Leo, A. Luches, A. Tepore: IEEE Trans. Nucl.Sci.NS-20, Number 3, 286 (1973)
68 C. Bovet, C. Pellegrini: Report UCRL-19892 Preprint, Lawrence Radiation Laboratory, Berkeley, Cal., USA, (1970); Part.Accelerators 2, 45 (1971)
69 L.J. Laslett, W.A. Perkins: Report ERAN-118 and UCRL-20143, Lawrence Radiation Laboratory, Berkeley, Cal., USA; Nucl.Instr.and Methods 97, 523 (1971)
70 C. Pellegrini, A.M. Sessler: Nucl.Instr.and Methods 84, 109 (1970)
71 S. van der Meer: Preprint ISR-PO-169-57, CERN, Geneva (1969)
72 E.A. Perelshtein, M.S. Perskii, V.F. Shevtsov: Report JINR-P9-6600, Dubna, USSR, (1972)
73 D. Keefe, G.R. Lambertson, L.J. Laslett, W.A. Perkins, J.M. Peterson, A.M. Sessler, R.W. Allison, Jr., W.W. Chupp, A.U. Luccio, J.B. Rechen: Phys.Rev.Lett.22, 558 (1969)
74 C. Andelfinger, W. Herrmann, A.U. Luccio, M. Ulrich: Report IPP 0/8, MPI für Plasmaphysik, (1971)
75 W. Walkinshaw: *A Spiral Ridged Bevatron*, A.E.R.E. Report Harwell (UK) (1956), unpublished
76 A.A. Garren, D.L. Judd, L. Smith, H.A. Willax: Nucl.Instr.and Methods 18, 19, 525 (1962)
77 L.J. Laslett: Report ERAN-71, Lawrence Radiation Laboratory, Berkeley, Cal., USA, (1970)
78 W.A. Perkins: Report ERAN-141, Lawrence Radiation Laboratory, Berkeley, Cal., USA, (1971)
79 L.J. Laslett: Report LBL-1709, Lawrence Berkeley Laboratory, Berkeley, Cal., USA, (1973); IEEE Trans.Nucl.Sci.NS-20, Number 3, 271 (1973)
80 K.R. Symon: Ref.[37], p.304 (1968)
81 I. Hofmann: unpublished material
82 C. Andelfinger, E. Buchelt, W. Dommaschk, J. Fink, W. Herrmann, I. Hofmann, D. Jacobi, P. Merkel, A. Schlüter, H.B. Schilling, U. Schumacher, M. Ulrich: Report IPP 0/30, MPI für Plasmaphysik, Garching, (1976)
83 L.J. Laslett: Proc.1963 Summer Study on Storage Rings, Accelerators and Experimentation at Super High Energies, Brookhaven National Laboratory, Upton, N.Y., BNL-7534 (1963) p.324
84 L.J. Laslett: Report ERAN-30, Lawrence Radiation Laboratory, Berkeley, Cal., USA, (1969)
85 L.J. Laslett: Report ERAN-200, Lawrence Berkeley Laboratory, Berkeley, Cal., USA, (1972)
86 I.N. Ivanov, M.L. Iovnovich, A.B. Kuznetsov, Yu.L. Obukhov, K.A. Reshetnikova, N.B. Rubin, V.P. Sarantsev, O.I. Yarkovoy: JINR Report P9-4132, Dubna, USSR, (1968)

87 E.D. Courant: Ann.Rev.Nucl.Sci.18, 435 (1968)
88 W.A. Perkins: Report ERAN-32, Lawrence Radiation Laboratory, Berkeley, Cal., USA, (1969)
89 I. Hofmann: Report IPP 0/16, MPI für Plasmaphysik, Garching, (1973)
90 D. Möhl, L.J. Laslett, A.M. Sessler: Report LBL-1062, Lawrence Berkeley Laboratory, Berkeley, Cal., USA, presented at the Symposium on Collective Methods of Acceleration, Dubna (USSR), (Sept.1972); Part.Accelerators 4, 159 (1973)
91 L.J. Laslett: Report ERAN-184, Lawrence Berkeley Laboratory, Berkeley, Cal., USA, (1972)
92 A.G. Bonch-Osmolovskii, G.V. Dolbilow, I.N. Ivanov, E.A. Perelshtein, V.P. Sarantsev, O.I. Yarkovoy: JINR-Report P9-4135, Dubna (USSR), (1968)
93 G.V. Dolbilov, I.N. Ivanov, E.A. Perelshtein, V.P. Sarantsev, V.F. Shertsov: JINR-Report P9-4737, Dubna (USSR), (1969)
94 C. Pellegrini, A.M. Sessler: Report ERAN-45, Lawrence Radiation Laboratory, Berkeley, Cal., USA, (1969); Nucl.Instr.and Methods 86, 273 (1970)
95 U. Schumacher: Report IPP 0/10, MPI für Plasmaphysik, Garching, (1972); Symp. on Collective Methods of Acceleration, Dubna (USSR), (Sept.1972)
96 C. Andelfinger, W. Herrmann, D. Jacobi, A.U. Luccio, W. Ott, U. Schumacher, M. Ulrich: IEEE Trans.Nucl.Sci.NS-20, Number 3, 276 (1973)
97 U. Schumacher, M. Ulrich: IInd Symp.on Collective Methods of Acceleration, Dubna (USSR), (Sept.1976); Proc.p.38 (1977)
98 C. Andelfinger, W. Dommaschk, I. Hofmann, P. Merkel, U. Schumacher, M. Ulrich: Proc.9th Int.Conf.on High-Energy Accelerators, Stanford (USA), (1974), p.218
99 C.E. Nielsen, A.M. Sessler, K.R. Symon: Int.Conf.on High Energy Accelerators, CERN, Geneva, (1959), p.239
100 A.A. Kolomenskii, A.N. Lebedev: Int.Conf.on High-Energy Accelerators, CERN, Geneva, (1959), p.115
101 J.C. Maxwell: *Scientific Papers*, Cambridge University Press, Vol.7, (1890), p.288
102 V.K. Neil, A.M. Sessler: Rev.Sci.Instr.36, 429 (1965)
R.J. Briggs, V.K. Neil: Plasma Phys.9, 209 (1967)
103 A.G. Ruggiero, V.G. Vaccaro: Report ISR-TH/68-33, CERN, Geneva, (1968)
104 D. Keefe: Proc.VIIIth Int.Conf.on High-Energy Accelerators, CERN, Geneva, (1971), p.397
105 K. Hübner, E. Keil, B. Zotter: Proc.VIIIth Int.Conf.on High-Energy Accelerators, CERN, Geneva, (1971), p.295
106 A.M. Sessler: IEEE Trans.Nucl.Sci.NS-20, Number 3, 854 (1973)
107 A. Faltens, G.R. Lambertson, J.M. Peterson, J.B. Rechen: Proc.IXth Int.Conf.on High-Energy Accelerators, Stanford (USA), (1974), p.226
108 E. Keil, W. Schnell: CERN Internal Report ISR-TH-RF-69-48, CERN, Geneva, (1969)
109 A.M. Sessler: IEEE Trans.Nucl.Sci.NS-18, Number 3, 1039 (1971)
110 J.D. Lawson: Report 76-09, CERN, Geneva, (1976)
111 A.M. Sessler: Int.Summer School of Applied Physics, Course on High-Intensity Relativistic Particle Beams, Erice, Sicely (June 1973), unpublished
112 A.M. Sessler, V.G. Vaccaro: CERN-Report 67-2, CERN, Geneva, (1967)
113 L.D. Landau: J.Phys.(USSR) 10, 25 (1946)
114 I.M. Kapchinskii, P.R. Zenkevich, Report ITEF-80, Inst.of Theoretical and Experimental Physics, Moscow, USSR, (1973)
115 D. Möhl: Report ERAN-178, Lawrence Berkeley Laboratory, Berkeley, Cal., USA, (1971)
116 B. Zotter: CERN-Report ISR-TH/69-35, CERN, Geneva, (1969)
117 A.C. Entis, A.A. Garren, L. Smith: IEEE Trans.Nucl.Sci.NS-18, Number 3, 1092, (1971)
118 A.C. Entis, A.A. Garren, D. Möhl: Report ERAN-174, Lawrence Berkeley Laboratory, Berkeley, Cal., USA, (1972)
A.G. Bonch-Osmolovskii: Report JINR-P9-6318, Dubna (USSR), (1972)
119 A. Faltens, L.J. Laslett: Symp.Coll.Meth.of Acceleration, Dubna (USSR), (Sept. 1972); Report LBL-1070, Berkeley, Cal., USA, (1972); Part.Accelerators 4, 151, (1973)
120 A. Faltens, E.C. Hartwig, D. Möhl, A.M. Sessler: Proc.8th Int.Conf.on Particle Accelerators, CERN, Geneva, (1971), p.338

121 B. Zotter, P. Bramham: CERN-Report ISR-TH/73-6, CERN, Geneva; IEEE Trans.Nucl. Sci.NS-20, No.3, 830 (1973)
122 H. Krauth: Bericht KFK 2105, Ges.für Kernforschung mbH, Karlsruhe, (1975)
123 C. Pellegrini, A.M. Sessler: Report ERAN-203, Lawrence Berkeley Laboratory, Berkeley, Cal., USA, (1973)
124 A.M. Sessler: Report ERAN-207, Lawrence Berkeley Laboratory, Berkeley, Cal., USA, (1973)
125 B.G. Shchinov, A.G. Bonch-Osmolovskii, V.G. Makhankov, V.N. Tsytovich: Plasma Physics 15, 211 (1973)
126 A.G. Bonch-Osmolovskii, E.A. Perelshtein: Izv.Rad.Phys.8, 1081, 1089 (1970); JINR Preprints P9-4424, P9-4425, Dubna/USSR (1969)
127 A.G. Bonch-Osmolovskii, E.A. Perelshtein, V.N. Tsytovich: JINR Preprint P9-4751, Dubna/USSR, (1969)
128 G.R. Lambertson, L.J. Laslett: Report ERAN-157, Lawrence Radiation Laboratory, Berkeley, Cal., USA, (1971)
129 A. Faltens: private communication
130 L.J. Laslett: Report ERAN-126, Lawrence Berkeley Laboratory, Berkeley, Cal., USA, (1971)
131 L.J. Laslett, V.K. Neil, A.M. Sessler: Rev.Sci.Instr.36, 436 (1965)
132 K. Hübner, V.G. Vaccaro: CERN-Report ISR-TH/70-44, CERN, Geneva, (1970)
133 W. Schnell, B. Zotter: CERN-Report ISR-GS-RF/76-26, CERN, Geneva, (1976)
134 K. Hübner, A.H. Ruggiero, V.G. Vaccaro: Proc.7th Int.Conf.on High-Energy Accelerators, Yerevan, USSR, (1969), p.343
135 F.J. Sacherer: Proc.9th Int.Conf.on High-Energy Accelerators, Stanford, USA, (1974), p.347
D. Möhl: Report ERAN-183, Lawrence Berkeley Laboratory, Berkeley, Cal., USA, (1971)
D. Möhl, A.M. Sessler: Proc.VIIIth Int.Conf.on High-Energy Accelerators, CERN, Geneva, (1971), p.334; Report LBL-42, Lawrence Berkeley Laboratory, Berkeley, Cal., USA, (1971)
136 A. Schlüter: private communication
137 B.V. Chirikov: Sov.Atomic Energy 19, 1149 (1965)
138 F.E. Mills: ref.[37], p.448
139 T.K. Fowler: ref.[37], p.457
140 V.I. Balbekov, A.A. Kolomenskii: Sov.Phys.-Techn.Phys.12, 1487 (1968)
141 D. Koshkarev, P. Zenkevich: Proc.VIIth Int.Conf.on High-Energy Accelerators, CERN, Geneva, (1971), p.496; Part.Accelerators 3, 1 (1972)
142 H.A. Grunder, G.R. Lambertson: Proc.VIII Int.Conf.on High-Energy Accelerators, CERN, Geneva, (1971), p.308
143 L.J. Laslett: LBL Internal Report ERAN-181, (1972); LBL Internal Report ERAN-177, Lawrence Berkeley Laboratory, Berkeley, Cal., USA, (1971)
144 D.M. LeVine: IEEE Trans.Nucl.Sci.NS-20, No.3, 327 (1973)
145 L.J. Laslett, A.M. Sessler, D. Möhl: Proc.III All-Union Nat'l.Part.Accelerator Conf., Moscow, USSR (1972); Nucl.Instr.Methods 121, 517 (1974)
146 W. Dommaschk, Report IPP 0/19, MPI für Plasmaphysik, Garching, (1973); Internal Report RPR-N30, unpublished
147 W. Dommaschk, I. Hofmann: to be published
148 W. Ott, W. Dommaschk, E. Springmann: Report IPP 0/23, MPI für Plasmaphysik, Garching, (1974)
149 I. Hofmann: Report IPP 0/35, MPI für Plasmaphysik, Garching, (1977)
150 L.J. Laslett: Report ERAN-51, Lawrence Radiation Laboratory, Berkeley, Cal., USA, (1970)
151 A.G. Bonch-Osmolovskii: JINR-P9-5299, Dubna, USSR, (1970)
152 P. Merkel: 4th Work Meeting on Coll.Ion Accelerators, Garching, (Febr.1971); Report IPP 0/4, MPI für Plasmaphysik, Garching, (1971)
153 L.J. Laslett: Report ERAN-144, Lawrence Radiation Laboratory, Berkeley, Cal., USA, (1971)
154 L.J. Laslett: Report ERAN-145, Lawrence Radiation Laboratory, Berkeley, Cal., USA, (1971)
155 L.J. Laslett, U. Schumacher: Proc.VIIIth Int.Conf.on High-Energy Accelerators, CERN, Geneva, (1971), p.468; Report UCRL-20855, Lawrence Berkeley Laboratory, Berkeley, Cal., USA, (1971)

156 L.J. Laslett, U. Schumacher: Report ERAN-149, (1971); Report ERAN-163, Lawrence Berkeley Laboratory, Berkeley, Cal., USA, (1971)
157 A.G. Bonch-Osmolovskii, K.A. Reshetnikova: JINR-Report P9-6136, Dubna, USSR, (1971)
158 Y.G. Globenko, D.G. Koshkarev: Symp.on Coll.Methods of Acceleration, Dubna, USSR, (1972)
159 A.U. Luccio: *Seminar Talk on Electron-Ion Instability,* Int.School of Applied Physics, Erice, Sicely (1973), unpublished
160 A.G. Bonch-Osmolovskii, V.A. Preizendorf, K.A. Reshetnikova: Report JINR-P9-6930, Dubna, USSR, (1973)
161 I. Hofmann: Internal Report RPR-N34, MPI für Plasmaphysik Garching, (1975) unpublished
162 A.G. Bonch-Osmolovskii, V.I. Danilov: Report JINR-P9-9886, Dubna, USSR, (1976)
163 D. Möhl, A.M. Sessler: Report ERAN-172, Lawrence Berkeley Laboratory, Berkeley, Cal., USA, (1971)
164 H.B. Schilling, E. Buchelt, G. Siller: Report IPP 0/22, MPI für Plasmaphysik, Garching, (1974)
165 A. Faltens: Report ERAN-63, Lawrence Berkeley Laboratory, Berkeley, Cal., USA, (1970)
166 W.T. Link: IEEE Trans.Nucl.Sci.NS-14, 777 (1967)
167 S.E. Graybill, S.V. Nablo: IEEE Trans.Nucl.Sci.NS-14, 782 (1967)
168 P. Champney, P. Spence: IEEE Trans.Nucl.Sci.NS-22, No.3, 970 (1975)
169 K.R. Prestwich: IEEE Trans.Nucl.Sci.NS-22, No.3, 975 (1975)
170 F.M. Charbonnier, J.P. Barbour, J.L. Brewster, W.P. Dyke, F.J. Grundhauser: IEEE Trans.Nucl.Sci.NS-14, 789 (1967)
171 W. Dommaschk: Part.Accelerators 5, 171 (1973)
172 C. Andelfinger, W. Ott: Report IPP 0/13, MPI für Plasmaphysik, Garching, (1972)
173 C. Andelfinger: Part.Accelerators 5, 105 (1973)
174 N.C. Christofilos, R.E. Hester, W.A.S. Lamb, D.D. Reagan, W.A. Sherwood, R.E. Wright: Proc.Int.Conf. on High-Energy Accelerators, Dubna, USSR, (1963), p.1073
175 J.W. Beal, N.C. Christofilos, R.E. Hester: IEEE Trans.Nucls.Sci.NS-16, No.3, 294 (1969)
176 R.T. Avery, G. Behrsing, W.W. Chupp, A. Faltens, E.C. Hartwig, H.P. Hernandez, C. MacDonald, J.R. Meneghetti, R.G. Nemetz, W. Popenuck, W. Salsig, D. Vanacek: IEEE Trans.Nucl.Sci.NS-18, 479 (1971)
177 V.D. Gitt, A.D. Kovalenko, P.I. Ryltsev, V.P. Sarantsev: JINR-Preprint P9-5601, Dubna, USSR, (1971)
178 R.E. Berg, H. Kim, M. Reiser: IEEE Trans.Nucl.Sci.NS-16, 1043 (1969)
179 M. Reiser: IEEE Trans.Nucl.Sci.NS-20, 310 (1973)
180 M. Friedman: Phys.Rev.Lett.24, 1098 (1970)
181 M.J. Rhee, G.T. Zorn, R.C. Placious, J.H. Sparrow: IEEE Trans.Nucl.Sci.NS-18, 468 (1971)
182 M.J. Rhee, W.W. Destler: Phys.Fluids 17, 1574 (1974)
183 W.W. Destler, P.K. Misra, M.J. Rhee: Phys.Fluids 18, 1820 (1975)
184 L.S. Barabash, I.A. Golutvin, G.V. Dolbilov, I.N. Ivanov, A.D. Kovalenko, V.G. Novikov, N.B. Rubin, E.A. Perelshtein, V.P. Sarantsev, V.A. Sviridov: Proc. IXth Int.Conf.on High-Energy Accelerators, Stanford, USA, (1974), p.318
185 P. Merkel: Report IPP 0/5, MPI für Plasmaphysik, Garching, (1971)
186 A.U. Luccio, U. Schumacher, E. Springmann: Report IPP 0/11, MPI für Plasmaphysik Garching, (1972)
187 A.U. Luccio, W. Herrmann: unpublished material
188 C. Bovet: Report ERAN-87, Lawrence Berkeley Laboratory, Berkeley, Cal., USA, (1970)
189 H. Toyama, H. Ishizuka: Report IPPJ-168, Inst.of Plasma Physics, Nagoya University, Nagoya, Japan, (1973); Part.Accelerators 5, 237 (1973)
190 H. Krauth: 4th Meeting on ERA, Garching, Febr.1971, Internal Report IPP 0/3, (1971), p.86
191 V. George, M.L. Iovnovich, V.G. Novikov, V.A. Preizendorf, N.B. Rubin, V.P. Sarantsev: Preprint JINR-P9-6555, Dubna, USSR, (1972)
192 R.H. Levy: ref.[37], p.318
193 G.S. Janes: Report ERAN-17, Lawrence Berkeley Laboratory, Berkeley, Cal., USA, (1968)

194 G.S. Janes, R.H. Levy, H.A. Bethe, B.T. Feld: Phys.Rev.145, 925 (1966)
195 A. Salop: Reports LBL-2047 (1973) and LBL-2440, Lawrence Berkeley Laboratory, Berkeley, Cal., USA (1973); Phys.Rev.A 8, 3032 (1973); Phys.Rev.A 9, 2496 (1974)
196 L.J. Laslett, B.S. Levine: Report ERAN-202, Lawrence Berkeley Laboratory, Berkeley, Cal., USA, (1972)
197 H.U. Siebert, D. Lehmann, G. Musiol, G. Zschornack: JINR-P9-10197, Dubna, USSR, (1976)
198 F.F. Rieke, W. Prepejchal: Phys.Rev.A 6, 1507 (1972)
199 G. Siller, H.B. Schilling, E. Buchelt: Report IPP 0/27, MPI für Plasmaphysik, Garching, (1975)
200 R.T. Avery, A. Faltens, E.C. Hartwig: IEEE Trans.Nucl.Sci.NS-18, No.3, 920 (1971)
201 J.B. Rechen, U. Schumacher: Report ERAN-152, Lawrence Berkeley Laboratory, Berkeley, Cal., USA, (1971)
202 G.R. Lambertson, D. Keefe, L.J. Laslett, W.A. Perkins, J.M. Peterson, J.B. Rechen: IEEE Trans.Nucl.Sci.NS-18, 501 (1971)
203 C. Andelfinger, W. Herrmann, M. Ulrich: Report IPP 0/9, MPI für Plasmaphysik, Garching, (1971)
204 J. Habanec, H. Guratzsch: JINR-Preprint P9-9141, Dubna, USSR, (1975)
205 U. Schumacher, M. Ulrich: Report IPP 0/32, MPI für Plasmaphysik, Garching, (1976)
206 U. Schumacher, C. Andelfinger, M. Ulrich: Phys.Letters 51A, 367 (1975)
207 U. Schumacher, C. Andelfinger, M. Ulrich: IEEE Trans.Nucl.Sci.NS-22, No.3, 989 (1975)
208 J. Schwinger: Phys.Rev.75, 1912 (1949)
209 R. Haensel, C. Kunz: Z.Angew.Physik 23, 276 (1967); DESY-Report 67/15, DESY, W.Germany (1967)
210 R.P. Godwin: Springer Tracts in Modern Physics 51 (1969)
211 D.H. Tomboulian, P.L. Hartmann: Phys.Rev.102, 1423 (1956)
212 L.J. Laslett: Report ERAN-142, Lawrence Berkeley Laboratory, Berkeley, Cal., USA, (1971)
213 J.W. Shearer, D.A. Nowak, E. Garelis, W.C. Condit: Proc.Int.Top.Conf.on Electron Beam Research and Technology, Albuquerque, USA, Vol.II, (1976), p.78
214 C. Andelfinger, W. Herrmann, A. Schlüter, U. Schumacher, M. Ulrich: IEEE Trans. Nucl.Sci.NS-18, 505 (1971)
215 W.W. Chupp, A. Faltens, E.C. Hartwig, D. Keefe, G.R. Lambertson, L.J. Laslett, W. Ott, J.M. Peterson, J.B. Rechen, A. Salop, R.W. Schmieder: Proc.IXth Int. Conf.on High-Energy Accelerators, Stanford, USA, 235 (1974)
216 V.P. Sarantsev, V.S. Alexandrov, L.N. Belyaev, L.S. Barabash, G.B. Dolbilov, A.K. Krasnych, V.I. Mironov, V.G. Novikov, G. Radonov, A.P. Sumbaev, S.I. Tutunikov, V.P. Phartushnyi, A.A. Phateev, A.S. Shtsheulin: JINR-Reports P9-10053 and P9-10054, Dubna, USSR; Proc.2nd Symp.Coll.Meth.of Acceleration, Dubna, USSR, Sept.1976, (1977), p.13
217 J.M. Peterson, J.B. Rechen: Report LBL-1385, Lawrence Berkeley Laboratory, Berkeley, Cal., USA, (1973); IEEE Trans.Nucl.Sci.NS-20, 790 (1973)
218 G.R. Lambertson, W.W. Chupp, A. Faltens, E.C. Hartwig, W.J. Herrmann, D. Keefe, L.J. Laslett, J.M. Peterson, J.B. Rechen, A. Salop: Part.Accelerators 5, 113 (1973)
219 U. Schumacher: Report 0/36, MPI für Plasmaphysik, Garching, (1977)
220 G.D. Ward: *Thesis*, Technical Report No.71-027, University of Maryland, USA, (1971)
221 G.D. Tsakiris, D.A. Hammer, R.C. Davidson: Bull.Amer.Phys.Soc.20, 1343 (1975)
222 H. Röhr, U. Schumacher, M. Ulrich: unpublished material
223 M.L. Iovnovich, V.P. Sarantsev, M.M. Fiks: Report JINR-P9-4850, Dubna, USSR, (1970)
224 H.U. Siebert, D. Lehmann, G. Musiol, G. Tschornak: JINR-Report P9-9366, Dubna, USSR, (1975)
225 R.W. Schmieder: Report LBL-2476, Lawrence Berkeley Laboratory, Berkeley, Cal., USA, (1973); ERAN-221, (1973); Phys.Lett.47A, 415 (1974)
226 H.G. Paretzke: Report GSF-Bericht S138, Ges.für Strahlen- und Umweltforschung, München, (1971)
227 R. Beaujean, W. Enge: Z.Physik 256, 416 (1972)

228 F.C. Young, J. Golden, C.A. Kapetanakos: NRL Memorandum Report 3391, Washington, D.C., USA, (1976); Rev.Sci.Instrum.48, 432, (1977)
229 V.P. Sarantsev, V.P. Rashevskij, A.K. Kaminskij, V.I. Mironov, V.P. Fartushnyj, A.P. Sergeev, V.G. Novikov, S.I. Tyutyunnikov, A.M. Kaminskaya: Report JINR-P9-5558, Dubna, USSR, (1971); JETP 60, 1980 (1971)
230 W.W. Destler, D.W. Hudgings, M.J. Rhee: to be published in IEEE Trans.Nucl.Sci. NS-24, No.3 (1977)
231 W.W. Destler, D.W. Hudgings, H. Kim, M. Reiser, M.J. Rhee, C.D. Striffler, G.T. Zorn: to be published in IEEE Trans.Nucl.Sci.NS-24, No.3 (1977)
232 H.S. Uhm: *Thesis*, Technical Report 77-002, University of Maryland, USA, (1976), H.S. Uhm, R.C. Davidson: to be published in Phys.Fluids (1977)
233 W.W. Destler, A. Greenwald, D.W. Hudgings, H. Kim, P.K. Misra, M.P. Reiser, M.J. Rhee, G.T. Zorn: Proc.IXth Int.Conf.on High-Energy Accelerators, Stanford, USA, (1974), p.230
234 J.G. Kalnins, H. Kim, J.G. Linhart: IEEE Trans.Nucl.Sci.NS-20, No.3, 324 (1973)
235 P. Merkel: Report IPP 0/24, MPI für Plasmaphysik, Garching, (1974); Part.Accelerators 7, 67 (1976)
236 W. Herrmann: Report IPP 0/25, MPI für Plasmaphysik, Garching, (1974); Part. Accelerators 7, 19 (1975)
237 P. Merkel: Technical Report No.76-076 PP-76-133, University of Maryland, USA, (1975); Part.Accelerators 8, 21 (1977)
238 U. Schumacher, I. Hofmann, P. Merkel, M. Reiser: Report IPP 0/31, MPI für Plasmaphysik, Garching, (1976); Part.Accelerators 7, 245 (1976)
239 R.C. Davidson, J.D. Lawson: Part.Accelerators 4, 1 (1972)
240 R.C. Davidson, S. Mahajan: Part.Accelerators 4, 53 (1972)
241 R.C. Davidson, C.D. Striffler: J.Plasma Phys.12, 353 (1974)
242 R.C. Davidson: *Theory of Nonneutral Plasmas*, W.A. Benjamin, Inc., Reading, Mass., USA, (1974)
243 J.W. Poukey, J.R. Freeman, M. Reiser: Part.Accelerators 6, 245 (1975)
244 V.P. Sarantsev: IEEE Trans.Nucl.Sci.NS-16, 15 (1969)
245 V.P. Sarantsev: Part.Accelerators 1, 145 (1970)
246 V.P. Sarantsev: Proc.8th Int.Conf.on High-Energy Accelerators, CERN, Geneva, (1971), p.391
247 V.P. Sarantsev: Symp.on Coll.Methods of Acceleration, Dubna, USSR (1972)
248 V.P. Sarantsev: Proc.IV All-Union National Accelerator Conf., Moscow, USSR, (1975), p.63
249 N.G. Anishchenko, A.S. Alekseev, N.I. Balalykin: JINR-P9-4722, Dubna, USSR, (1969)
250 V.K. Plotnikov: Int.Symp.on Coll.Meth.of Accleration, Dubna, USSR (1972)
251 I.V. Chuvilo, I.M. Kapchinsky, V.K. Plotnikov, R.M. Vengrov: Proc.IXth Int. Conf.on High-Energy Accelerators, Stanford, USA, (1974), p.314
252 R.M. Vengrov, A.A. Drozdovskii, I.M. Kapchinskii, B.V. Kurakin, N.Ya. Popova, V.K. Plotnikov, I.V. Chuvilo: Proc.IV All-Union National Accelerator Conf., Moscow, USSR, Vol.I, (1975), p.91
253 V.P. Grigoryev, A.N. Didenko, Yu.P. Usov, G.P. Fomenko, E.G. Furman, V.V. Tsygankov, V.L. Chakhlov, Yu.G. Yushkov: Proc.Int.Top.Conf.on Electron Beam Res. and Technology, Albuquerque, N.M., USA, Vol.II, (1976), p.152
254 S. Kawasaki, A. Miyahara, K. Huke, H. Ishizuka, G. Horikoshi, H.M. Saad, Y. Kubota: IEEE Trans.Nucl.Sci.NS-20, No.3, 280 (1973)
255 S. Kawasaki, N. Kobayashi, Y. Kubota, A. Miyahara: IEEE Trans.Nucl.Sci.NS-22, No.3, 992 (1975)
256 D. Keefe: Proc.7th Int.Conf.on Particle Accelerators, Yerevan, USSR, (1969), p.447
257 D. Keefe, W.W. Chupp, A.A. Garren, G.R. Lambertson, L.J. Laslett, A.U. Luccio, W.A. Perkins, J.M. Peterson, J.B. Rechen, A.M. Sessler: Nucl.Instr.Methods 93, 541 (1971)
258 D. Keefe: Proc.IVth All-Union Conf.on Particle Accelerators, Moscow, USSR, Vol.I, (1975), p.109
259 D. Keefe: Report LBL-5536, Lawrence Berkeley Laboratory, Berkeley, Cal., USA, (1976)
260 R.A. Beck, A. Nakach: 4th Meeting on ERA, Garching, Report IPP 0/3, (1971), p.6

261 H. Bruck: Symp.on Coll.Meth.of Acceleration, Dubna, USSR (Sept.1972)
262 H. Bruck: 4th Meeting on ERA, Garching, Report IPP 0/3, (1971), p.105
263 P. Kappe: Report KFK Ext.Ber.3/69/30, Karlsruhe, (1969)
264 W. Zernial: 4th Meeting on ERA, Garching, Report IPP 0/3, (1971), p.1
265 C. Dustmann, W. Heinz, H. Hermann, P. Kappe, H. Krauth, L. Steinbock, W. Zernial: Proc.8th Int.Conf.on High-Energy Accelerators, CERN, Geneva, (1971), p.408
266 C. Dustmann, W. Heinz, H. Krauth, L. Steinbock, W. Zernial: Symp.on Coll.Meth. of Acceleration, Dubna, USSR (1972)
267 C. Dustmann, H. Krauth, L. Steinbock, W. Zernial: IEEE Trans.Nucl.Sci.NS-20, No.3, 283 (1973)
268 C. Dustmann, W. Heinz, H. Krauth, L. Steinbock, W. Zernial: Proc.IXth Int.Conf. on High-Energy Accelerators, Stanford, USA, (1974), p.250
269 L. Steinbock: Proc.8th Int.Conf.on High-Energy Accelerators, CERN, Geneva, (1971), p.478
270 C. Andelfinger, W. Herrmann, P. Merkel, W. Ott, A. Schlüter, U. Schumacher, M. Ulrich, H. Welter: Contr.3rd Work Meeting on ERA, Report IEKP 3/69-30, KFZ Karlsruhe, (1969)
271 C. Andelfinger, W. Herrmann, A.U. Luccio, P. Merkel, W. Ott, U. Schumacher, M. Ulrich: Proc.4th Work Meeting on ERA, MPI für Plasmaphysik, Garching, Report IPP 0/3, (1971), p.8
272 A. Schlüter: Proc.VIIIth Int.Conf.on High-Energy Accelerators, CERN, Geneva, (1971), p.402
273 C. Andelfinger, W. Dommaschk, J. Fink, R. Griek, W. Herrmann, I. Hofmann, D. Jacobi, P. Merkel, W. Ott, A. Schlüter, U. Schumacher, M. Ulrich: Proc. IVth All-Union Conf.on Part.Accelerators, Moscow, USSR, Nov.1974, Vol.I, (1975), p.71
274 U. Schumacher, M. Ulrich, W. Dommaschk, I. Hofmann, P. Merkel: Proc.First Int.Conf.on Electron Beam Res.and Technology, Albuquerque, USA, Nov.1975, Vol. II, (1976), p.385
275 J. Fink, W. Herrmann, W. Ott, J.M. Peterson: Proc.IXth Int.Conf.on High-Energy Accelerators, Stanford, USA, (1974), p.223
276 W. Herrmann: Report IPP 0/15, MPI für Plasmaphysik, Garching, (1973); Proc. 2nd Symp.on Coll.Methods of Acceleration, Dubna, USSR, Sept.1976, (1977), p.60
277 C. Andelfinger, E. Buchelt, W. Dommaschk, J. Fink, W. Herrmann, I. Hofmann, A.U. Luccio, P. Merkel, D. Jacobi, H.-B. Schilling, A. Schlüter, U. Schumacher, M. Ulrich: IEEE Trans.Nucl.Sci.NS-24, No.3, 1622 (1977)
278 H. Ebert: Physikalisches Taschenbuch, 5.Aufl. (Vieweg, Braunschweig, 1976)
279 M. Reiser: private communication;
D.W. Hudgings, R.A. Meger, C.D. Striffler, W.W. Destler, H. Kim, M. Reiser, M.J. Rhee: Physics Publication No.78-121, University of Maryland, USA, (1977), to be published
280 J.M. Hauptmann, L.J. Laslett, W.W. Chupp, D. Keefe: Proc.9th Int.Conf.on High-Energy Accelerators, Stanford, USA, (1974), p.240